AUDEL®

Mathematics for Mechanical Technicians and Technologists

Principles
Formulas
Problem Solving

by JOHN D. BIES

Rex Miller, *Consulting Editor*

Macmillan Publishing Company
New York
Collier Macmillan Publishers
London

By the same author:

Sheet Metal Work

Copyright © 1986 by Macmillan Publishing Company, a division of Macmillan, Inc.

All rights reserved. No part of this book may be reproduced or transmitted in any form or by any means, electronic or mechanical, including photocopying, recording or by any information storage and retrieval system, without permission in writing from the Publisher.

Macmillan Publishing Company
866 Third Avenue, New York, N.Y. 10022
Collier Macmillan Canada, Inc.

Library of Congress Cataloging-in-Publication Data

Bies, John D., 1946–
  Mathematics for mechanical technicians and technologists.

  "An Audel book."
  Includes index.
  1. Engineering mathematics. I. Title.
TA330.B5 1986 620′.0042 86-5209
ISBN 0-02-510620-1

Macmillan books are available at special discounts for bulk purchases for sales promotions, premiums, fund-raising, or educational use. For details, contact:

        Special Sales Director
        Macmillan Publishing Company
        866 Third Avenue
        New York, N.Y. 10022

10 9 8 7 6 5 4 3 2 1

Printed in the United States of America

While every precaution has been taken in the preparation of this book, the Publisher assumes no responsibility for errors or omissions. Neither is any liability assumed for damages resulting from the use of the information contained herein.

Contents

Preface   *v*

PART I   Basic Mathematics and Formulas

Chapter 1   Arithmetic and Algebra   3
Powers and Roots—Logarithms—Algebra

Chapter 2   Area and Volume   15
Geometric Formulas

Chapter 3   Triangulation: Right and Oblique Planes   58
Right Angle Trigonometry—Oblique Angle Trigonometry—Other Trigonometric Formulas

PART II   Mechanical Technology Formulas and Calculations

Chapter 4   Basic Concepts of Mechanics   83
Basic Definitions and Unit Systems—Force Systems—Friction—Levers

Chapter 5   Complex Mechanics   125
Wheels and Pulleys—Center of Gravity and Moments of Inertia—Velocity and Acceleration—Work and Power—Centrifugal Force

Chapter 6   Strength of Materials   168
Fundamentals of Strength of Materials—Simple Shear and Stress Formulas—Torsion and Torque—Strength of Compression Members

Chapter 7 Fluidics — 199
Basic Properties of Fluids—Static Fluid Concepts and Calculation—Dynamic Hydraulic Systems—Pneumatic Systems

Chapter 8 Cams and Gears — 227
Cams—Gears

Chapter 9 Machine Elements — 260
Screws, Rivets, and Splines—Springs and Bearings

Chapter 10 Machining Operations — 290
Cutting Speeds and Feeds—Basic Machining Operations—Power and Force Calculations in Machining

PART III Industrial Management Formulas and Calculations

Chapter 11 Management Controls — 319
Industrial Organization—Span, Production, and Inventory Controls—Quality Assurance, Manufacturing Controls, and Budget Cost Controls

Chapter 12 Economics in Machining — 338
Machining Costs—Cost and Production Rate Equations—Formulas for Optimization

Chapter 13 Facility and Human Resources Management — 354
Physical Facilities—Human Resources Management

Answers to Exercises — 371

Appendixes — 385

Index — 415

Preface

The use of mathematics, formulas, and quantitative analysis is the backbone of industrial technology. A competent technician and technologist must not only have a sound understanding of mathematical theory, but must also be able to apply these principles to solving practical industrial and mechanical problems. The critical need of modern industry is for technical personnel competent in mathematics and the ability to select and use appropriate formulas for decision making. Unfortunately, the number of individuals skilled in mathematics is limited. Furthermore, those with mathematical skills often have little idea of what formulas are available for problem solving and when and where they should be used.

Those that recognize their shortcomings will strive to obtain the knowledge necessary to become effective employees. This can be obtained by attending formal classes in community colleges and universities, industrial seminars and workshops, and/or by reading in professional journals and books.

As a consultant to business and industry, the author of this book became aware of the dearth of reference materials that would help the technician and technologist obtain information about mathematical formulas to help them solve everyday problems. Formulas used within the field of mechanical technology are scattered throughout many different textbooks, references, periodicals, manuals, and reports, making it difficult for individuals to obtain needed information quickly.

This book is a response to the need for a text that not only deals with mathematical formulas, but also assembles them in one volume for easy reference. Hence, it is written for use as a reference by technical personnel, as text material in a classroom setting, or for the individual who seeks self-improvement via home study.

Mathematics for Mechanical Technicians and Technologists is designed to serve as a review and introduction to the use of mathematical formulas as a problem-solving tool. The content in each chapter progresses from simple to increasingly more complex concepts and formulas. Formulas are presented, explained, and used in example problems, and the principles and concepts necessary for their correct applications are also provided.

This handbook is divided into three major parts. The first is a basic review of mathematics and conversions that is needed for understanding how formulas are manipulated for decision making. The major topics covered are arithmetic and algebraic calculations, geometry and related formulas, basic trigonometry, and the use of other basic formulas employed in problem solving. Part II is the main portion of the book. Presented here are formulas and information of major concern to mechanical technology problems, such as basic and complex mechanics, strength of materials, fluidics, cams and gears, machine elements, and machining operations. Part III provides information to technical personnel who are in, or desire to be in, management positions. Topics such as management controls, economics in machining, and facility and human resources management are presented.

PART I

Basic Mathematics and Formulas

CHAPTER I

Arithmetic and Algebra

- **Powers and Roots**   • **Logarithms**   • **Algebra**

The basic requirement for using formulas in the field of mechanical technology is a working knowledge of arithmetic and algebra. For the purposes of this book it is assumed that the reader possesses a working knowledge of the basic mathematical rules and computational techniques upon which all other formulas are calculated. This would include, for example, adding, subtracting, multiplying, and dividing fractions, decimals, and numbers with dissimilar signs; converting fractions to decimals; and working with reciprocals. Our review begins with a discussion of powers and roots, then moves on to logarithms, and concludes with a review of algebraic rules and operations.

Powers and Roots

Powers are also referred to as *exponents*, which are notations as to how many times a digit or term is to be taken as a factor. As an example, the expression 5^2 is read as "five squared," which is the same thing as $5 \times 5 = 25$, and 6^3 is read as "six cubed" or $6 \times 6 \times 6 = 216$. Likewise, 5^6 is equivalent to 5 times itself six times,

4 • ARITHMETIC AND ALGEBRA

or $5 \times 5 \times 5 \times 5 \times 5 \times 5 = 15,625$. The two most common exponents found in technical formulas are the square and cube (see Appendix B).

In some cases, one will be required to make calculations with zero and negative exponents. Any value, except for zero, that is raised to the power of zero will always be equal to one. Thus, $3^0 = 1$ and $47^0 = 1$. In addition, any number raised to a negative power will be equal to its reciprocal. Hence, $3^{-3} = 1/3^3 = 1/27 = 0.037037$.

Notation by powers of ten is used to show the position of the decimal point, and is often referred to as the *scientific notation*. This notation is used to give a product of two factors, one of which is a given value or number, while the other is a positive or negative power of ten.

In scientific notation, the expression 3.1416×10^4 means 3.1416 with the decimal point moved four places to the *right*, or 31,416. The expression 3.1416×10^{-7} means 3.1416 with the decimal point moved seven places to the left, or 0.000 000 314 16. (When writing decimals that exceed five digits, it is best to place a space after every third digit.) This notation system should be used only when dealing with very large or very small quantities.

The ability to work with exponents in various formulas is critical to the solution of many problems. Therefore, one should be knowledgeable about the rules associated with exponent operations. The first rule states that when multiplying the same base number with exponents, the exponents are added so that their sum will become the new exponent for that base. Hence, $(3^2)(3)(3^5) = 3^{2+1+5} = 3^8$. Conversely, when dividing two values with the same base number, the exponent of the divisor will be subtracted from the exponent of the dividend. For example, $12^6 \div 12^3 = 12^{6-3} = 12^3$. Finally, numbers raised to the power of one will *not* change the value of the base number. Thus, $7^1 = 7$.

The roots of numbers are the opposite of powers. The symbol used to indicate a root is $\sqrt{}$ and is called a *radical*, while the number in the symbol is the *radicand*. The radical, as presented, is used to denote the *square root* of a number; that is, a number that is multiplied by itself (squared) will be equal to the radicand. Thus, $\sqrt{25} = 5$, $\sqrt{81} = 9$, $\sqrt{3} = 1.7321$, and so on. If four figures to the right of the decimal of the root are sufficient, it is recommended that the solution be obtained from mathematical tables (see Appendix B).

An *index* is sometimes used in combination with the radical sign. The index is used to calculate roots greater than the square root. For example, $\sqrt[3]{27}$ (the *cube* root of 27) equals 3 (i.e., $3 \times 3 \times 3 = 27$). The value of the index is used to determine the root of the radicand. Thus, $2 \times 2 \times 2 \times 2 = 16$; 2 is the fourth root of 16.

Exercises

Solve for the following numerical answers:

1. $4^3 =$
2. $16^4 =$
3. $(7^3)(7^2) =$

Simplify the following expressions:

4. $3^4 - 3^2 =$
5. $(\frac{2}{3})^3 =$
6. $(5^{-3})(5^4)(5^7) =$

Convert the following into their actual values:

7. $36.134 \times 10^{-7} =$
8. $4.74 \times 10^9 =$
9. $\sqrt{16} =$
10. $\sqrt[3]{64} =$

Logarithms

Mathematical formulas frequently involve tiresome arithmetical computations. To simplify them, logarithms are often used. The term *logarithm* is derived from the Greek words *logos*, meaning "a ratio," and *arithmos*, meaning "number." Basically, a logarithm is an exponent of power to which a base number (usually 10) must be raised in order to produce a given number.

Another way of expressing 10^2, or 100, is to say that the logarithm of 100 to the base 10 is 2. Hence, since $8^3 = 64$, the logarithm of 64 to the base 8 is 3. Within the context of this section, we will

6 • ARITHMETIC AND ALGEBRA

deal only with logarithms to the base 10. Such logarithms are referred to as *common logarithms*.

Common Logarithms

The common logarithm (abbreviated as log) of 1 (10^0) is 0, the common logarithm of 10 (10^1) is 1, while the common logarithm of 100 (10^2) is 2. All numbers between 10 and 100 must therefore be calculated by 10 raised to some power *between* 1 and 2. Thus, any number can be expressed as a raised power of 10: 78 can be expressed as $10^{1.8920946}$, 6777 can be written as $10^{3.8310375}$, and so on. Hence, the common logarithm of a number is an expression of the power to which the number 10 must be raised in order to calculate that number.

Logarithms are made up of two numbers. The first is the *integral* or *characteristic*, which is the whole number to the left of the decimal point. The second is the *mantissa*, which is the decimal part of the number to the right of the decimal point.

To identify the characteristic of decimals, we refer back to the scientific notation system, where negative exponents are used to show the position of the decimal point. Keeping this in mind, note the characteristics of the following decimal numbers:

$$\log 0.1 \ \ \ = -1, \bar{1}, \text{ or } 9-10$$
$$\log 0.01 \ \ = -2, \bar{2}, \text{ or } 8-10$$
$$\log 0.001 \ = -3, \bar{3}, \text{ or } 7-10$$
$$\log 0.0001 = -4, \bar{4}, \text{ or } 6-10$$

The mantissas of logarithms have been precalculated in logarithmic tables, which are commonly available.

The problem of combining a negative characteristic with a positive mantissa, such as -2 and $+.60314$, respectively, is overcome by using one of the following expressions: $2.60314 - 10$, $\bar{8}.60314$, or $0.60314 - 8$

Exercises

Write the characteristics of the logarithms for the following numbers:

11. 745

12. 34,678

13. 4
14. 0.15
15. 0.0000034
16. 245.037

Using Logarithmic Tables

The use of logarithmic tables is essential, especially when electronic calculators are not available. Four- and five-place tables are commonly available. Special tables are available to eight places where extreme accuracy is required.

To find the log of any (positive) number is quite simple. An inspection of the logarithmic table shows that the number (N) increases from 100 to 1000, which can also be used for the numbers 1 to 99. For example:

$$\log .141 = 9.14922 - 10$$
$$\log 1.41 = 0.14922$$
$$\log 14.1 = 1.14922$$
$$\log 141 = 2.14922$$

To find the log of 37.456, for example, we first find that it falls between 1.57345 (log 37.45) and 1.57356 (log 37.46). The difference here is 0.00011, and our number is 6/10 between the two values, or 0.000066. By adding this amount to 1.57345, we obtain 1.573516, rounded off to 1.57352, for the log of 37.456.

To find a number (N) corresponding to a given log, we employ the opposite procedure just described. For example, consider the following:

$$\log = N$$
$$0.98682 = 9.701$$
$$0.98682 - 4 = 0.0009701$$
$$0.98682 + 4 = 97{,}010$$

Exercises

Find the log for the following numbers:

17. 0.1255

8 • ARITHMETIC AND ALGEBRA

18. 1,247
19. 23.457
20. 0.000987

Find N for the following log values:

21. 0.58557
22. 0.28198 + 6
23. 0.52535 − 7

Logarithm Operations

Before one can effectively use logarithms for mathematical calculations, four basic principles or properties must be known:

1. The log of a product will be equal to the sum of the logs of all its factors. This can be represented by:

$$\log (AB) = \log A + \log B$$

2. The log of a quotient is equal to the log of its numerator minus the log of its denominator. This can be represented by:

$$\log (A/B) = \log A - \log B$$

3. The log of the nth power of a number will be equal to n times the log of that number. This is represented as:

$$\log (A^n) = (n)(\log A)$$

4. The log of the nth root of a number will be equal to the log of that number divided by n. This is represented as:

$$\log \sqrt[n]{A} = (\log A)/n$$

Bearing in mind these four basic properties of logarithms, let us now consider their applications. Here are some examples:

EXAMPLE 1

Find $a = (6.234)(0.0003493)(56.2)$.

Solution:

$$\begin{aligned}
\log 6.234 &= 0.78718 \\
\log 0.0003493 &= 0.54320 - 4 \\
\underline{\log 56.2 } &= \underline{0.74974 + 1} \\
\text{Sum of logs} &= 2.08014 - 3 = 0.08014 - 1 \\
a &= 0.12238
\end{aligned}$$

EXAMPLE 2

Find $b = 7.036/0.0004612$.

Solution:

$$\begin{aligned}
\log 7.036 &= 0.84733 \\
\underline{\log 0.0004612 } &= \underline{0.66389 - 4} \\
\text{Difference between logs} &= 0.18344 + 4 \\
b &= 15{,}256
\end{aligned}$$

EXAMPLE 3

Find $c = (0.0274)^3$

Solution:

$$\begin{aligned}
\log 0.0274 &= 0.43775 - 2 \\
(0.43775 - 2)(3) &= 1.31325 - 6 = 0.31325 - 5 \\
c &= 2.0571 \times 10^{-5}
\end{aligned}$$

EXAMPLE 4

Find $d = \sqrt[3]{3.482}$

Solution:

$$\begin{aligned}
\log 3.482 &= 0.54183 \\
(0.54183)/3 &= 0.18061 \\
d &= 1.5153
\end{aligned}$$

Exercises

Using logarithms, find the correct values, to four significant digits, for the following:

24. $\sqrt{0.5329}$

25. $\sqrt[4]{7}$

26. $(\sqrt[3]{0.00245} \times 0.3149 \times 72340)/51 \times (-.7249)^2$

27. $[(3.7)(-5.46)(3.14)]/[(9.45)(0.301)]$

28. $(40.027/3.1892)^{2/3}$

Algebra

Algebra is an extension of arithmetic, with both numbers and letters used in the expression of values. The vast majority of formulas used in business and industry are algebraic expression.

All rules of arithmetic also apply to algebra. The multiplication sign ($\times$) used in arithmetic, however, should not be used in algebra because of potential confusion with the variable x. Parentheses or a multiplication dot will be used instead in multiplication operations. If two separate variables, such as p and s, are presented as ps, this also denotes the multiplication of the two letters.

Algebraic Expressions

Algebra is used where the numerical values of all quantities are not known. Algebraic expressions are used either to simplify complex calculations or when they are the only means for solving a problem.

Upper- and lowercase letters of the alphabet are normally used as algebraic symbols. For example, suppose we have two unmeasured variables (A and B) that are added together, subtracted, multiplied, or divided. Without measures (numerical quantities) arithmetic cannot be used. However, with algebra we can express these operations as follows:

$$A + B = C$$
$$A - B = D$$
$$AB = E$$
$$\frac{A}{B} = F$$

As in arithmetic, parentheses, brackets, and braces are used to indicate the order in which calculations are to be performed. Thus:

$$\frac{(A-B)D}{C-E} = F$$

means that we first subtract B from A, then multiply that difference by D, which is then divided by C. That quotient is then subtracted by E to give the value of F.

In algebraic formulas Greek letters are often used as substitutes for the English alphabet. When such a situation is encountered, there should be no confusion, for the Greek letter performs the same function as other letters: It stands for a given value. In some cases, certain Greek letters represent specific constants (e.g., π (pi) = 3.141592654 or 22/7). The Greek alphabet is presented in Table 1-1.

Table 1-1. The Greek Alphabet

Letter	Uppercase Symbol	Lowercase Symbol
Alpha	A	α
Beta	B	β
Gama	Γ	γ
Delta	Δ	δ
Epsilon	E	ϵ
Zeta	Z	ζ
Eta	H	η
Theta	Θ	θ
Iota	I	ι
Kappa	K	κ
Lambda	Λ	λ
Mu	M	μ
Nu	N	ν
Xi	Ξ	ξ
Omicron	O	o
Pi	Π	π
Rho	P	ρ
Sigma	Σ	σ
Tau	T	τ
Upsilon	Y	υ
Phi	Φ	ϕ
Chi	X	χ
Psi	Ψ	ψ
Omega	Ω	ω

Exponents are used in algebraic expressions to multiply a quantity by itself a given number of times. Exponents are used according to the following arrangements:

12 • ARITHMETIC AND ALGEBRA

$$x^1 = (x)$$
$$x^2 = (x)(x)$$
$$x^3 = (x)(x)(x)$$
$$x^4 = (x)(x)(x)(x)$$
$$x^5 = (x)(x)(x)(x)(x)$$
$$x^6 = (x)(x)(x)(x)(x)(x)$$

In addition to numerical exponents, alphabetical exponents are also used, such as x^n. This is read as "x to the nth power," and is used when the value of the exponent is unknown.

When multiplying a factor raised to a power (e.g., x^n) by the same factor raised to another power (e.g., x^p), the product will be the sum of the original exponents. In our example, $x^{n \times p} = x^{n+p}$. Likewise, $x^{4 \times 7} = x^{11}$.

Another possible situation is to multiply a raised quantity (e.g., y^k) by itself p times. This would be expressed as $(y^k)^p$. Likewise, $(y^2)^8 = y^{16}$.

Reciprocal quantities are expressed algebraically by negative exponents:

$$C^{-1} = 1/C$$
$$C^{-3} = 1/C^3$$
$$C^{-12} = 1/C^{12}$$

When combining quantities raised to negative powers, the same rules apply as in arithmetic:

$$C^6 C^{-4} = C^{6-4} = C^2$$
$$(C^{-2})^{-2} = C^{(-2)(-2)} = C^4$$
$$(C^3)^{-4} = C^{(3)(-4)} = C^{-12}$$

It should also be remembered that any quantity raised to the zero power is always equal to 1. Hence, $B^0 = 1$ and $B^2 B^{-2} = 1$.

Fractional or decimal exponents are often quite useful. Perhaps the simplest example is the square root of the quantity Y, which is expressed as $\sqrt{Y}$. We can also present this quantity as an exponential expression by the fraction ½ or the decimal 0.5. Thus, the square root of Y is equal to $Y^{1/2}$ or $Y^{0.5}$.

Other root expressions can also be expressed with exponents:

$$Y^{1/3} = Y^{0.333} = \sqrt[3]{Y}$$
$$Y^{1/7} = Y^{0.143} \quad \sqrt[7]{Y}$$
$$Y^{3/8} = Y^{1.75}$$
$$Y^{1/n} = \sqrt[n]{Y}$$

These exponents can be manipulated in calculations as follows:

$$(V^4)^{1/2} = V^{(1/2)(4)} = V^2$$
$$(V^{1/4})^{-7} = V^{(1/4)(-7)} = V^{-7/4}$$
$$(V^{1.8})^{-4} = V^{(-4)(1.8)} = V^{-7.2}$$
$$(V^{0.8})^{0.7} = V^{0.56}$$

Exercises

Complete the following calculations:

29. $(5)(xy) =$

30. $(2x)(4b)(3) =$

31. $(5y)(ay) + 3 =$

32. $\dfrac{2x}{4y} =$

33. $\dfrac{24xyz}{6xy} =$

34. $(a)(a)(a)(a) =$

35. $\dfrac{c^A}{c^B} =$

36. $27^{2/3} =$

37. $(4c^4)^0 =$

Algebraic Operations

Though the basic rules of arithmetic apply to algebra, there are some differences, since both letters and numbers appear in algebraic formulas. The problem here is that numbers have specific values, while letters represent unknown quantities.

The basic rules of adding and subtracting algebraic formulas are that only like variables can be added together: $A + A = 2A$, $B + 3B + B + 2B = 7B$, etc. However, $B + C$ does not equal BC ($B + C = B + C$). The same rules apply to subtracting algebraic expressions: $A - A = 0$, $3B - B = 2B$, $2C - 3C - 4C = -5C$. When several letters are used in algebraic expressions, the same procedures are employed. Consider the problem of adding the expression

$(3w + 2h + 4dt - 2p - 4xy + 10z)$ to $(4w + 4h - 6dt + 2xy + 2.5z - 7g)$. To solve this problem, the following procedure would be used:

$$\begin{array}{r}3w + 2h + 4dt - 2p - 4xy + 10z\\ 4w + 4h - 6dt + 2xy + 2.5z - 7g\\ \hline 7w + 6h - 2dt - 2p - 2xy + 12.5z - 7g\end{array}$$

In multiplication, the same principles used in arithmetic are applied here. Each term in the multiplicand algebraic expression is multiplied by each term in the multiplier. An example of this is in the problem $(2a - 2)(a + 3)$:

$$\begin{array}{r}2a - 2\\ a + 3\\ 2a^2 - 2a\\ 6a - 6\\ \hline 2a^2 + 4a - 6\end{array}$$

With division, the same concept is applied—only in reverse. Consider the problem $(3a^2 + 6a - 9)/(a + 3)$:

$$\begin{array}{r}3a - 3\\ a + 3 \overline{\smash{)}3a^2 + 6a - 9}\\ 3a^2 + 9a\\ \hline -3a - 9\\ \underline{-3a - 9}\end{array}$$

Exercises

Complete the following problems:

38. $2x + 3y + x + z + 2y =$

39. $5a + 4b - 2a + 2b - 6b =$

40. $3x + 2y - (x - 2y) =$

41. $(2a - 2)(a + 3) =$

42. $(4a^2 - 3a + 3)(4a - 3) =$

43. $\dfrac{(4z^3 + 6z^2 + 1)}{(2z - 1)} =$

CHAPTER 2

Area and Volume

- **Geometric Formulas**

One of the most basic problems encountered within the field of mechanical technology is the calculation of surface area and solid volume. Area and volume problems are critical in determining storage capacity, shipping weights, and product material costs. The formulas used for these calculations are based on the principles, laws, and theories of geometry.

Geometry is the field of mathematics that addresses the properties and relationships of lines, surfaces, and their measurement. Geometry involves an understanding of the differences between geometric shapes and their relationship to one another. It is used in a variety of fields including plumbing, electronics, surveying, machining, and construction. In addition, problems dealing with volume, area, capacity, and weights require the use of geometric principles. (Here again, a knowledge of basic geometry is presupposed. This includes a knowledge of the various angles and plane figures.)

Applied geometric principles employ relationships between points and lines. It should be remembered that two points will determine the length and direction of a line. In turn, a plane can be formed by three or more points, a line and a point, or two or more lines (a plane formed by two lines will not have a definite size or shape). When a third dimension is added to plane figures, it is possible to generate a geometric solid. Solids, then, represent real-world objects that are often encountered in business, industry, and the workplace. In all, there are six basic geometric solids: cubes, prisms, cylinders, cones, spheres, and pyramids.

Most measurements of plane figures will deal with either linear (length, width, and height) or surface area measures. With solids, however, an additional measurement must be determined: volume. While linear measures are always given in terms of the units of measure—inches (in.), foot (ft), meter (m), etc.—and area measures are specified in terms of units squared—in.^{2} (sq. in.), ft^2, m^2, etc.—volume measures are given in cubed units—in.^{3} (cu. in.), ft^3 (cu. ft.), m^3, etc.

Geometric Formulas

Areas and Dimensions of Plane Figures

Determining areas and dimensions of plane figures is central to many problems faced in various fields of technology. The following paragraphs provide formulas that can be used when dealing with plane or flat surfaces. (When possible, the notations used in the formulas will correspond to related illustrations. When this is not possible, the notation will be described or defined within the paragraph itself.)

SQUARE

To find the area (A) of a square (Figure 2-1), one of two formulas can be used. The first makes use of the dimension of the square's side (s), while the second incorporates the diagonal dimension (d):

$$A = s^2 \text{ or}$$
$$A = \frac{1}{2}d^2$$

In addition, when the area of a square is known, it is possible to calculate the side and diagonal dimensions by using the following formulas:

$$s = 0.7071d \text{ or}$$
$$s = \sqrt{A}$$
$$d = 1.414s \text{ or}$$
$$d = 1.414\sqrt{A}$$

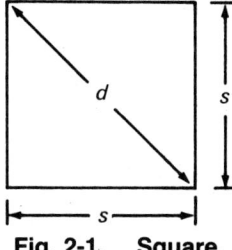

Fig. 2-1. Square.

Consider the problem where we have to determine the area of two squares. The first has a side 25 cm long, and the second has a diagonal dimension of 1.75 in. To calculate the areas of these two problems, the following procedures would be used:

$$A = s^2$$
$$= 25^2$$
$$= 625 \text{ cm}^2$$

$$A = \frac{1}{2}d^2$$
$$= (\frac{1}{2})(1.75)^2$$
$$= \frac{3.0625}{2}$$
$$= 1.53 \text{ in}^2$$

If the area of a square is 1024 m² (square meters), and we must find the length of the side and the diagonal, the following procedures should be used:

$$s = \sqrt{A}$$
$$= \sqrt{1024}$$
$$= 32 \text{ m}$$

$$d = 1.414\sqrt{A}$$
$$= 1.414\sqrt{1024}$$
$$= (1.414)(32)$$
$$= 45.248 \text{ m}$$

RECTANGLE

As in finding the area of a square, rectangles make use of either

18 • AREA AND VOLUME

diagonal or side dimensions (Figure 2-2). The formulas used to calculate the area of rectangles are:

$$A = ab$$
$$A = a\sqrt{d^2 - a^2}$$
$$A = b\sqrt{d^2 - b^2}$$

Furthermore, to determine the sides and diagonal of a rectangle with certain known values (i.e., area and one linear measurement), the following formulas are used:

$$a = \frac{A}{b} = \sqrt{d^2 - b^2}$$
$$b = \frac{A}{a} = \sqrt{d^2 - a^2}$$
$$d = \sqrt{a^2 + b^2}$$

Bearing all these formulas in mind, it is possible to calculate the area of a rectangle with sides that are 15 and 37 cm long by the following procedure:

$$A = ab$$
$$= (15)(37)$$
$$= 555 \text{ cm}^2$$

To find the side (b) and diagonal (d) of a rectangle whose other side (a) is 17.5 in., and that has an area 328.125 in.², the following procedures are used:

$$b = \frac{A}{a} = \frac{328.125}{17.5} = 18.75 \text{ in.}$$
$$d = \sqrt{a^2 + b^2} = \sqrt{17.5^2 + 18.75^2} = \sqrt{657.8125} = 25.648 \text{ in.}$$

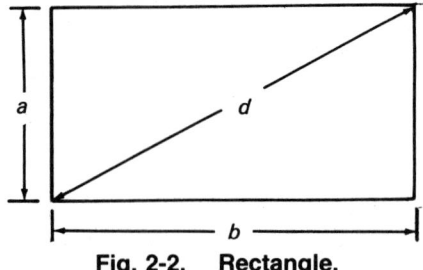

Fig. 2-2.   Rectangle.

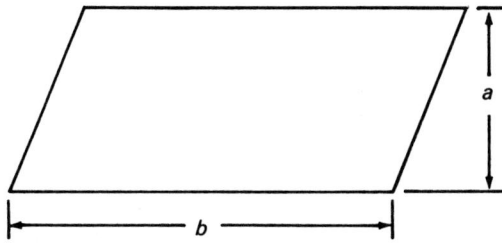

Fig. 2-3. Parallelogram.

PARALLELOGRAM

The area of a parallelogram is calculated by multiplying the dimension base side of the quadrilateral by its height (see Figure 2-3). The area and dimensional formulas used here are:

$$A = ab$$
$$a = \frac{A}{b}$$
$$b = \frac{A}{a}$$

Consider a parallelogram whose base is 16 m and height is 5.5 m. The area would be calculated as follows:

$$\begin{aligned}A &= ab \\ &= (16)(5.5) \\ &= 88 \text{ m}^2\end{aligned}$$

If the area of a parallelogram is 248 in.^{2}, and it has a height of 8.25 in., the base could be calculated by using the formula:

$$\begin{aligned}b &= \frac{A}{a} \\ &= \frac{248}{8.25} \\ &= 30.061 \text{ in.}\end{aligned}$$

RIGHT TRIANGLE

The area of a right triangle (one with a 90° angle) is determined by using the length of its base (c) and its height (b) sides (Figure 2-4):

20 • AREA AND VOLUME

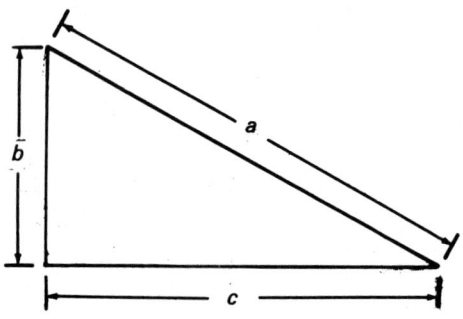

Fig. 2-4.   Right triangle.

$$A = \frac{bc}{2}$$
$$a = \sqrt{b^2 + c^2}$$
$$b = \sqrt{a^2 - c^2}$$
$$c = \sqrt{a^2 - b^2}$$

An example of how the area formula would be used is to find the area of a right triangle whose base and height sides are 34.75 cm and 14.45 cm, respectively:

$$A = \frac{bc}{2}$$
$$= \frac{(34.75)(14.45)}{2}$$
$$= 251.07 \text{ cm}^2$$

To find side a of a right triangle whose base is 12 in. and height is 16 in., the following procedure is used:

$$a = \sqrt{b^2 + c^2}$$
$$= \sqrt{16^2 + 12^2}$$
$$= \sqrt{400} = 20 \text{ in.}$$

If the hypotenuse (a) of a right triangle is equal to 16.7 m and its height is 4 m, the base measurement would be calculated as:

$$c = \sqrt{a^2 - b^2}$$
$$= \sqrt{16.7^2 - 4^2}$$
$$= \sqrt{294.89} = 17.17 \text{ m}$$

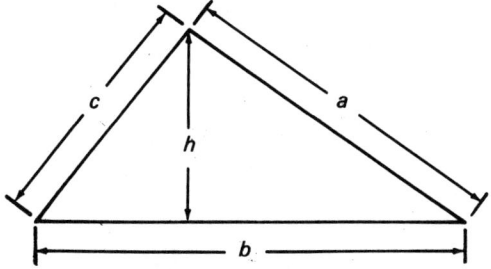

Fig. 2-5. Acute triangle.

ACUTE TRIANGLE

Since acute triangles (those with all angles at less than 90°) do not have a right angle, height (h) is measured from the vertex of the triangle and 90° to its base (Figure 2-5). The basic formulas used to calculate area are:

$$A = \frac{bh}{2}$$

$$A = \frac{b}{2}\sqrt{a^2 - [(a^2 + b^2 + c^2)/2b]^2}$$

To find the area of an acute triangle with sides of $a = 10$ in., $b = 9$ in., and $c = 8$ in., the following sequence of calculations is used:

$$\begin{aligned}
A &= \frac{b}{2}\sqrt{a^2 - [(a^2 + b^2 - c^2)/2b]^2} \\
&= \frac{9}{2}\sqrt{10^2 - [(10^2 + 9^2 - 8^2)/(2)(9)]^2} \\
&= 4.5\sqrt{100 - [(100 + 81 - 64)/18]^2} \\
&= 4.5\sqrt{100 - 42.25} \\
&= (4.5)(7.6) \\
&= 34.20 \text{ in.}^2
\end{aligned}$$

OBTUSE TRIANGLE

As in area calculations in acute triangles, calculations in obtuse triangles (those with an angle greater than 90°) make use of a measured height dimension (Figure 2-6). These area formulas, however, are

22 • AREA AND VOLUME

somewhat different due to the nature of a plane figure. The area formulas for obtuse triangles are:

$$A = \frac{bh}{2}$$
$$A = \frac{b}{2}\sqrt{a^2 - [(c^2 - a^2 - b^2)/2b]^2}$$

To employ the first formula, consider an obtuse triangle with a base length of 54 mm and a height of 124 mm:

$$A = \frac{bh}{2}$$
$$= \frac{(54)(124)}{2}$$
$$= 3348 \text{ mm}^2$$

For an obtuse triangle where $a = 6$ in., $b = 5$ in., and $c = 9$ in., its area would be found using the following procedures:

$$A = \frac{b}{2}\sqrt{a^2 - [(c^2 - a^2 - b^2)/2b]^2}$$
$$= \frac{5}{2}\sqrt{6^2 - [(9^2 - 6^2 - 5^2)/(2)(5)]^2}$$
$$= 2.5\sqrt{36 - [(81 - 36 - 25)/10]^2}$$
$$= 2.5\sqrt{36 - 4}$$
$$= 2.5\sqrt{32}$$
$$= 14.142 \text{ in.}^2$$

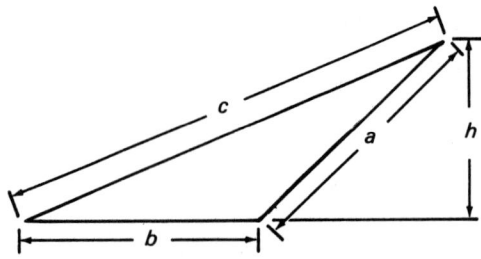

Fig. 2-6.   Obtuse triangle.

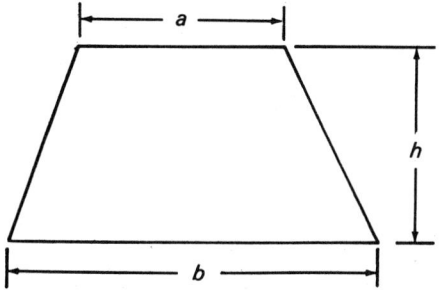

Fig. 2-7. Trapezoid.

TRAPEZOID

A trapezoid is a quadrilateral with only two sides parallel to each other. As illustrated in Figure 2-7, calculating the area of a trapezoid involves the use of the lengths of the parallel sides (a and b) plus the distance (h) between them:

$$A = \frac{(a+b)h}{2}$$

To use this formula, consider the trapezoid where side $a = 30$ cm, side $b = 45$ cm, and the distance between the two sides is 39 cm:

$$\begin{aligned} A &= \frac{(a+b)h}{2} \\ &= \frac{(30+45)39}{2} \\ &= \frac{2925}{2} \\ &= 1462.5 \text{ cm}^2 \end{aligned}$$

TRAPEZIUM

A trapezium is a quadrilateral where none of the sides is parallel to each other. Required for calculating a trapezium's area are two height dimensions, plus the differences in length between opposite sides (see Figure 2-8). It should be noted that the area of a trapezium can also be calculated by drawing a diagonal and computing the area

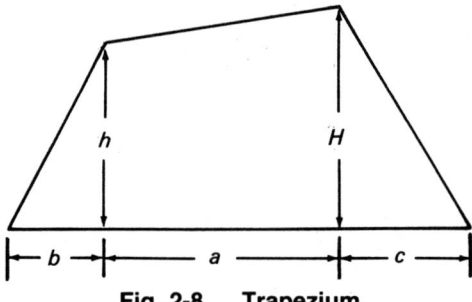

Fig. 2-8. Trapezium.

of each triangle formed. The basic formula used to calculate the area of a trapezium is:

$$A = \frac{(H+h)a + bh + cH}{2}$$

An example of how this formula would be used is for a trapezium where $a = 11$, $b = 3$, $c = 4$, $h = 9$, and $H = 13$ meters. The calculation would procedure as follows:

$$\begin{aligned}
A &= \frac{(H+h)a + bh + cH}{2} \\
&= \frac{(13+9)11 + [(3)(9)] + [(4)(13)]}{2} \\
&= \frac{321}{2} \\
&= 160.5 \text{ m}^2
\end{aligned}$$

REGULAR HEXAGON

To calculate the area of a regular hexagon (a polygon with six equal sides and angles), it is necessary to know the radius of either its circumscribed circle (R) or its inscribed circle (r) (see Figure 2-9). With this in mind it is possible to note the following relationships:

$$\begin{aligned}
R &= s \\
s &= 1.155r \\
r &= 0.866s \text{ or} \\
r &= 0.866R
\end{aligned}$$

To calculate the area of a regular hexagon, several formulas are

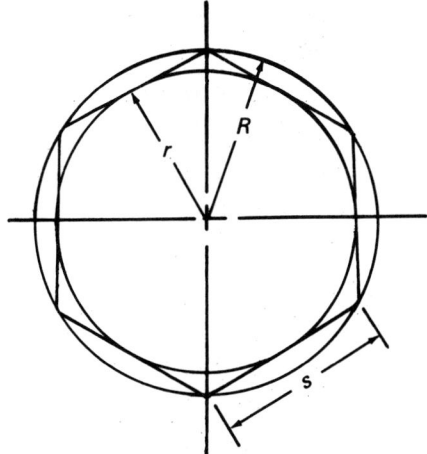

Fig. 2-9. Regular hexagon.

available; the selection of the formula used is determined by the amount and type of data provided. These formulas are:

$$A = 2.598s^2 \text{ or}$$
$$A = 2.598R^2 \text{ or}$$
$$A = 3.464r^2$$

A typical problem for a regular hexagon would be in the example where the length of each side (s) is equal to 12 in., and it is required that we find the area of the figure and the radii of its inscribed and circumscribed circles. The following procedures would be used:

$$A = 2.598(12)^2$$
$$= (2.598)(144)$$
$$= 374.112 \text{ in.}^2$$

$$r = 0.866s$$
$$= (0.866)(12)$$
$$= 10.392 \text{ in.}$$

$$R = s$$
$$= 12 \text{ in.}$$

REGULAR OCTAGON

Another regular polygon frequently encountered is the octagon (eight sides and angles). Similar to the hexagon, this plane figure makes use of side and radii dimensions (Figure 2-10). The relationship of these measures are:

$$R = 1.307s \text{ or}$$
$$R = 1.082r$$
$$r = 1.207s \text{ or}$$
$$r = 0.924R$$
$$s = 0.765R \text{ or}$$
$$s = 0.828r$$

The formulas used to calculate the area of a regular octagon are:

$$A = 4.828s^2$$
$$A = 2.828R^2, \text{ or}$$
$$A = 3.314r^2$$

As an example, consider the regular octagon that has an inscribed circle of 2.5 m. Find its area, the length of its sides, and the radius of the circumscribed circle. The solution to these problems are:

$$A = 3.314r^2$$
$$= (3.314)(2.5)^2$$
$$= (3.314)(6.25)$$
$$= 20.7125 \text{ m}^2$$

$$s = 0.828r$$
$$= (0.828)(2.5)$$
$$= 2.07 \text{ m}$$

$$R = 1.082r$$
$$= (1.082)(2.5)$$
$$= 2.705 \text{ m}$$

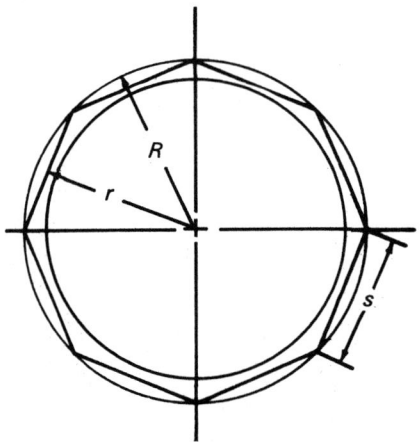

Fig. 2-10. Regular octagon.

REGULAR POLYGONS

In addition to hexagons and octagons, other regular polygons are encountered on the job. Examples of these are pentagons (five sides), heptagon (seven sides), nonagons (nine sides), and decagons (ten sides). Figure 2-11 illustrates a regular polygon with elements used in calculating dimensional and area measures.

Note that angular measures are shown by the Greek letters alpha (α) and beta (β). In addition to these elements, the number of sides (n) is also used in regular-polygon formulas. Basic measurement formulas used here are:

$$\alpha = \frac{360°}{n}$$
$$\beta = 180° - \alpha$$
$$R = \sqrt{r^2 + (s^2/4)}$$
$$r = \sqrt{R^2 - (s^2/4)}$$
$$s = 2\sqrt{R^2 - r^2}$$

The formulas used for calculating the area of regular polygons are:

$$A = \frac{nsr}{2} \text{ and}$$
$$A = \frac{ns}{2} \sqrt{R^2 - (s^2/4)}$$

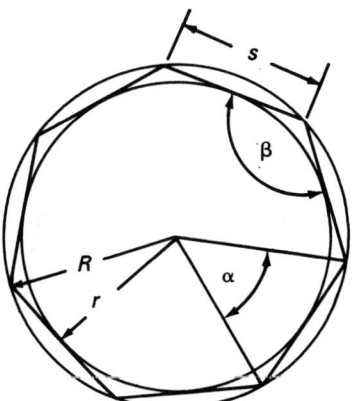

Fig. 2-11. Regular polygon.

28 • AREA AND VOLUME

To find the area of a polygon that has 12 sides 4.2 in. in length and is inscribed in a circle with a diameter of 16 in., the following procedure would be used:

$$A = \frac{ns}{2}\sqrt{R^2 - (s^2/4)}$$
$$= \frac{(12)(4.2)}{2}\sqrt{8^2 - (4.2^2/4)}$$
$$= 25.2\sqrt{59.59}$$
$$= (25.2)(7.719)$$
$$= 194.53 \text{ in.}^2$$

CIRCLE

Formulas for circle specifications are important in a broad section of technical fields. Linear measurements here are specified in terms of the distance about the circumference (C) of the circle, its diameter (d), and radius (r) (see Figure 2-12). In the fields of civil engineering and surveying it is frequently required to calculate the length of an arc whose center angle is known—this is based on using diameter dimensions. The basic dimension formulas used for circles are:

$$C = 2\pi r = \pi d$$
$$r = \frac{C}{6.2832}$$
$$d = \frac{C}{\pi}$$

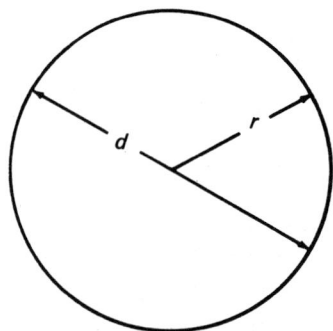

Fig. 2-12.   Circle.

The length of an arc where the center angle is 1° will be equal to $0.008727d$. In addition, the length of an arc where the center angle is $n°$ will be equal to $0.008727nd$.

Formulas for the area of a circle and related relationships are:

$$A = \pi r^2$$
$$r = \sqrt{A/\pi}$$
$$r = 0.564\sqrt{A}$$
$$d = \sqrt{A/0.7854}$$
$$d = 1.128\sqrt{A}$$

An example of a typical circle problem would be to find the area whose diameter is 16 m:

$$\begin{aligned} A &= \pi r^2 \\ &= (3.14159)(8^2) \\ &= (3.14159)(64) \\ &= 201.06 \text{ m} \end{aligned}$$

To find the diameter of a circle whose area is 26.8 in.^{2}, the following procedures would be used:

$$\begin{aligned} d &= 1.128\sqrt{A} \\ &= 1.128\sqrt{26.8} \\ &= (1.128)(5.177) \\ &= 5.840 \text{ in.} \end{aligned}$$

CIRCULAR SECTOR

A circular sector is any part of a circle that is bounded by any two radii (r) and the arc (l) included between them (Figure 2-13). Cal-

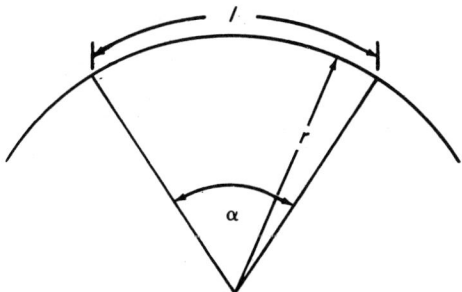

Fig. 2-13. Circular sector.

culations made of circular sectors may use either angular or nonangular measures. Note that where angle measures are required, Greek letters are used to indicate angle size. The basic formulas and relationships for circular sectors are:

$$A = \frac{rl}{2} = 0.00872\alpha r^2$$

$$l = \frac{(r\alpha\pi)}{180} = 0.01745r = \frac{2A}{r}$$

$$\alpha = \frac{(57.296)(l)}{r}$$

$$r = \frac{2A}{l} = \frac{(57.296)(l)}{\alpha}$$

An example of a circular-sector problem would be to find the area and the arc length of a sector that has a radius of 35 cm and an angle (α) of 72°. The calculations used to solve this problem are:

$$A = 0.008727 r^2$$
$$= (0.008727)(72)(35)$$
$$= 21.992 \text{ cm}^2$$

$$l = 0.01745 r$$
$$= (0.01745)(35)(72)$$
$$= 43.974 \text{ cm}$$

CIRCULAR SEGMENT

A circular segment is that part of a circle that is marked off or made separate by a straight line (Figure 2-14). In addition to mathematical

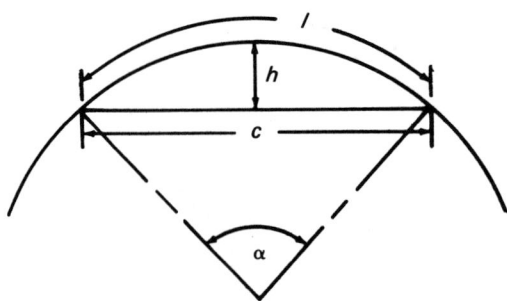

Fig. 2-14. Circular segment.

manipulations, circular-segment areas and measurements can also be determined by use of mathematical tables such as presented in Appendix C. The basic formulas used in making circular-segment calculations are:

$$A = \frac{rl - c(r-h)}{2}$$
$$c = 2\sqrt{h(2r-h)}$$
$$r = \frac{c^2 + 4h^2}{8h}$$
$$l = 0.01745r\alpha$$
$$h = r - (.5) - \sqrt{4r^2 - c^2} = r[1 - \cos(\alpha/2)]$$
$$\alpha = \frac{(57.296)(l)}{r}$$

In a typical problem the radius of the circular segment is 16 in. and its height is 2 in. Determine the length of the chord (c) of the segment:

$$\begin{aligned}c &= 2\sqrt{h(2r-h)} \\ &= 2\sqrt{2[(2)(16) - 2]} \\ &= 2\sqrt{60} \\ &= 15.492\end{aligned}$$

Another problem would be to find the area of a circular segment where the radius is 10 m, the length of the chord is 16.48 m, the length of the arc is 19.37 m, and the height of the segment is 4.336 m:

$$A = \frac{rl - c(r-h)}{2} = (10)(19.37) - \frac{(16.48)(10 - 4.336)}{2}$$
$$= \frac{193.7 - 93.34}{2} = 50.18 \text{ m}^2$$

FILLET OR SPANDREL

A fillet or spandrel is the triangular space formed within a right angle by an arc (Figure 2-15). There are three possible formulas available for calculating the area of a fillet:

$$A = r^2 - \frac{\pi r^2}{4}$$
$$A = 0.215r^2$$
$$A = 0.1075c^2$$

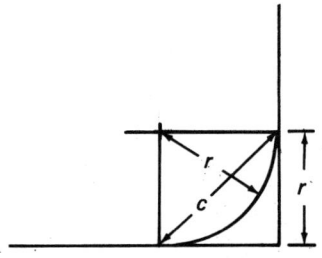

Fig. 2-15.   Fillet or spandrel.

To calculate the area of a spandrel that has a radius of 0.235 in. the following procedure is used:

$$A = 0.215r^2$$
$$= (0.215)(0.235^2)$$
$$= (0.215)(0.0552)$$
$$= 0.0119 \text{ in.}^2$$

To find the area of a fillet that has a chord of 3.45 cm, the following method would be employed:

$$A = 0.1075c^2$$
$$= (0.1075)(3.45^2)$$
$$= (0.1075)(11.9025)$$
$$= 1.280 \text{ cm}^2$$

ELLIPSE

The formulas used to calculate specifications for ellipses are *approximations* rather than exact values. As shown in Figure 2-16, the formulas for perimeter (P) or circumference are based on the two axes (major axis and minor axis) of the figure:

$$A = \pi ab$$
$$P = \pi\sqrt{2(a^2+b^2)} \text{ or}$$
$$P = \pi\sqrt{2(a^2+b^2)-(a-b)^2/2.2}$$

It should be noted that the second perimeter formula is recommended where greater accuracy is needed.

To find the area and perimeter of an ellipse with a major di-

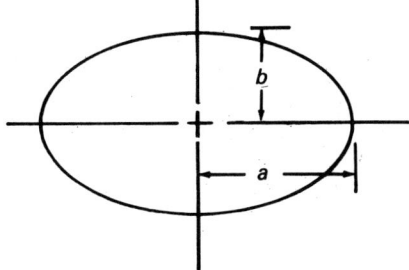

Fig. 2-16. Ellipse.

ameter of 20 in. and a minor diameter of 15 in., the following calculations would be made:

$$A = \pi ab$$
$$= (3.14159)(10)(7.5)$$
$$= 235.62 \text{ sq. in.}$$

$$P = \pi\sqrt{2(a^2 + b^2)}$$
$$= 3.14159\sqrt{2(10^2 + 7.5^2)}$$
$$= 3.14159\sqrt{312.5}$$
$$= (3.14159)(17.678)$$
$$= 55.537 \text{ in.}$$

or for greater accuracy:

$$P = \pi\sqrt{2(a^2 + b^2) - (a - b)^2/2.2}$$
$$= 3.14159\sqrt{2(10^2 + 7.5^2) - (10 - 7.5)^2/2.2}$$
$$= 3.14159\sqrt{312.5 - 2.841}$$
$$= (3.14159)(17.597)$$
$$= 55.283 \text{ in.}$$

PARABOLA

A parabola is a curve formed by a plane that cuts through the side of a cone (Figure 2-17). The formula for calculating the area of a portion of a parabola is:

$$A = \frac{4xy}{3}$$

AREA AND VOLUME

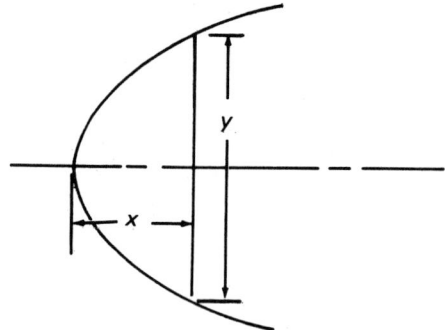

Fig. 2-17.   Parabola.

For example, in a parobola where the x dimension is 84 mm and the y dimension is 54 mm, the area of that portion of the parabola would be calculated in the following manner:

$$A = \frac{4xy}{3}$$
$$= \frac{(4)(84)(54)}{3}$$
$$= 6048 \text{ mm}^2$$

Exercises

Calculate the area for the following squares:

1. What is the area of a square whose sides are 5.75 m? What is its diagonal measurement?

2. What is the area of a square whose diagonal is 14 in.? What is the length of its sides?

3. Calculate the area of a rectangle whose sides are 3.5 and 7.25 in. What is this rectangle's diagonal measurement?

4. If the area of a rectangle is 126 cm² and one of its sides is 8 cm, what is the length of its other side and diagonal?

5. Calculate the area of a parallelogram whose base is 24 ft and height is 8 ft.

6. What is the area of a right triangle whose sides are equal to 8 and 14 in? What is the length of its third side (hypotenuse)?

7. Calculate the area for the two acute triangles shown in Figure 2-18.

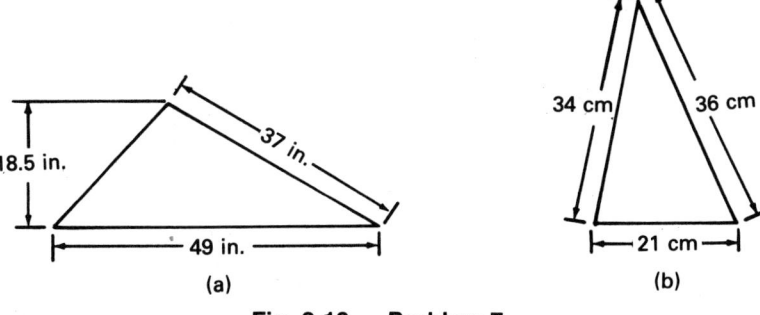

Fig. 2-18. Problem 7.

8. Calculate the area of the two obtuse triangles shown in Figure 2-19.

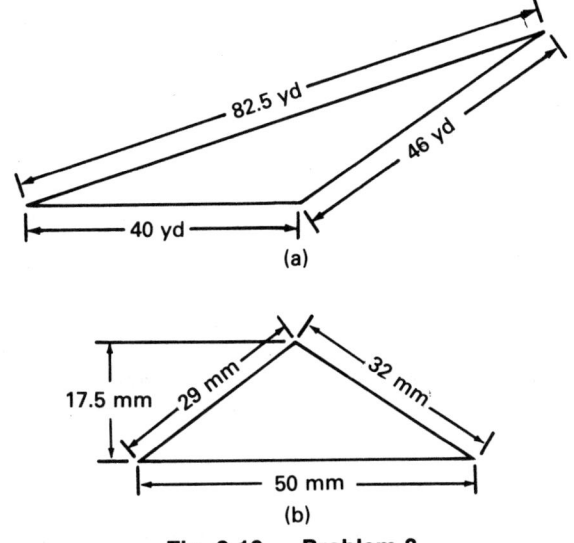

Fig. 2-19. Problem 8.

9. What is the area of a trapezoid whose two parallel sides are 8 dm apart and have lengths of 12.75 dm and 24 dm?

36 • AREA AND VOLUME

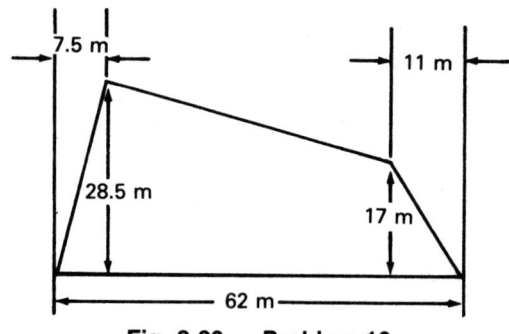

Fig. 2-20. Problem 10.

10. What is the area for the trapezium illustrated in Figure 2-20?

11. What are the areas of three regular hexagons that have circumscribed diameters of 18, 64, and 124 in.? What is the length of the sides of each of the hexagons?

12. Calculate the area of a regular octagon that is inscribed in a circle with a diameter of 24.5 m. Find the length of its sides.

13. What is the area of a 17-sided regular polygon whose sides are 3.75 cm long and that is circumscribed about a circle with a radius of 16 cm?

14. What is the area and circumference of three circles with diameters of 6, 8, and 11.64 in.?

15. If the area of a circle is 124 m^2, what is the length of its diameter?

16. What is the area of a circular sector whose interior angle is 42° and radius is 16 m? What is the length of its arc?

17. Calculate the interior angle of a circular sector that has a radius of 24.64 in. and whose arc has a length of 100 in.

18. What is the interior angle of a circular segment whose radius is 18 ft and arc length is 12 ft?

19. Find the area of a circular segment with a radius of 2.4 in., an arc length of 4 in., a chord length of 6.5 in., and a height of 3. in.

Geometric Formulas • 37

20. Find the area of a fillet that has a radius of 0.128 in. What is the area of a spandrel that has a chord length of 3.5 ft?
21. Calculate the areas of two ellipses with major and minor axes of 12 and 8.5 in. and 87 and 64 cm, respectively.
22. Find the areas of the shaded areas in the two parabolas shown in Figure 2-21.

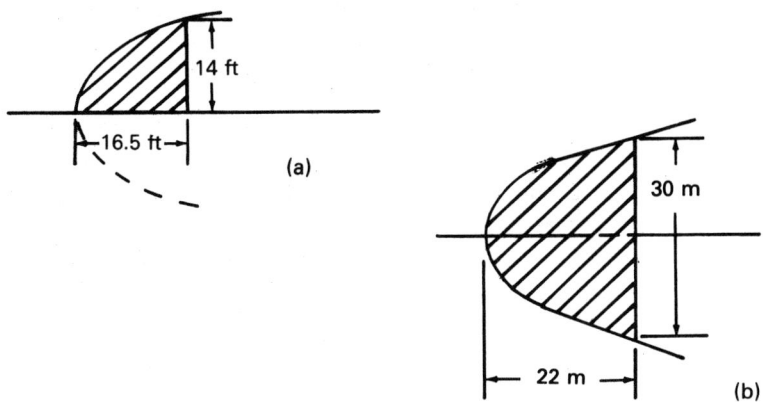

Fig. 2-21.   Problem 22.

Volumes of Solid Figures

Often volumes must be calculated for determining product costs, weights, treatment, and other related factors. This section presents the formulas for figuring the volumes (V) of common solid figures. Again, where possible the notations used in these formulas will correspond to related illustrations.

CUBE

The volume of a cube is calculated by cubing its side measurement (Figure 2-22). The two basic formulas used here are:

$$V = s^3$$
$$s = \sqrt[3]{V}$$

AREA AND VOLUME

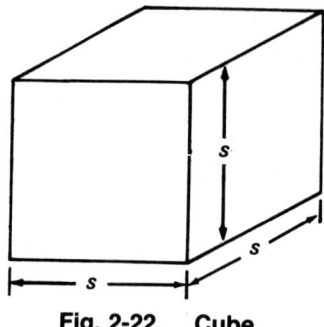

Fig. 2-22. Cube.

To find the volume of a cube whose side is 21.45 cm, the following procedure is used:

$$V = s^3$$
$$= (21.45)^3$$
$$= 9869.199 \text{ cm}^3$$

SQUARE PRISM

To calculate the volume of a square prism, all three side measurements must be used—length, height, and width (Figure 2-23). Volume and related formulas for square prisms are:

$$V = abc$$
$$a = \frac{V}{bc}$$
$$b = \frac{V}{ac}$$
$$c = \frac{V}{ab}$$

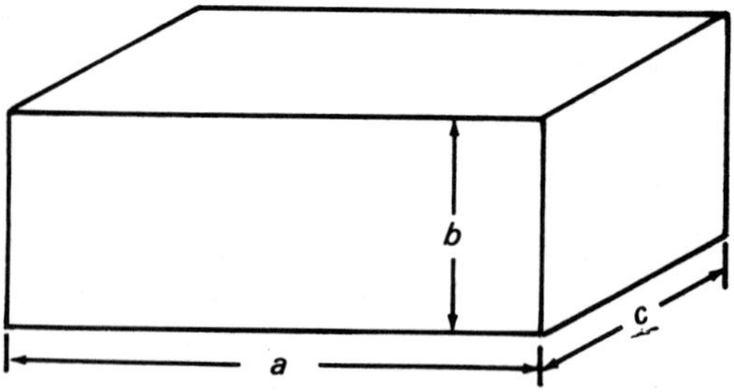

Fig. 2-23. Square prism.

To find the volume of a square prism whose sides are 16, 8, and 4 in., the following method is employed:

$$V = abc$$
$$= (16)(8)(4)$$
$$= 512 \text{ in.}^3$$

PRISM

The volume of a prism (Figure 2-24) requires that the area of the ends be calculated first. Using our illustration as an example, the pentagon's area must be found before the prism's volume can be calculated. Once this is determined, the following formula used:

$$V = hA$$

Note that the height (h) measure is taken *perpendicular* to the end surfaces.

To calculate the volume of a prism that has a regular hexagon base with a side of 8 cm and is 32 cm high, the following procedure would be used:

Step 1: Area of hexagon $= 2.598s^2$
$$= (2.598)(8^2)$$
$$= 166.272 \text{ cm}^2$$

Step 2: $V = hA$
$$= (32)(166.272)$$
$$= 5320.704 \text{ cm}^3$$

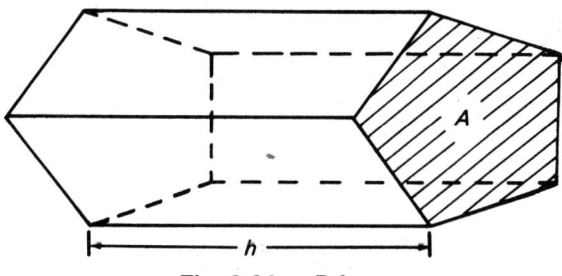

Fig. 2-24. Prism.

PYRAMID

As in calculating the volume of a prism, the area of a pyramid's base must be known before the volume can be calculated. (Figure 2-25). The first volume formula to be given is for a pyramid with a polygon base; the second is for a pyramid with a regular polygon base where the base's area is incorporated into the calculation itself:

$$V = \frac{hA}{3}$$

$$V = \frac{nsrh}{6}$$

$$V = \frac{nsh}{6}\sqrt{R^2 - s^2/4}$$

In the second and third formulas, n = the number of sides that are found in the regular polygon, s = the length of the sides, r = the radius of the inscribed circle, and R = the radius of the circumscribed circle.

An example of a pyramid problem would be to find the volume of a pyramid that has a height of 11 ft and a parallelogram base with sides of 4 and 6 ft, respectively:

$$V = \frac{hA}{3} = \frac{(11)(4)(6)}{3}$$

$$= \frac{264}{3} = 88 \text{ ft.}^3$$

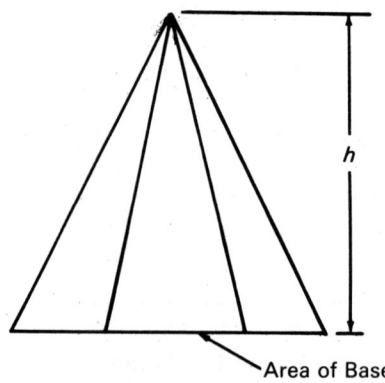

Fig. 2-25.   Pyramid.

FRUSTUM OF A PYRAMID

The frustum is that part of the pyramid that remains after the top has been cut off parallel to the base. To solve for its volume, two areas must be accounted for, the top surface and the base. These are noted as A_1 and A_2 in Figure 2-26. The formula used for this problem is:

$$V = \frac{h(A_1 + A_2 + \sqrt{A_1 A_2})}{3}$$

For example, find the volume of a pyramid that has a regular hexagon base with an area of 415.68 cm, a top surface area of 103.92 cm, and a height of 34 cm:

$$\begin{aligned}
V &= \frac{h(A_1 + A_2 + \sqrt{A_1 A_2}}{3} \\
&= \frac{34[103.92 + 415.68 + \sqrt{(103.92)(415.68)}\,]}{3} \\
&= \frac{34(519.6 + \sqrt{43197.466})}{3} \\
&= \frac{34(519.6 + 207.84)}{3} \\
&= \frac{24732.96}{3} \\
&= 8244.32 \text{ cm}^3
\end{aligned}$$

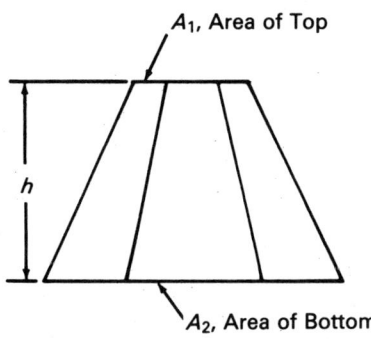

Fig. 2-26.   Frustum of pyramid.

42 • AREA AND VOLUME

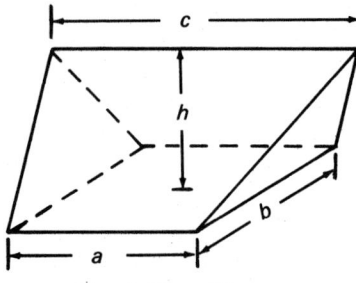

Fig. 2-27. Wedge.

WEDGE

A common solid that is frequently overlooked in volume problems is the wedge (Figure 2-27), yet it is frequently encountered in the workplace. Volume measures here are primarily used for weight and materials needed for production. The basic formula used for calculating the volume of a wedge is:

$$V = \frac{(2a+c)bh}{6}$$

A problem using this formula would be to find the volume of a wedge where $a = 14$ in., $b = 10$ in., $c = 20$ in., and the height is 16 in. The volume would be found by:

$$\begin{aligned} V &= \frac{(2a+c)bh}{6} \\ &= \frac{[(2)(14)+20](10)(16)}{6} \\ &= \frac{7680}{6} \\ &= 1280 \text{ in}^3 \end{aligned}$$

CYLINDER

Two basic formulas are used when working with cylinders (Figure 2-28). The first is used to find the cylinder's volume (V), and the second is used to find the surface area (S) of the cylinder itself:

$$V = \pi r^2 h \text{ or}$$
$$V = \frac{\pi d^2 h}{4}$$
$$S = 2\pi r h \text{ or}$$
$$S = \pi d h$$

To find the total surface area of the cylinder plus its end surfaces, the following formula is used:

$$A = 2\pi r(r+h) \text{ or}$$
$$A = \pi d\left(\frac{d}{2} + h\right)$$

To find the volume and the area of the cylinder surface when the diameter of the cylinder is 25.4 cm and its length (or height) is 267 cm, the following calculations would be made:

$$V = \frac{\pi d^2 h}{4}$$
$$= \frac{(3.14159)(25.4^2)(267)}{4}$$
$$= \frac{541,163.13}{4}$$
$$= 135,290.78 \text{ cm}^3$$

$$S = \pi d h$$
$$= (3.14159)(25.4)(267)$$
$$= 21,305.64 \text{ cm}^2$$

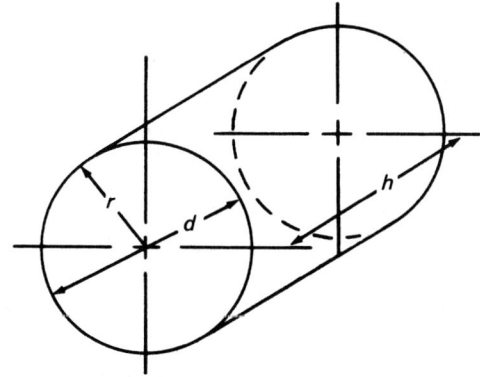

Fig. 2-28. Cylinder.

CONE

As for cylinders, the two basic formulas used with cones are for finding the volume and the area of the conical surface (A) (see Figure 2-29).

$$V = \frac{r^2 h}{3} \text{ or}$$

$$V = \frac{\pi d^2 h}{12}$$

$$A = \pi r \sqrt{r^2 + h^2} \text{ or}$$
$$A = \pi r s \text{ or}$$
$$A = \frac{\pi d s}{2}$$

$$S = \sqrt{r^2 + h^2} \text{ or}$$
$$S = \sqrt{(d^2/4) + h^2}$$

To find the volume and conical-surface area of a cone with a circle base diameter of 3 in. and a height of 8 in., the following procedures would be used:

$$V = \frac{\pi d^2 h}{12}$$
$$= \frac{(3.14159)(3^2)(8)}{12}$$
$$= \frac{226.19448}{12}$$
$$= 18.850 \text{ in.}^3$$

$$A = \pi r \sqrt{r^2 + h^2}$$
$$= 3.14159 \sqrt{1.5^2 + 8^2}$$
$$= 3.14159 \sqrt{66.25}$$
$$= (3.14159)(8.13941)$$
$$= 25.571 \text{ in.}^2$$

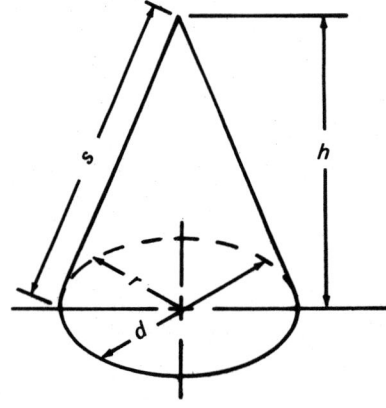

Fig. 2-29. Cone.

FRUSTUM OF A CONE

The volume calculations for the frustum of a cone require that the base and top surface diameter be known (Figure 2-30). Again, both volume and conical-surface area formulas are available:

$$V = \frac{\pi h(R^2 + Rr + r^2)}{3} \text{ or}$$

$$V = \frac{\pi h(D^2 + Dd + d^2)}{12}$$

$$A = \pi s(R + r) \text{ or}$$
$$A = \frac{\pi s(D + d)}{2}$$

$$a = R - r$$
$$s = \sqrt{a^2 + h^2} \text{ or}$$
$$s = \sqrt{(R-r)^2 + h^2}$$

To find the volume of a frustum of a cone where the base diameter is 18 m, the smaller diameter is 8 m, and the frustum's height is 9.75 m, the following calculations are made:

$$V = \frac{\pi h(D^2 + Dd + d^2)}{12}$$
$$= \frac{[(3.14159)(9.75)][(18^2 + (18)(8) + 8^2)]}{12}$$
$$= \frac{(30.6305)(532)}{12}$$
$$= 1357.952 \text{ m}^3$$

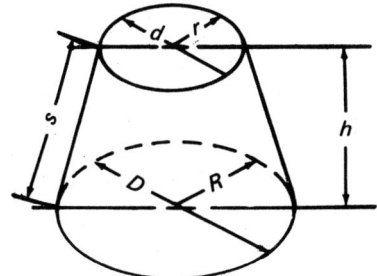

Fig. 2-30.   Frustum of a cone.

46 • AREA AND VOLUME

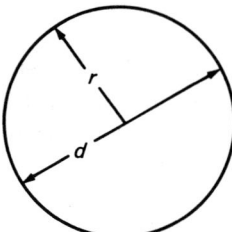

Fig. 2-31. Sphere.

SPHERE

The two basic formulas most often called upon when working with spheres are for volume and area (see Figure 2-31). Here, areas pertain to the surface area of the sphere:

$$V = \frac{4\pi r^3}{3} \text{ or}$$

$$V = \frac{\pi d^3}{6}$$

$$A = 4\pi r^2 \text{ or}$$
$$A = \pi d^2$$

$$r = \sqrt[3]{3V/4\pi} \text{ or}$$
$$r = 0.6204 \sqrt[3]{V}$$

To find the volume and surface area of a sphere that has a diameter of 16.75 in., the following method would be used:

$$V = \frac{\pi d^3}{6}$$
$$= \frac{(3.14159)(16.75^3)}{6}$$
$$= \frac{14,763.657}{6}$$
$$= 2460.61 \text{ in.}^3$$

$$A = \pi d^2$$
$$= (3.14159)(16.75^2)$$
$$= 881.41 \text{ in.}^2$$

SPHERICAL SECTOR

A spherical sector is that part of a sphere that is defined by its radii and the spherical arc included within them. In appearance (Figure 2-32) the spherical sector resembles an inverted cone with a domed top. Formulas used with this solid figure are:

$$V = \frac{2\pi r^2 h}{3}$$

$$A = \pi r \left(2h + \frac{c}{2}\right)$$

$$c = 2\sqrt{h(2r - h)}$$

To find the volume, chord, and surface area for a spherical sector that has a radius of 4.5 in. and a sector height of 2.2 in., the following procedures would be used:

$$V = \frac{2\pi r^2 h}{3}$$
$$= \frac{(2)(3.14159)(4.5^2)(2.2)}{3}$$
$$= \frac{279.916}{3}$$
$$= 93.305 \text{ in.}^3$$

$$c = 2\sqrt{h(2r - h)}$$
$$= 2\sqrt{2.2[(2)(4.5) - 2.2]}$$
$$= 2\sqrt{14.96}$$
$$= 7.736 \text{ in.}$$

$$A = \pi r\left(2h + \frac{c}{2}\right)$$
$$= [(3.1459)(4.5)]\left[(2)(2.2) + \frac{7.736}{2}\right]$$
$$= (14.137)(8.268)$$
$$= 116.885 \text{ in.}^2$$

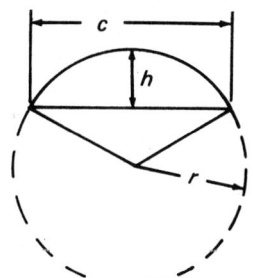

Fig. 2-32. Spherical sector.

48 • AREA AND VOLUME

SPHERICAL SEGMENT

As shown in Figure 2-33, a spherical segment is that portion of a sphere that has been cut away by a plane. The segment will have a circular-flat base with a domed top. The basic formulas used here are:

$$V = \pi h^2 \left(r - \frac{h}{3} \right) \text{ or}$$

$$V = \pi h \left(\frac{c^2}{8} + \frac{h^2}{6} \right)$$

$$A = 2\pi r h \text{ or}$$

$$A = \pi \left(\frac{c^2}{4} + h^2 \right)$$

$$c = 2\sqrt{h(2r-h)}$$

$$r = \frac{c^2 + 4h^2}{8h}$$

To find the volume, radius, and surface area of a spherical segment that has a height of 40 cm and a chord length of 125 cm, the following procedures would be used:

$$V = \pi h \left(\frac{c^2}{8} + \frac{h^2}{6} \right)$$
$$= (3.14159)(40) \left(\frac{125^2}{8} + \frac{40^2}{6} \right)$$
$$= (125.664)(2219.792)$$
$$= 278,974.942 \text{ cm}^3$$

$$r = \frac{c^2 + 4h^2}{8h}$$
$$= \frac{125^2 + (4)(40^2)}{(8)(40)}$$
$$= \frac{15,625 + 6,400}{320}$$
$$= 68.828 \text{ cm}$$

$$A = 2\pi r h$$
$$= (2)(3.14159)(68.828)(40)$$
$$= 17,298.349 \text{ cm}^2$$

Geometric Formulas • 49

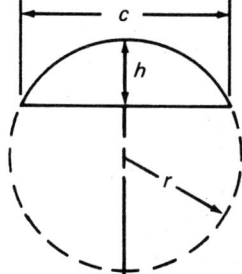

Fig. 2-33. Spherical segment.

SPHERICAL ZONE

Similar to a spherical segment, the spherical zone is made from a sphere that has been cut by two parallel planes (Figure 2-34). The top and bottom surfaces of this solid will appear as true circles. The basic formulas for spherical zone calculations are:

$$V = \left(\frac{\pi}{6}\right)(h)\left(\frac{3c_1^2}{4} + \frac{3c_2^2}{4} + h^2\right)$$

$$A = 2\pi r h$$

$$r = \sqrt{\frac{c_2^2}{4} + [(c_2^2 - c_1^2 - 4h^2)/8h]^2}$$

If the height of a spherical zone is 1.5 m, with the top surface diameter equal to 3 m and the bottom surface diameter 4 m, the spherical zone's volume would be calculated in the following manner:

$$V = \left(\frac{\pi}{6}\right)(h)\left(\frac{3c_1^2}{4} + \frac{3c_2^2}{4} + h^2\right)$$
$$= \left(\frac{3.14159}{6}\right)(1.5)\left[\frac{3(3^2)}{4} + \frac{3(4^2)}{4} + 1.5^2\right]$$
$$= (0.52360)(1.5)(6.75 + 12 + 2.25)$$
$$- 16.494 \text{ m}^3$$

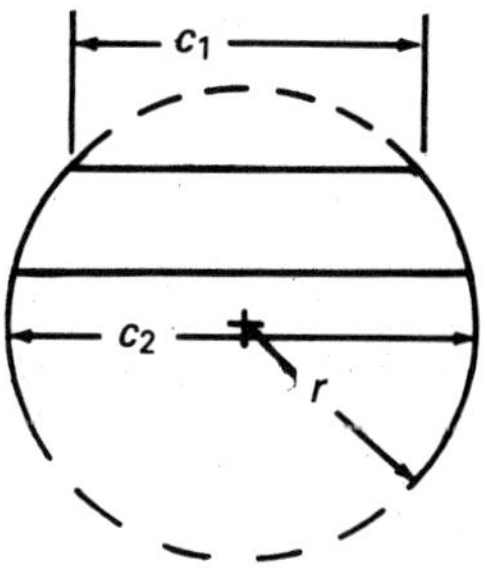

Fig. 2-34. Spherical zone.

SPHERICAL WEDGE

Both volume and area formulas used for spherical wedges (Figure 2-35) employ the central angle of the shape. The two formulas used here are:

$$V = \left(\frac{\alpha}{360}\right)\left(\frac{4\pi r^3}{3}\right)$$

$$A = \left(\frac{\alpha}{360}\right)(4\pi r^2)$$

To calculate the volume and area for a spherical wedge that has a diameter of 120 mm and a central angle of 38°, the following procedures would be used:

$$V = \left(\frac{\alpha}{360}\right)\left(\frac{4\pi r^3}{3}\right)$$
$$= \left(\frac{38}{360}\right)\left[\frac{(4)(3.14159)(60^3)}{3}\right]$$
$$= (0.10556)(904,777.92)$$
$$= 95,508.357 \text{ mm}^3$$

$$A = \left(\frac{\alpha}{360}\right)(4\pi r^2)$$
$$= \left[\frac{38}{360}\right][(4)(3.14159)(60^2)]$$
$$= (0.10556)(45,238.896)$$
$$= 4,775.418 \text{ mm}^2$$

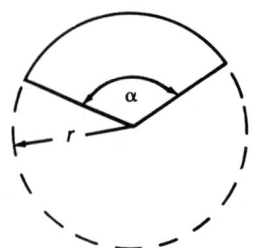

Fig. 2-35. Spherical wedge.

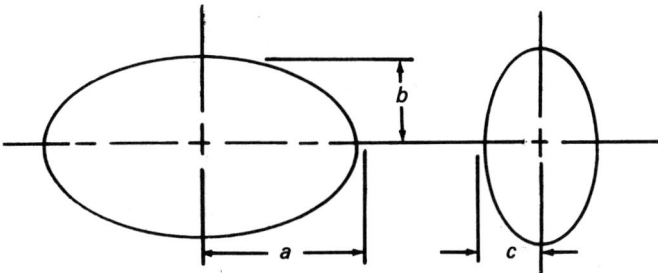

Fig. 2-36. Ellipsoid.

ELLIPSOID

The three-dimensional version of an ellipse is the ellipsoid (Figure 2-36). A unique feature of this solid is that all its sections are either ellipses or circles. As in ellipse calculations, ellipsoid formulas employ major and minor axis dimensions. The formula commonly used here is:

$$V = \left(\frac{4\pi}{3}\right)(abc)$$

As an example, find the total volume for an ellipsoid where $a = 4$ in., $b = 3.5$ in., and $c = 3$ in.:

$$\begin{aligned} V &= \left(\frac{4\pi}{3}\right)(abc) \\ &= \frac{(4)(3.14159)}{3}(4)(3.5)(3) \\ &= (4.1888)(42) \\ &= 175.930 \text{ in.}^3 \end{aligned}$$

PARABOLOID

A paraboloid is a three-dimensional solid that is formed by a parabola revolving about its axis (Figure 2-37). Both volume and surface-area formulas are used with this figure. These, plus a related formula, are presented here (note that p is a mathematical relationship that is used in the area formula):

AREA AND VOLUME

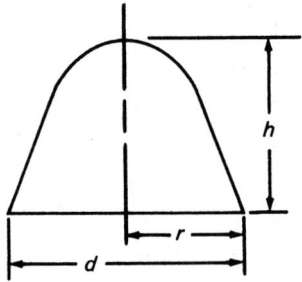

Fig. 2-37. Paraboloid.

$$A = \frac{2\pi}{3p}\sqrt{(d^2/4+p^2)^3 - p^3}$$

$$V = \frac{\pi r^2 h}{2}$$

$$p = \frac{d^2}{8h}$$

The volume of a paraboloid with a height of 30 ft and a diameter of 12.5 ft is calculated in the following manner:

$$V = \frac{\pi r^2 h}{2}$$
$$= \frac{(3.14159)(6.25^2)(30)}{2}$$
$$= \frac{3,681.551}{2}$$
$$= 1840.776 \text{ ft}^3$$

PARABOLOID SEGMENT

When a paraboloid is cut by a plane that is parallel to its base, a paraboloid segment is formed (see Figure 2-38). The formulas used to calculate the volume for this solid are:

$$V = \frac{h\pi(R^2 + r^2)}{2} \text{ or}$$
$$V = \frac{h\pi(D^2 + d^2)}{8}$$

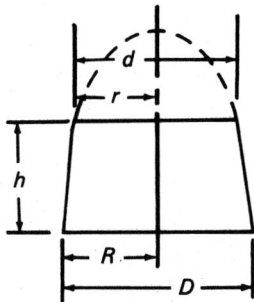

Fig. 2-38. Paraboloidal segment.

The volume for a paraboloid segment whose height is 17 cm and larger and smaller diameters are 15 and 8.2 cm, respectively, is calculated in the following manner:

$$V = \frac{h\pi(D^2 + d^2)}{8}$$
$$= \frac{(17)(3.14159)(15^2 + 8.2^2)}{8}$$
$$= (6.6759)(292.24)$$
$$= 1950.965 \text{ cm}^3$$

TORUS

In the field of geometry, a torus (also known as a tore) is a solid that is generated by the revolution of a conic section about an axis in a plane other than its diameter. The resulting shape, then, resembles a donut (Figure 2-39). Both area and volume formulas are frequently used for this solid:

$$V = 2\pi^2 R r^2 \text{ or}$$
$$V = \frac{D d^2 \pi^2}{4}$$

$$A = 4\pi^2 R r \text{ or}$$
$$A = \pi^2 D d$$

54 • AREA AND VOLUME

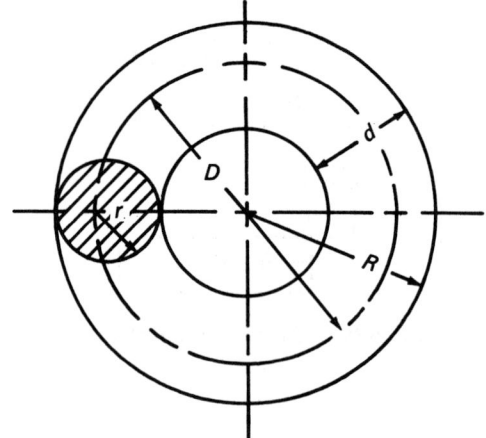

Fig. 2-39.   Torus.

An example of a problem employing these formulas would be for a torus in which $d = 2$ in. and $D = 12$ in.:

$$V = \frac{Dd^2\pi^2}{4}$$
$$= \frac{(12)(2^2)(3.14159^2)}{4}$$
$$= \frac{473.7402}{4}$$
$$= 118.445 \text{ in.}^3$$

$$A = \pi^2 Dd$$
$$= (3.14159^2)(12)(2)$$
$$= 236.870 \text{ in.}^2$$

Exercises

23. Compute the volume of three cubes that have sides of 3, 24, and 18.74 in., respectively.

24. What is the length of the sides of a cube that has a total volume of 2744 ft^3?

Geometric Formulas • 55

25. Calculate the volume of a square prism that has a length of 21 cm, a width of 8.2 cm, and a height of 14.75 cm.
26. Calculate the volume of an octagonal prism where one of the octagon's sides is equal to 80 mm and the total height of the solid is 240 mm.
27. Find the volume of a regular polygon pyramid with a base having seven sides 0.98 in. long, an inscribed circle radius of 1.25 in., and a height of 4.35 in.
28. What is the volume for a regular polygon pyramid that has a base area of 18.754 ft^2 and a height of 3.2 ft?
29. Find the volume of a frustum of a square pyramid that has a height of 12 cm, a base with all sides equal to 4 cm, and a top side with all sides equal to 3.45 cm.
30. Find the volume of the wedge shape illustrated in Figure 2-40. If this wedge were made out of an alloy that weighed 240 grams per cubic millimeter (g/mm^3), what would be the total weight of this wedge?

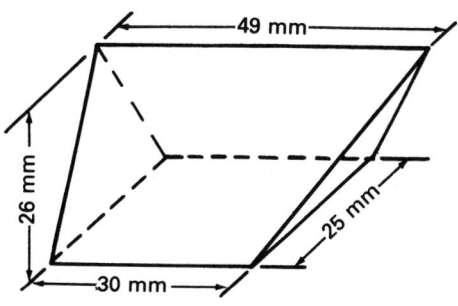

Fig. 2-40.   Problem 30.

31. How much resin could a cylindrical container hold if its diameter was 24 in. and it had a height of 48 in.? To construct this cylinder, how much sheet material would be needed?
32. What is the area for a cone that has a base diameter of 1.12 m and a height of 2 m? What would be the length of its side?
33. Calculate the total volume of natural gas that can be held by the container illustrated in Figure 2-41.

56 • AREA AND VOLUME

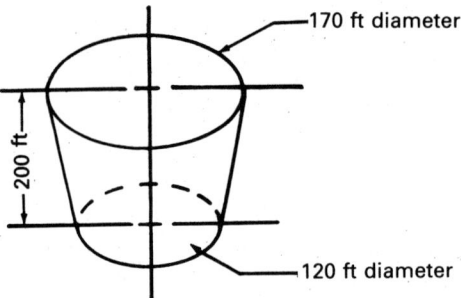

Fig. 2-41. Problem 33.

34. What is the total surface area of a globe that has a diameter of 18 in.? What is its total volume?

35. Calculate the total volume of the storage container illustrated in Figure 2-42.

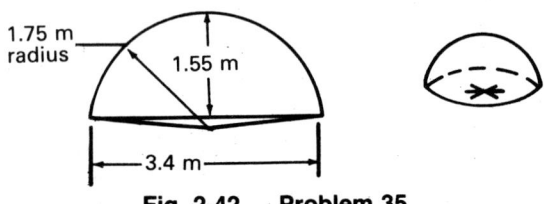

Fig. 2-42. Problem 35.

36. How much sheet material would be required to make the bowl shown in Figure 2-43? What is its maximum holding capacity?

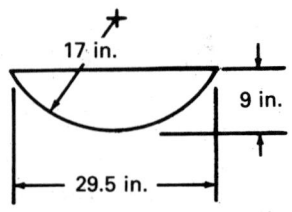

Fig. 2-43. Problem 36.

37. Find the total volume of a spherical zone having a top and bottom cross-sectional length of 14 and 23 cm, respectively, and a height of 6.4 cm. What would be the diameter of the sphere from which this solid was cut?

Geometric Formulas • 57

38. Calculate the total volume for three spherical wedges with central angles of 23°, 45°, and 68° and radii of 1.2 ft, 24 m, and 34 yd, respectively.

39. Find the total capacity of the ellipsoid tank illustrated in Figure 2-44.

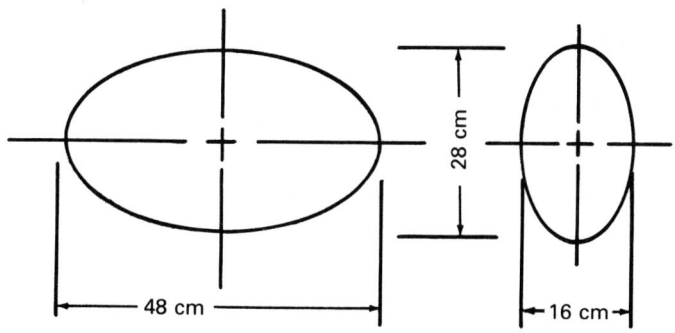

Fig. 2-44.   Problem 39.

40. Find the total surface area for a paraboloid container that has a height of 2.3 ft and base diameter of 1.75 ft. What is its total surface area?

41. Calculate the total surface area and volume of the torus-shaped unit illustrated in Figure 2-45.

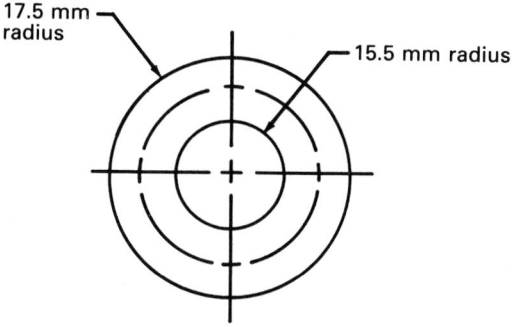

Fig. 2-45.   Problem 41.

CHAPTER 3

Triangulation: Right and Oblique Planes

- Right Angle Trigonometry
- Oblique Angle Trigonometry
- Other Trigonometric Formulas

Problems involving right and oblique triangular planes are found in a wide variety of technologies. Perhaps the field most closely allied to triangulation problems is that of surveying. Surveying principles are not only used for the plotting of landmasses and mapping but also for the exact placement of machinery and measuring equipment. In fact, the installation of many laser measuring units can be accomplished only by the use of surveying principles.

Triangulation formulas are based on the use of trigonometry. The term *trigonometry* is derived from the two Greek words, *trigonon*, meaning "three-angle" or "triangle," and *metria*, which literally means "measurement." Thus, trigonometry deals with ratios and measurements associated with triangles.

The mathematics of trigonometry is generally divided into two major areas: plane and spherical. *Plane trigonometry* deals with calculations on flat "plane" surfaces, while *spherical trigonometry* deals with ratios and measurements of triangles on curved or spherical planes. Since plane trigonometry is what is most frequently utilized in most areas of technology (except for some surveying and engineering situations), spherical trigonometry will not be covered in this book.

Plane trigonometry is usually subdivided into two subcategories: *right angle trigonometry* and *oblique angle trigonometry*.

Right Angle Trigonometry

Trigonometry is based on certain types of ratios known as *trigonometric functions*. The term *function* is used to describe relationships such as: a circle is a function of its radius, the surface area of a sphere is a function of its diameter, and the volume of a cube is a function of its sides. Since the triangle is the basis for trigonometric measures, we employ the functions of its angles as the basis for all calculations.

Trigonometric Functions

In trigonometry, six functions are given in relationship to a right triangle (Figure 3-1). When drawing a 90° triangle, it is common practice to label the right angle of this triangle C and its opposite side, known as the *hypotenuse*, as c. The remaining angles are then labeled A and B, and their opposite sides, a and b. The six basic trigonometric functions are:

sine, abbreviated as sin
cosine, abbreviated as cos
tangent, abbreviated as tan
cotangent, abbreviated as cot
secant, abbreviated as sec
cosecant, abbreviated as csc

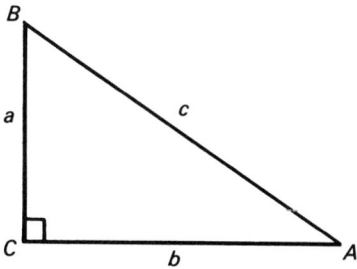

Fig. 3-1. Labeling of a right triangle.

The six functions are ratios that involve two of the three sides of the triangle ABC and are given to describe a given angle. An example of how these ratios are expressed for a right triangle's angles are:

$$\sin A = \text{opposite side} : \text{hypotenuse} = \frac{a}{c}$$

$$\sin B = \text{opposite side} : \text{hypotenuse} = \frac{b}{c}$$

$$\cos A = \text{adjacent side} : \text{hypotenuse} = \frac{b}{c}$$

$$\cos B = \text{adjacent side} : \text{hypotenuse} = \frac{a}{c}$$

$$\tan A = \text{opposite side} : \text{adjacent side} = \frac{a}{b}$$

$$\tan B = \text{opposite side} : \text{adjacent side} = \frac{b}{a}$$

$$\cot A = \text{adjacent side} : \text{opposite side} = \frac{b}{a}$$

$$\cot B = \text{adjacent side} : \text{opposite side} = \frac{a}{b}$$

$$\sec A = \text{hypotenuse} : \text{adjacent side} = \frac{c}{b}$$

$$\sec B = \text{hypotenuse} : \text{adjacent side} = \frac{c}{a}$$

$$\csc A = \text{hypotenuse} : \text{opposite side} = \frac{c}{a}$$

$$\csc B = \text{hypotenuse} : \text{opposite side} = \frac{c}{b}$$

Specific numerical values for all angles are assigned to trigonometric functions and are normally given in the form of a table. These values can also be determined by the use of a scientific/engineering calculator. To find the function of angles that are expressed to the nearest second, the same interpolation procedure is followed as was presented in logarithms.

As an example, suppose we were to find the tangent for the angle 34°42'. Using a trigonometric function table, find 34°42' and look under the tangent column, where the value of the tangent is found to be 0.69243.

As an illustration of the use of interpolation, let's take the tangent of B, which equals 0.3727. If a functions table was only available to the nearest 10 minutes, and we wanted to solve to the nearest minute, then the following procedure would be used:

$$\tan 20°20' = 0.3706$$
$$\tan B = 0.3727$$
$$\tan 20°30' = 0.3739$$

Since the value of B falls between 20°20' and 20°30', we note that the total difference between the 20' and 30' value is 0.0033, which is expressed as 33. In addition, the difference between the tangent of B and the 20' value is 0.0021, or 21. To find angle B, we note that it is 21/33, or 7/11, of the way from 20°20' to 20°30'. 7/11 = 0.636, or 6'; therefore, $B = 20°6'$.

Care must be taken when working with the trigonometric functions table. After careful review of this table, you will note that angle sizes are given at the top and bottom of each column. It is therefore possible to have the same function for two different angle sizes. Note that the value 0.69243 is not only the tangent of 34°42', but also the cotangent for 55°18'. This illustrates that in any right triangle the following values will be equal:

$$\sin A = \cos B$$
$$\cos A = \sin B$$
$$\tan A = \cot B$$
$$\cot A = \tan B$$
$$\sec A = \csc B$$
$$\csc A = \sec B$$

Using Trigonometric Functions

In solving problems that deal with right triangles, it is advantageous to remember that there are other relationships that can be used to check as well as solve problems. As presented in Chapter 2, you will remember that in a right triangle:

62 • TRIANGULATION: RIGHT AND OBLIQUE PLANES

$$c^2 = a^2 + b^2$$
$$a = \sqrt{(c-b)(c+b)}$$
$$b = \sqrt{(c-a)(c+a)}$$

Furthermore, since the sum of all the angles of any triangle will equal 180°, the following relationships can be found:

$$A + B + C = 180°$$
$$A + B = 90°$$

An example of a typical problem encountered in right angle trigonometry is for the triangle illustrated in Figure 3-2. Here, triangle ABC has two known quantities: $B = 37°23'$ and $a = 1.357$ cm. To find the angle size of A and the length of sides b and c, the following procedures are used:

$$\begin{aligned} A &= 90° - B \\ &= 90° - 37°23' \\ &= 52°37' \end{aligned}$$

To find the length of sides b and c, it is necessary to make use of trigonometric functions, since only one side and one angle are known:

$$\tan B = \frac{b}{a}$$
$$\begin{aligned} b &= (a)(\tan B) \\ &= (1.357)(\tan 37°23') \\ &= (1.357)(0.76410) \\ &= 1.0369 \text{ cm} \end{aligned}$$

$$\cos B = \frac{a}{c}$$
$$\begin{aligned} c &= (a)(\cos B) \\ &= (1.357)(\cos 37°23') \\ &= (1.357)(0.79459) \\ &= 1.0783 \text{ cm} \end{aligned}$$

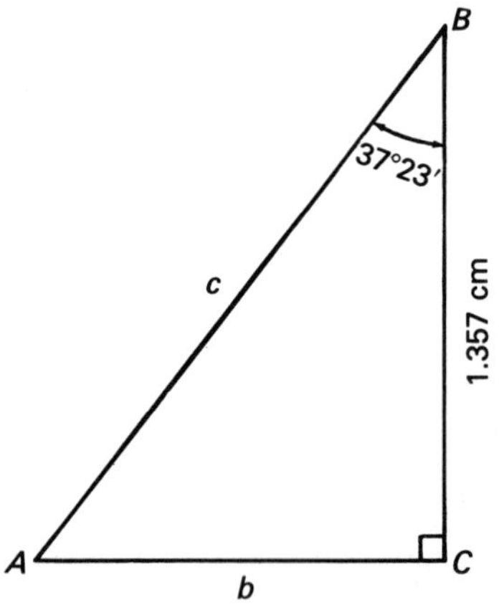

Fig. 3-2. Right triangle problem.

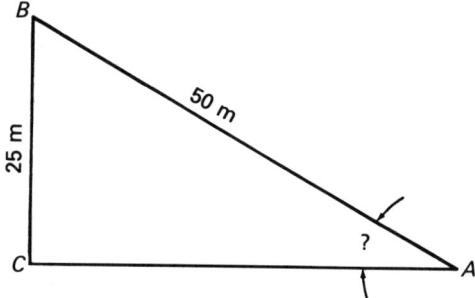

Fig. 3-3. Finding unknown angle.

Another problem would be to find the size of angle A for the triangle shown in Figure 3-3. Here, only two sides are known: opposite and hypotenuse. Because of the givens, it is possible to use either the sine or cosecant function. Examples of how each one would be calculated are:

$$\sin A = \frac{a}{c}$$
$$= \frac{25}{50}$$
$$= 0.5000$$

Therefore:

$$A = 30°$$

$$\csc A = \frac{c}{a}$$
$$= \frac{50}{25}$$
$$= 2.000$$

Therefore:

$$A = 30°$$

Exercises

Solve for the missing values for right triangle ABC in the following problems.

1. Find the length of side a when $b = 6$ in. and $c = 11$ in.

2. What is the length of the hypotenuse for triangle ABC when angle B is equal to 72°14′ and side a is equal to 14 m?

3. What is the size of angle A if side $a = 7$ ft and side $b = 11.5$ ft.?

4. If side $c = 38$ mm and side $a = 14.57$ mm, what size are angles A and B?

5. If side $b = 101$ in. and side $c = 320$ in., find the size of angles A and B.

6. If angle $A = 27°11′$ and side $c = 34$ cm, what is the size of angle B and the lengths of sides a and b?

7. Find angle A and B for a right triangle where $a = 12$ cm and $b = 20.2$ cm. What is the length of side c?

8. Using the tangent function, find side a when $b = 32$ in. and angle $A = 35°$.

9. A surveyor is standing 120 m from the base of a column and notes that the angle of elevation to the top of the column is 50°. If his sighting is made at a height of 5.667 m from the ground, what is the height of the column?

10. Calculate the angle of inclination (rise) for a stairway that has a tread of 8.5 in and a rise of 6.75 in.

11. The major dimensions for a truss are given in Figure 3-4. Calculate the size of ∠CHG and ∠HCG.

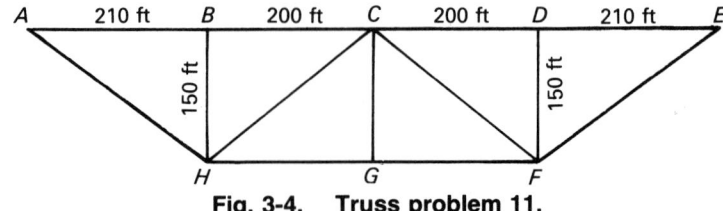

Fig. 3-4.   Truss problem 11.

12. While on a site-surveying job, a surveyor measured a distance of 1.2 mi along a sighting line labeled as AC. If the angle between sighting lines AB and BC was 54°34′, what would be the calculated distance across the body of water shown in Figure 3-5?

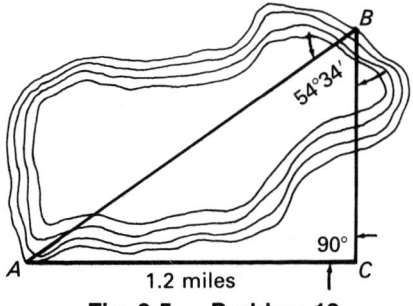

Fig. 3-5. Problem 12.

13. During a surveying job, measures were taken as shown in Figure 3-6. Calculate the size of angle φ.

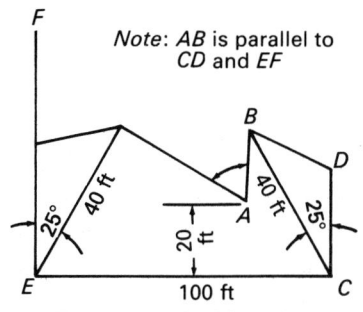

Fig. 3-6. Problem 13.

14. Using Figure 3-7, find the length for diameter D. The dimensions given are in millimeters.

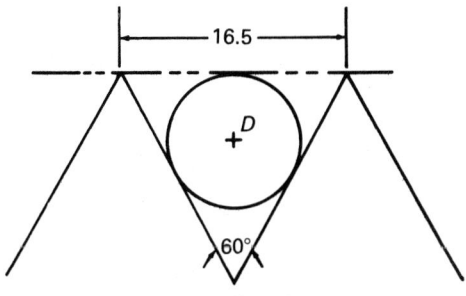

Fig. 3-7. Problem 14.

66 • **TRIANGULATION: RIGHT AND OBLIQUE PLANES**

15. Calculate the size of dimension Y in the drawing illustrated in Figure 3-8.

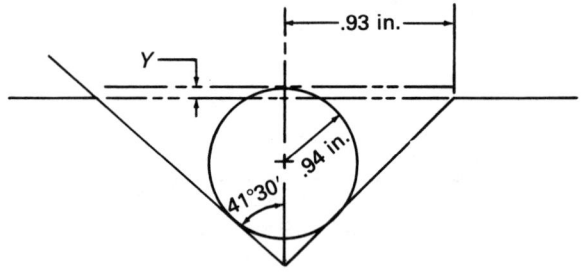

Fig. 3-8.   Problem 15.

Oblique Angle Trigonometry

There are probably more oblique angle problems than right angle problems encountered in everyday situations. Distances between parts, lengths of sides, and angles between datum points must be known to lay out parts and determine processing procedures. To solve oblique triangle problems requires the understanding and use of special mathematical laws and procedures.

In some cases, it is possible to solve oblique triangle problems by constructing a perpendicular. An example of this is illustrated in Figure 3-9. Here, oblique triangle KLM has several knowns: $\angle K = 23°$, side $KL = 144.5$ m, and side $LM = 138.75$ m. To find $\angle L$, $\angle M$, and side MK, the following procedures can be used:

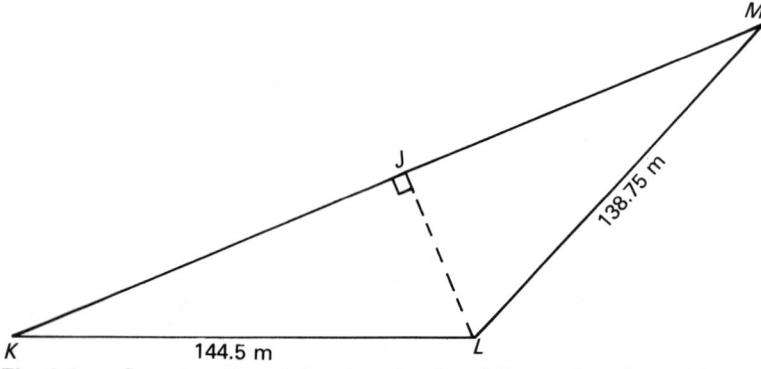

Fig. 3-9.   Constructing right triangles for oblique triangle problems.

1. Altitude *LJ* is drawn from $\angle L$ so that it is perpendicular to side *KM*. To solve the length of *LJ*, the following calculations are made:

$$LJ = (KL)(\sin K)$$
$$= (144.5)(\sin 23°)$$
$$= (144.5)(0.39073)$$
$$= 56.46 \text{ m}$$

2. With side *LJ* known, it is now possible to find $\angle M$ as follows:

$$\sin M = \frac{LJ}{LM}$$
$$= \frac{56.46}{138.75}$$
$$= 0.40692$$

Therefore:

$$\angle M = 24°00'41.5''*$$

3. To solve for the size of $\angle L$, we find the sum of angles *K* and *M* and subtract it from 180°:

$$\angle L = 180° - (\angle K + \angle M)$$
$$= 180° - (23° + 24°00'41.5'')$$
$$= 180° - 47°00'41.5''$$
$$= 132°59'18.5''$$

4. Knowing the length of altitude *LJ*, it is also possible to solve for side *KM*, where $KM = KJ + JM$. *KJ* and *JM* are solved in the following manner:

---

*Since sin *M* was between 24° and 24°1', it was necessary to interpolate the exact value of 0.40692. Since sin 24° = 0.40674, sin *M* = 0.40692, and sin 24°1' = 0.40700, the following calculations were made:

$$\begin{array}{rr} 0.40700 & 0.40692 \\ -0.40674 & -0.40674 \\ \hline -0.00026 & -0.00018 \end{array}$$

$$\frac{.00018}{.00026} = .6923$$
$$(.6923)(60'') = 41.5''$$

68 • TRIANGULATION: RIGHT AND OBLIQUE PLANES

$$MJ = \sqrt{(LM - LJ)(LM + LJ)}$$
$$= \sqrt{(138.75 - 56.45)(138.75 + 56.45)}$$
$$= \sqrt{(82.3)(195.2)}$$
$$= \sqrt{16,064.96}$$
$$= 126.75$$

$$KJ = \sqrt{(KL - LJ)(KL + LJ)}$$
$$= \sqrt{(144.5 - 56.45)(144.5 + 56.45)}$$
$$= \sqrt{(88.05)(200.95)}$$
$$= \sqrt{17,693.647}$$
$$= 133.02$$

$$KM = MJ + KJ$$
$$= 126.75 + 133.02$$
$$= 259.77 \text{ m}$$

As can be seen by this procedure, right triangle trigonometry can be used to solve for oblique triangle problems, though it is often lengthy and somewhat cumbersome. To facilitate this procedure, and streamline the process, a series of laws and formulas based on these laws have been developed to solve for oblique triangle problems.

Trigonometric Laws

LAW OF SINES

The Law of Sines is based on the fact that the longest side of a triangle is always opposite the largest angle, and the shortest side is opposite the smallest angle (Figure 3-10). The Law of Sines states:

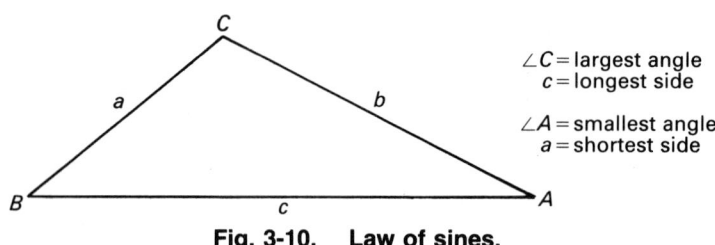

Fig. 3-10. Law of sines.

Oblique Angle Trigonometry • 69

In any triangle, the ratio between the lengths of any two sides will be equal to the ratio between the sines of the angles opposite those sides.

The Law of Sines is normally stated in terms of the following trigonometric proportions:

$$\frac{a}{\sin A} = \frac{b}{\sin B} = \frac{c}{\sin C}$$

When using the Law of Sines to solve for missing values, three of four measures must be known. As an example, consider the oblique triangle illustrated in Figure 3-11. In triangle DEF, side $DE = 14.25$ ft, side $EF = 12.33$ ft, and angle $D = 24°20'$. Using the Law of Sines to solve for the size of angle F and side DF, the procedure is as follows:

1. Determine the Law of Sines proportion that will use a known side and a known opposite angle (i.e., side EF and angle D). To find angle F, the following calculations are made:

$$\frac{\sin F}{DE} = \frac{\sin D}{EF}$$

$$\frac{\sin F}{14.25} = \frac{\sin 24°20'}{12.33}$$

$$\sin F = (14.25)(\sin 24°20')/12.33 = \frac{(14.25)(0.41204)}{12.33} = 0.47620$$

Therefore:

$$\angle F = 28°26'15.5''$$

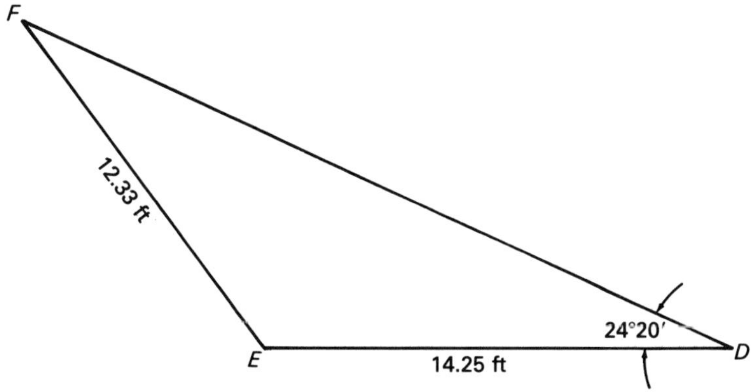

Fig. 3-11. Oblique triangle problem.

2. With the value of angles F and D known, we can now calculate the size of angle E ($\angle D + \angle E + \angle F = 180°$) as being 127°13′44.5″. Solving for side DF, we use the following ratios:

$$\frac{DF}{\sin E} = \frac{EF}{\sin D}$$

$$\frac{DF}{\sin 127°13'44.5''} = \frac{12.33}{\sin 24°20'}$$

$$DF = \frac{(\sin 127°13'44.5'')(12.33)}{\sin 24°20'}$$

$$= \frac{(0.79622)(12.33)}{0.41204}$$

$$= 23.826 \text{ ft}$$

LAW OF COSINES

The Law of Cosines is used to calculate the length of an oblique triangle's leg when the opposite angle and two other sides are known. The relationship that exists under the Law of Cosines may be stated as:

The square of any side of a triangle will be equal to the sum of the squares of the other two sides minus the product of those two sides multiplied by the cosine of the included angle.

The basic formulas used to solve the length of unknown sides are:

$$a^2 = b^2 + c^2 - 2bc \cos A$$
$$b^2 = a^2 + c^2 - 2ac \cos B$$
$$c^2 = a^2 + b^2 - 2ab \cos C$$

When solving for the angles of oblique triangles, these formulas are transposed to:

$$\cos A = \frac{b^2 + c^2 - a^2}{2bc}$$
$$\cos B = \frac{a^2 + c^2 - b^2}{2ac}$$
$$\cos C = \frac{a^2 + b^2 - c^2}{2ab}$$

It should be noted that the Law of Cosines should only be used when smaller numbers are involved and when it is impossible to make uses of less complex formulas. Because of the plus and minus signs included in the Law of Cosines, it is not suitable for logarithmic computations.

For an example of how this law is used, let us take triangle ABC with known values of $A = 34°$, $c = 5$ cm, and $b = 6$ cm. To find side a and angle B, the following procedures would be used:

1. Since no two ratios of the sine formula contain three parts of the triangle, we cannot use the Law of Sines. To find side a, we will implement the following calculations:

$$\begin{aligned} a^2 &= b^2 + c^2 - 2bc \cos A \\ &= 6^2 + 5^2 - 2(6)(5)(\cos 34°) \\ &= 36 + 25 - 49.7424 \\ &= 11.2576 \end{aligned}$$

Therefore:
$$a = 3.355 \text{ cm}$$

2. To complete the problem, we calculate the value of angle B by:

$$\begin{aligned} \cos B &= \frac{c^2 + a^2 - b^2}{2ca} \\ &= \frac{5^2 + 3.355^2 - 6^2}{2(5)(3.355)} \\ &= \frac{30.256}{33.55} \\ &= 0.90316 \end{aligned}$$

Therefore:
$$\angle B = 25°25'24.3''$$

LAW OF TANGENTS

Because of the complexities associated with the Law of Tangents, it is often used with large numbers and in combination with logarithms. Basically, the Law of Tangents states:

In any triangle, the tangent of half the difference of two angles will be proportional to the tangent of half their sum as the difference of the opposite sides is to their sum.

72 • TRIANGULATION: RIGHT AND OBLIQUE PLANES

The three formulas used to describe the Law of Tangents are:

$$\frac{a-b}{a+b} = \frac{\tan 1/2(A-B)}{\tan 1/2(A+B)}$$

$$\frac{a-c}{a+c} = \frac{\tan 1/2(A-C)}{\tan 1/2(A+C)}$$

$$\frac{b-c}{b+c} = \frac{\tan 1/2(B-C)}{\tan 1/2(B+C)}$$

(*Note:* Negative numbers should be avoided. Using the first formula as an example, if b is greater than a, simply interchange A and B and a and b throughout the formula.)

For an example of how this law would be used, we will use oblique triangle ABC with angle $B = 85°$, $a = 3.59$ yd, and $c = 4.85$ yd. To find the size of angles A and C, the following formula would be selected:

$$\frac{c-a}{c+a} = \frac{\tan 1/2(C-A)}{\tan 1/2(C+A)}$$

Then:

$$\frac{(c-a)}{(c+a)} = \frac{4.85 - 3.59}{4.85 + 3.59}$$

$$= \frac{1.26}{8.44}$$

$$C + A = 180° - B$$
$$= 95°$$

Therefore:

$$\frac{(C+A)}{2} = 47°30'$$

Then:

$$\frac{1.26}{8.44} = \frac{\tan 1/2(C-A)}{\tan 47°30'}$$

$$\tan 1/2(C-A) = \frac{(1.26)(\tan 47°30')}{8.44}$$

$$= \frac{(1.26)(1.0913)}{8.44}$$

$$= 0.1629$$

$$= 9°15'8''$$

Therefore:

$$C = 9°15'8'' + 47°30'$$
$$= 56°45'8''$$
$$A = 47°30' - 9°15'8''$$
$$= 38°14'52''$$

Solutions to Problems

Problems that deal with oblique triangles are divided into four categories. These categories are based upon the amount and type of information available. Each problem type may be solved by the Law of Sines, the Law of Cosines, and/or the Law of Tangents. Table 3-1 presents a summary of these four types of oblique triangle problems and recommended solution(s).

Exercises

Using the Law of Sines, solve for the missing angle(s) and side(s) of the following oblique triangles:

16. In triangle ABC: $\angle A = 63°$, $\angle C = 49°$, and $c = 3$ cm.
17. In triangle DEF: $\angle D = 24°$, $f = 8$ m, and $d = 6$ m.
18. In triangle ABC: $\angle C = 25°27'$, $a = 6.15$ ft, and $c = 4.25$ ft.
19. In triangle ABC: $\angle A = 30°$, $a = 2.45$ km, $c = 4.9$ km.

Table 3-1. Oblique Triangle Problems

Problem Category	Solution
1. One side and two angles given	Law of Sines
2. Two sides and an opposite angle are given	Law of Sines—this is a somewhat ambiguous problem
3. Two sides and an included angle are given	Law of Tangents for finding the two unknown angles; the Law of Sines can be used to find the third side. If only the third side is to be determined, then the Law of Cosines is used
4. Three sides are given	Law of Cosines

Using the Law of Cosines, solve for the missing angle(s) and side(s) of the following oblique triangles:

20. In triangle ABC: $\angle A = 67°$, $b = 3$ yd, and $c = 2$ yd.

21. In triangle DEF: $\angle E = 67°$, $e = 3$ m, and $f = 2.5$ m.

22. In triangle ABC: $\angle C = 24°$, $b = 3$ cm, and $c = 2.5$ cm.

23. In triangle FGH: $f = 3$ mi, $g = 4$ mi, and $h = 6$ mi.

24. In triangle XYZ: $\angle X = 37°12'$, $\angle Z = 43°29'$, and $x = 21.5$ in.

25. In triangle LMN: $\angle L = 14°30'$, $l = 8.75$ mm, and $m = 24.25$ mm.

26. In triangle PQR: $p = 4$ dm, $q = 9$ dm, and $r = 7$ dm.

27. In triangle FGH: $\angle F = 29°56'$, $\angle G = 82°13'$, and $h = 5.4$ in.

Using the Law of Tangents, solve for the missing angle(s) and side(s) of the following oblique triangles:

28. In triangle PQR: $r = 47.18$ in., $q = 39.96$ in., and $\angle P = 67°31'40''$.

29. In triangle ABC: $\angle C = 32°18'50''$, $b = 9.137$ cm, and $a = 3.028$ cm.

30. In triangle XYZ: $\angle Y = 123°12'20''$, $x = 0.5366$ km, and $z = 0.7815$ km.

Other Trigonometric Formulas

There are a number of other formulas that have been derived from the principles of right and oblique triangle trigonometry. These formulas are based on ratios that were established in trigonometric functions. The following section is a brief presentation of other trigonometric formulas.

Trigonometric Identities

There are times when trigonometric identities should be understood and known, such as in cases where certain formulas demand that a specific form of trigonometric function be used over another (e.g., sine of an angle instead of some other function) but that particular function is not known. Therefore, to convert a given value into the required form, trigonometric identities need to be employed.

FUNDAMENTAL IDENTITIES

Fundamental trigonometric identities address the six basic functions used in triangle problems. These identities are:

$$\csc A = \frac{1}{\sin A}$$
$$\sec A = \frac{1}{\cos A}$$
$$\tan A = \frac{1}{\cot A}$$
$$\tan A = \frac{\sin A}{\cos A}$$
$$\cot A = \frac{\cos A}{\sin A}$$

$$\sin^2 A + \cos^2 A = 1$$
$$\sec^2 A = \tan^2 A + 1$$
$$\csc^2 A = \cot^2 A + 1$$

COMPLEMENTARY ANGLES

Complementary angles are angles whose sum equals 90°. This type of relationship is expressed as trigonometric identities in the following expressions:

$$\sin A = \cos (90° - A)$$
$$\cos A = \sin (90° - A)$$
$$\tan A = \cot (90° - A)$$
$$\cot A = \tan (90° - A)$$
$$\sec A = \csc (90° - A)$$
$$\csc A = \sec (90° - A)$$

NEGATIVE ANGLES

There are times when negative angles are encountered, especially in navigation, surveying, and electronics. Before these identities are given, it is necessary to review certain facts about negative angles. First, if the moving leg of an angle revolves in a *clockwise* direction, then the angle generated will be a *negative angle*.

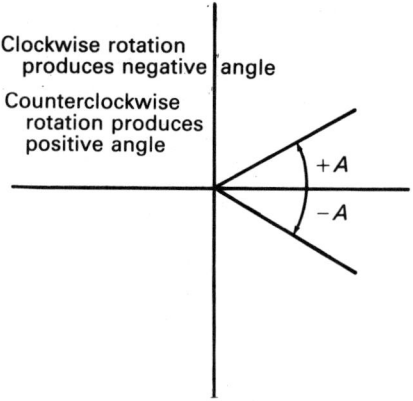

Fig. 3-12.   Negative angle.

Figure 3-12 shows the positive ($+A$) and negative ($-A$) values of two identical angles. The negative angle identities resulting in this type of relationship are given as follows:

$$\sin(-A) = -\sin A$$
$$\cos(-A) = \cos A$$
$$\tan(-A) = -\tan A$$
$$\cot(-A) = -\cot A$$
$$\sec(-A) = \sec A$$
$$\csc(-A) = -\csc A$$

MULTIPLE ANGLES

Many trigonometric problems involve the use of more than one angle for their solution. Furthermore, the basic theories of trigonometry are "angle oriented" and rely heavily on the use of multiple angles. For these reasons, it only seems logical that there are a fairly large number of identities for multiple angles. These are:

$$\sin(A \pm B) = \sin A \cos B \pm \cos A \sin B$$
$$\cos(A \pm B) = \cos A \cos B \pm \sin A \sin B$$
$$\tan(A \pm B) = \frac{\tan A \pm \tan B}{1 \pm \tan A \tan B}$$

$$\sin 2A = 2 \sin A \cos A$$
$$\cos 2A = \cos^2 A - \sin^2 A$$
$$\tan 2A = \frac{2 \tan A}{1 - \tan^2 A}$$

$$\sin x + \sin y = \frac{2 \sin (x+y)}{2 \cos (x-y)/2}$$
$$\sin x - \sin y = \frac{2 \cos (x+y)}{2 \sin (x-y)/2}$$
$$\cos x + \cos y = \frac{2 \cos (x+y)}{2 \cos (x-y)/2}$$
$$\cos x - \cos y = \frac{-2 \sin (x+y)}{2}$$

Half-Angle Formulas

Half-angles are sometimes used in solving surveying and navigational problems. Half-angles in trigonometric triangles will always be acute. The basic identities for half-angles are:

$$\frac{\sin A}{2} = \pm \sqrt{(1 - \cos A)/2}$$
$$\frac{\cos A}{2} = \pm \sqrt{(1 + \cos A)/2}$$
$$\frac{\tan A}{2} = \pm \sqrt{(\sin A)/(1 + \cos A)}$$

When working with half-angle and related formulas, a value known as the *semiperimeter*, or *s*, is used. This is equal to half of the sum total of the legs of a triangle, or $(a+b+c)/2$ (see Figure 3-13). The formulas for the half-angles of triangles when *s* is known are:

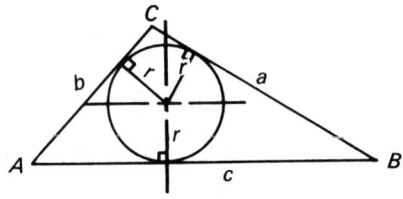

Fig. 3-13. Geometric proof where s=(a+b+c)/2

TRIANGULATION: RIGHT AND OBLIQUE PLANES

$$\frac{\tan A}{2} = \sqrt{(s-b)(s-a)/(s)(s-c)}$$

$$\frac{\tan B}{2} = \sqrt{(s-c)(s-a)/(s)(s-b)}$$

$$\frac{\tan C}{2} = \sqrt{(s-a)(s-b)/(s)(s-c)}$$

$$\frac{\sin A}{2} = \sqrt{(s-b)(s-c)/bc}$$

$$\frac{\sin B}{2} = \sqrt{(s-a)(s-c)/ac}$$

$$\frac{\sin C}{2} = \sqrt{(s-a)(s-b)/ab}$$

$$\frac{\cos A}{2} = \sqrt{(s)(s-a)/bc}$$

$$\frac{\cos B}{2} = \sqrt{(s)(s-b)/ac}$$

$$\frac{\cos C}{2} = \sqrt{(s)(s-c)/ab}$$

In applying half-angle formulas it is possible to find the angles in any order. For example, in solving for angle A in triangle ABC, where $a = 5.8$ m, $b = 4.5$ m, and $c = 6.2$ m, the following procedures are used:

$$s = \frac{a+b+c}{2}$$
$$= \frac{5.8+4.5+6.2}{2}$$
$$= \frac{16.5}{2}$$
$$= 8.25$$

$$\frac{\sin A}{2} = \sqrt{(s-b)(s-c)/bc}$$
$$= \sqrt{(8.25-4.5)(8.25-6.2)/(4.5)(6.2)}$$
$$= \sqrt{7.6875/27.9}$$
$$= \sqrt{0.27554}$$
$$= 0.52492$$

Therefore:

$$\frac{A}{2} = 31°39'46''$$
$$A = 63°19'32''$$

Haversines

Haversines are sometimes encountered in navigational problems and are used in the solution of oblique triangles. By definition, the haversine of an angle is equal to half of the difference of the cosine of that angle from 1. Mathematically, it is presented as:

$$\text{hav } A = \frac{(1 - \cos A)}{2}$$

Related Formulas

Using the semiperimeter value for a triangle, it is also possible to solve for two other related measures. The first is for the area of a triangle, where the area of triangle $ABC = \sqrt{s(s-a)(s-b)(s-c)}$. To find the area of triangle DEF, where $d = 7$ in., $e = 12.5$ in., and $f = 9.75$ in., the following procedures would be used:

$$\begin{aligned}
\text{Area} &= \sqrt{s(s-d)(s-e)(s-f)} \\
&= \sqrt{14.625(14.625-7)(14.625-12.5)(14.625-9.75)} \\
&= \sqrt{1155.232} \\
&= 33.989 \text{ in.}^2
\end{aligned}$$

The second formula is used to find the radius (r) of the circle inscribed in a triangle. This radius is sometimes referred to as the *apothem* of a triangle. The apothem of triangle $ABC = \sqrt{s(s-a)(s-b)(s-c)}s$. To find the radius of an inscribed circle for triangle LMN, where $l = 40$ mm, $m = 70$ mm, and $n = 110$ mm, the following calculations are made:

$$r = \sqrt{\frac{s(s-l)(s-m)(s-n)}{s}} = \sqrt{\frac{110(110-40)(110-70)(110-110)}{110}}$$

$$= \sqrt{2800} = 52.915 \text{ mm}$$

Exercises

Write the following expressions in another form:

31. $\tan(A+B)$
32. $\tan(P-Q)$
33. $\sin(X-Y)$
34. $\sin 37° \cos 22° + \cos 37° \sin 22°$

Solve for the following:

35. Find the value of $\sin N/2$ if N is in the third quadrant and $\cos N = -3/7$. (Note: Only tangents are positive in third quadrant)
36. Express the following as a function of a positive acute angle: $\cos(-102°)$, $\cot(-304°)$, $\tan(-96°)$, and $\cos(-313°)$.
37. Find the haversines for the following angles: $30°$, $27°42'$, and $78°33'$.
38. Find the semiperimeter of a triangle whose sides are 5, 16, and 22 in.
39. Using the semiperimeter, calculate the area of a triangle where $a = 32$ cm, $b = 96$ cm, and $c = 112$ cm.
40. Determine the apothem of a triangle where $a = 810$ m, $b = 360$ m, and $c = 990$ m?

PART II

Mechanical Technology Formulas and Calculations

CHAPTER 4

Basic Concepts of Mechanics

- **Basic Definitions and Unit Systems**
- **Force Systems**
- **Friction**
- **Levers**

The field of mechanics deals with the effects of forces on bodies that result in or prevent motion. *Statics*, a branch of mechanics, is a study of bodies that are in equilibrium. Statics has to do with forces that act on a body and causes it to remain at rest or to move at a uniform rate of velocity. A second branch of mechanics, known as *dynamics*, deals with bodies that are not at equilibrium and examines forces that cause bodies to move at nonuniform velocity. Of the two, dynamics is the more complex and characteristic of real-world situations.

Dynamics can be further broken down into two subgroupings. The first, *kinetics*, deals with forces that act on a body and the motions that they cause. The second subgroup of dynamics, *kinematics*, is the study of only the motions of bodies, without reference to the forces that affect them. This chapter will present the basic concepts and formulas used in the field of mechanics.

Basic Definitions and Unit Systems

Mechanics is simply the application of a few principles of physics, and involves the solving of problems by either graphical or mathematical procedures. All quantities in the field of mechanics are divided into two groups: scalar and vector. *Scalar* quantities are measures such as time, volume, and density. *Vector* measures, by comparison, are force velocity, acceleration, moment, and displacement.

Scalar and vector measures are used in many technological and engineering fields to solve problems. Before one can proceed with the solving of these problems, however, it is first necessary to have an understanding of a few basic definitions and of the system of units employed in mechanics.

Definitions

Force is the action of one body on another and is simply viewed as the pushing or pulling of a body. Forces can take many forms, including mechanical, magnetic, gravitational, and electrical.

Matter is anything that takes up space. It can be a gas, a solid, or a liquid, and can come in any size, including molecules, atoms, and electrons.

Inertia is a property of matter. This property is defined as the characteristic of a body to resist changes in its state of rest or uniform motion.

Mass is the amount of matter in a body and is a measure of the inertia of a body. Mass is usually expressed in kilograms.

Work is the product of force and the distance through which the force acts. Traditionally, work has been expressed in such relationships as foot-pounds or kilogram-meters. The SI (Système Internationale d'Unités) metric unit for work, however, is the joule.

Power is an expression used to describe the rate at which work is performed and is calculated as the product of force and distance divided by time. It has been noted in terms such as foot-pounds per minute or kilogram-meters per second, but is expressed as the watt (1J/s) in SI metric units.

Horsepower (hp) is the traditional unit of power in engineering. One hp is equal to 746 watts (w), 33,000 ft-lb/min, or 550 ft-lb/s. In electrical work, the horsepower unit is not used; in its place, the kilowatt (kw) is substituted. The value of 1 hp = 0.746 kw. The metric SI unit here is the watt instead of horsepower.

The *torque* or *moment* of a force is the tendency of that force to turn or rotate a body about a given axis. The common expressions used to describe torque are pound-feet and kilogram-meters. The SI metric units used here is the newton-meter.

Momentum is the product of mass and velocity.

Velocity is a distance-time ratio expressed in measures such as feet per second, millimeters per second, and meters per second.

Acceleration is a measure of the change of velocity over a period of time and is expressed in measures such as feet per second squared, or feet per second per second. The SI metric unit used for acceleration is the meter per second squared.

Impulse is the product of force and time.

Unit Systems

Two unit systems are used in mechanics: *absolute* and *gravitational*. The basic units used in the absolute unit system are length, time, and mass. From these units it is possible to derive the measure of force. With absolute units it will be possible to use either the centimeter-gram-second (cgs) or meter-kilogram-second (MKS) systems. The MKSA, or meter-kilogram-second-ampere, is used to unite MKS with electromagnetic units.

The units used with the gravitational system are length, time, and force. From these three gravitational units, mass is derived. Mass is derived from the formula $M = W/g$, where W is the weight of the object and g is the acceleration due to gravity (i.e., 32.2 ft./s^2 or 9.81 m/s^2).

An internationally agreed-upon standard for defining a unit of force is the pound force. This standard has a value of 32.1740 ft/s^2 and is denoted by the symbol g_o. Therefore, whenever the term *mass* is used in a mechanics formula, the standard pound force, or $g_o = 32.1740$ ft/s^2, is assumed. In the MKS system, this force is the newton (N), where 1 N = 0.22481 lb.

Force Systems

Two basic approaches are used to solve force systems: graphical and algebraic. The first is quick, simple, and useful where extreme accuracy is not required. Presented in this section are procedures used to describe the composition of forces and to find their resolution.

Definitions

Resultants are single forces that produce the same effect upon a body as two or more other forces acting together on that body.

Components are different or separate forces that can be combined or act together.

The *composition of forces* is the determination of the resultant of two or more forces.

The *resolution of forces* is the finding of two or more components of a given force.

Concurrent forces are those that have lines of action that have a common point of intersection.

Nonconcurrent forces are those that have lines of action that do not have a common point of intersection.

Parallel forces are nonconcurrent and have lines of action that are parallel to one another.

Colinear forces are two forces that have the same line of action.

Couple forces are two forces that have equal magnitudes, are parallel, and act in opposite directions.

Coplanar forces are all in the same plane.

Noncoplanar forces are not in the same plane.

Skewed forces are nonconcurrent, nonparallel, and noncoplanar.

Graphical Solutions

As already mentioned, many mechanics' problems can be solved by graphical methods. Though it is not a "mathematical method" per se, graphics is based on mathematical concepts and theories. Therefore, it is necessary to be aware of these procedures. In all, there are five basic categories of graphical force problems. These are parallelogram of forces, polygon of forces, adding and subtracting forces, parallel forces, and moment of a force.

PARALLELOGRAM OF FORCES

Figure 4-1 is an illustration of the two forces AB and AC. To find the resultant of AB and AC, we must be able to describe the magnitude and direction of the two forces as they act together. These forces are graphically drawn as two adjacent sides of a parallelogram.

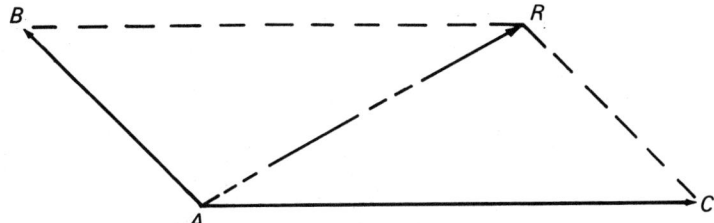

Fig. 4-1. Parallelogram of forces.

The resultant of these forces will be specified as the diagonal of the parallelogram, or AR. Thus, AR is the graphical representation of the magnitude and direction of forces AB and AC.

An adaptation of the parallelogram-of-forces problem is where the two forces do not originate at the same point. Figure 4-2 is an example of this situation. Here, forces D and E begin at two separate points. To find their resultant, their lines of action are extended until they meet at P. From P, the magnitude of D and E are drawn so that they represent two adjacent sides of a parallelogram. To find the direction and magnitude of D and E, the same graphical procedures are used as in Figure 4-1.

Another, slightly different, parallelogram-of-forces problem involves more than two forces that originate from one point (Figure 4-3). To find this system's resultant, first solve for the resultant of the two forces AB and AC. Their resultant, AR_1, then becomes a force to act along with force AD. Thus, AR_1 and AD are two adjacent sides of a parallelogram, the resultant of which is AR. AR, then, is the direction and magnitude of forces AB, AC, and AD. If additional forces were found, the same procedures would be used until all forces were accounted for.

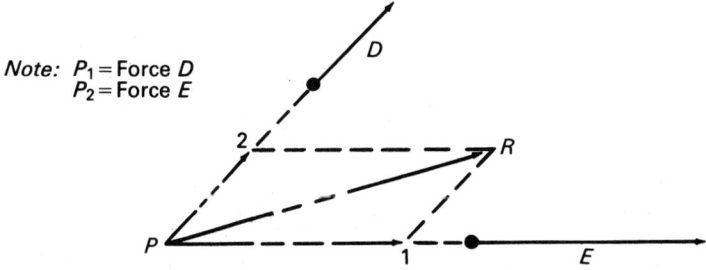

Note: $P_1 =$ Force D
$P_2 =$ Force E

Fig. 4-2. Adaptation of parallelogram-of-forces problem.

88 • **BASIC CONCEPTS OF MECHANICS**

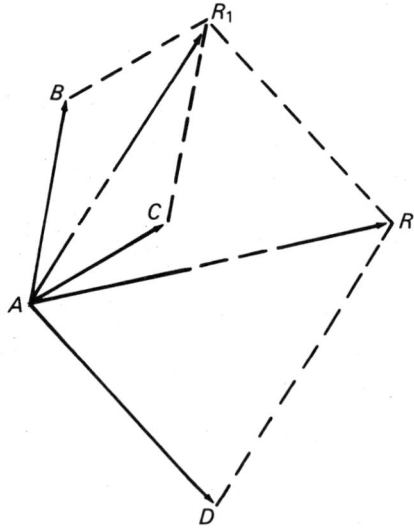

Fig. 4-3. Parallelogram of forces with multiple forces.

POLYGON OF FORCES

Cases that involve multiple forces originating from a common point can incorporate a graphical technique that constructs a polygon for finding their resultant. Shown in Figure 4-4 are six forces that orig-

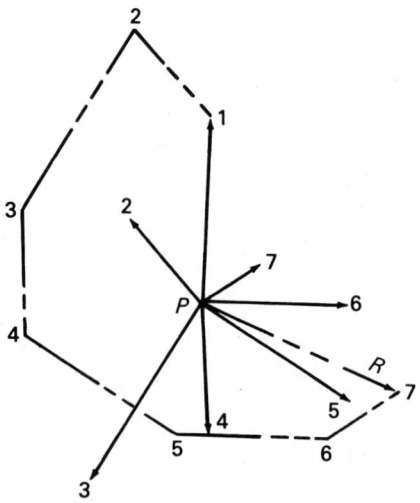

Fig. 4-4. Polygon of forces.

inate from a common point, P, and are all in the same plane. To find their resultant (R at point 7), begin by constructing the magnitude and direction of force P2 at the end of force P1. Then construct force P3 at the end of constructed force P2. Continue this procedure until all forces are accounted for. The resultant of these forces will be the line drawn from point P to the end of the last constructed force line.

ADDING AND SUBTRACTING FORCES

The resultant of two or more forces that originate from the same point and act in the same direction can be determined by either addition or subtraction. If all forces are going in the same direction, addition is used; forces going in opposite directions are subtracted.

Shown in Figure 4-5 are two illustrations of this concept. Note that the numerical values used for addition and subtraction are equal to the magnitude of the appropriate forces.

PARALLEL FORCES

There are two basic situations that can arise with parallel forces. The first is shown in Figure 4-6. Here the two parallel forces are acting in the same direction. Their magnitude, then, will be equal to their sum. The location of the resultant, however, is determined first by drawing an adjoining line, AC. The location of the resultant will be inverse to the magnitude of the two parallel forces. In other words, $AB:CE = CD:AD$.

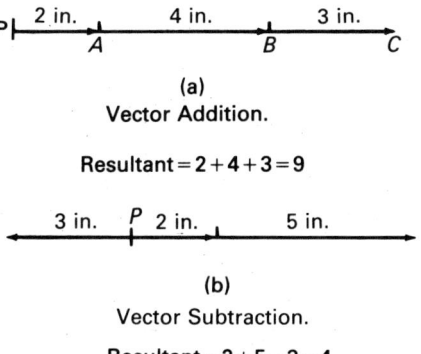

(a)
Vector Addition.

Resultant = 2 + 4 + 3 = 9

(b)
Vector Subtraction.

Resultant = 2 + 5 − 3 = 4

Fig. 4-5. Vector addition and subtraction.

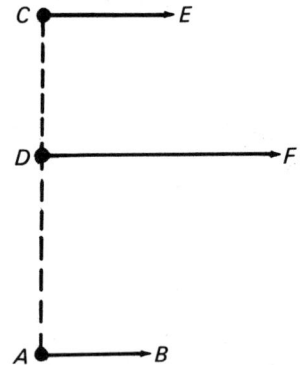

Fig. 4-6. Sum of parallel forces.

Where parallel forces in the same plane act in opposite directions, their magnitude will be equal to the difference of these forces and the direction of the resultant will be in the same direction as the greater of two forces, The location of the resultant will again begin on a line drawn between the two forces and be situated according to the proportion $AB:CD = CE:AE$ (see Figure 4-7).

MOMENT OF FORCE

The last graphical procedure to be considered is the moment of a force with respect to a point. This is determined by multiplying the

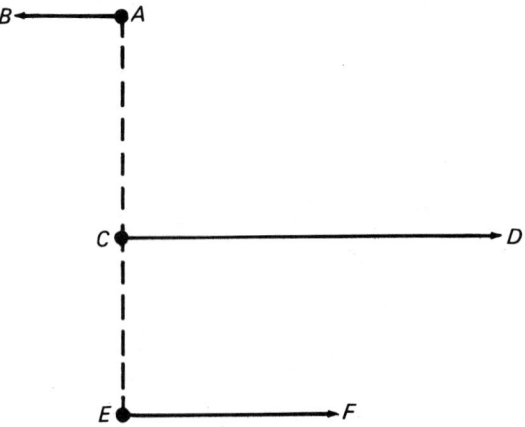

Fig. 4-7. Difference between parallel forces.

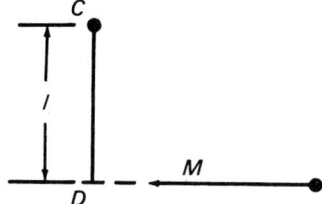

Fig. 4-8. Moment of a force.

force by the perpendicular distance from the point to the direction of the force. In our example, Figure 4-8, the moment of force M in relation to point C will be equal to $(M)(l)$, where l is the length of line CD.

Exercises

1. Find and locate the resultant of the four force systems shown in Figure 4-9.

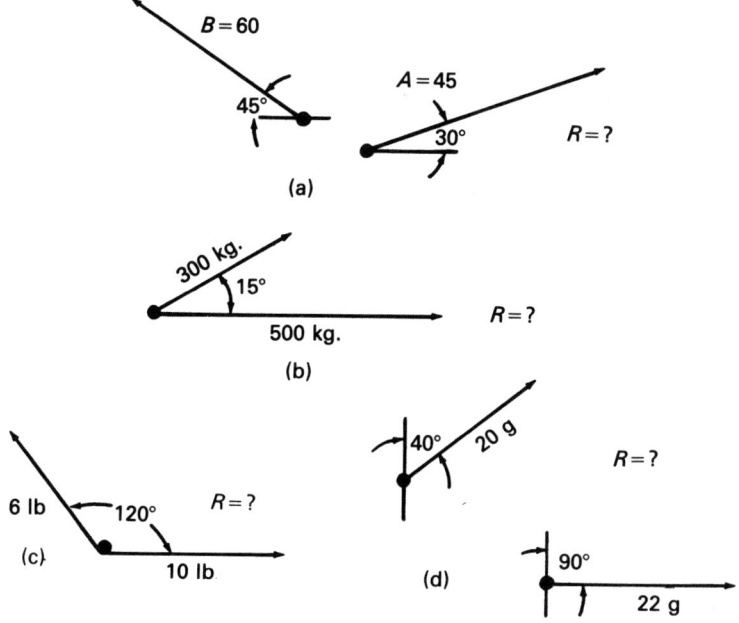

Fig. 4-9. Problem 1.

92 • BASIC CONCEPTS OF MECHANICS

2. What is the magnitude and location of the resultant of two forces, A and B, if A is 400 kg down and B is 300 kg down and 4 m to the right of A.

3. Find and locate the resultant of the two force systems shown in Figure 4-10.

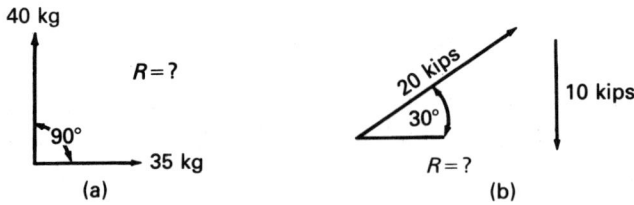

Fig. 4-10.   Problem 3.

4. Find and locate the result of the multiple forces shown in the three systems in Figure 4-11.

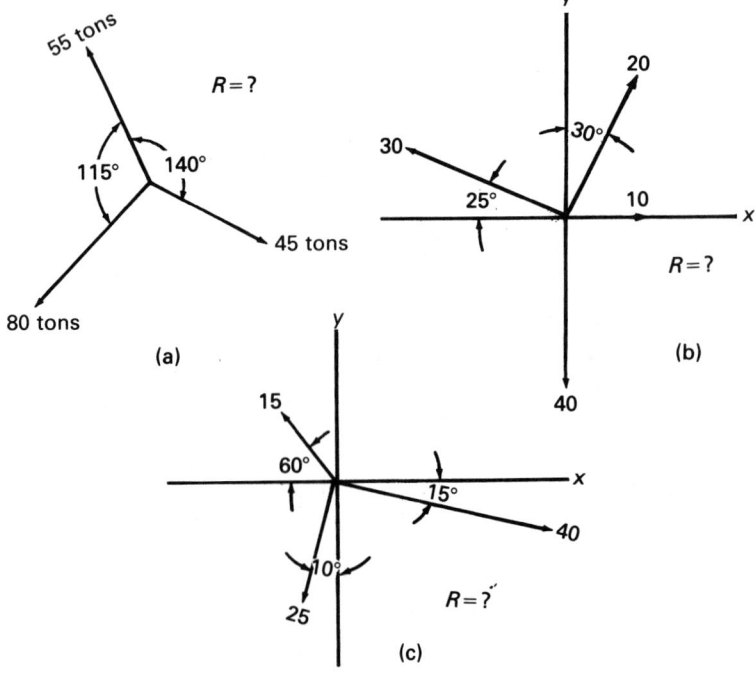

Fig. 4-11.   Problem 4.

5. Determine the magnitude of the moment of force for the two systems illustrated in Figure 4-12.

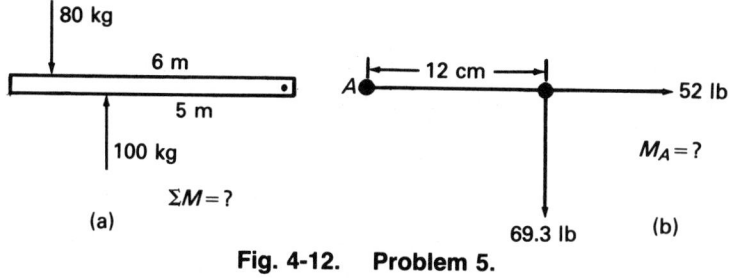

Fig. 4-12. Problem 5.

Mathematical Solutions: Coplanar Forces

There are a number of basic formulas that are used in the solution of force systems that are in the same plane. These formulas and procedures are based on traditional mathematical laws and principles that are applied to mechanics problems. In this section is a presentation of formulas used in solving coplanar force system problems.

CONCURRENT FORCES

There are three subcategories of problems that are involved in concurrent force systems. The first is the calculation of two concurrent components for a single force. A force system arrangement for this is shown in Figure 4-13. Here one must find the components F_1 and F_2 for the force F. If F_1 and F_2 are not at right angles to each other or to F, then the following formulas are used:

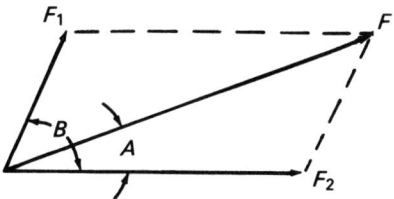

Fig. 4-13. Finding two concurrent components of a single force.

94 • BASIC CONCEPTS OF MECHANICS

$$F_1 = \frac{(F)(\sin A)}{\sin B}$$

$$F_2 = \frac{[F][(\sin(B-A))]}{\sin B}$$

In situations where F_1 and F_2 are at right angles to each other (Figure 4-14), the following formulas are used:

$$F_1 = (F)(\sin A)$$
$$F_2 = (F)(\cos A)$$

For example, take the case where $F = 500$ kg, angle $A = 30°$, and angle $B = 110°$. To find the two components for F, the following procedure would be used:

$$F_1 = \frac{(F)(\sin A)}{\sin B}$$

$$= \frac{(500)(\sin 30°)}{\sin 110°}$$

$$= \frac{500)(0.50000)}{0.93969}$$

$$= \frac{250}{0.93969}$$

$$= 266.045 \text{ kg}$$

$$F_2 = \frac{[F][\sin(B-A)]}{\sin B}$$

$$= \frac{[500][\sin(110-30)]}{\sin 110°}$$

$$= \frac{[500][\sin 80°]}{\sin 110°}$$

$$= \frac{[500][0.98481]}{0.93969}$$

$$= \frac{492.405}{0.93969}$$

$$= 524.008 \text{ kg}$$

In another type of force system problem the values of two concurrent forces are known, and one must find their resultant. Again,

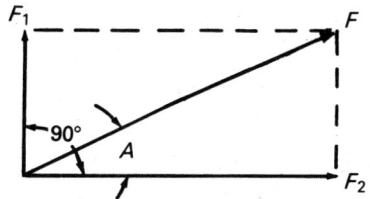

Fig. 4-14. Component forces forming a right angle.

there are two sets of formulas: the first is where the two concurrent forces are not at right angles to each other and the second is where they are (Figure 4-15). The formulas used to calculate the resultant for concurrent forces not at right angles, and its angle, are:

$$F_R = \frac{(F_1)(\sin B)}{\sin A} \text{ or}$$

$$F_R = \frac{(F_2)(\sin B)}{\sin (B-A)} \text{ or}$$

$$F_R = \sqrt{F_1^2 + F_2^2 + 2F_1 F_2 \cos B}$$

$$\tan A = \frac{(F_1 \sin B)}{(F_1 \cos B + F_2)}$$

When the concurrent forces are at 90° to each other, then the following formulas may be used:

$$F_R = \frac{F_2}{\cos A} \text{ or}$$

$$F_R = \frac{F_1}{\sin A} \text{ or}$$

$$F_R = \sqrt{F_1^2 + F_2^2}$$

$$\tan A = \frac{F_1}{F_2}$$

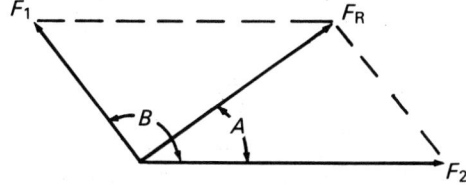

Fig. 4-15. Finding resultant of two concurrent forces.

96 • BASIC CONCEPTS OF MECHANICS

As an example, let's find the resultant of two forces of 200 lb and 750 lb that are at 84° to each other. The procedures used here are:

$$F_R = \sqrt{F_1^2 + F_2^2 + 2FF_2 \cos B}$$

$$= \sqrt{200^2 + 750^2 + (2)(200)(750) \cos 84°}$$

$$= \sqrt{40,000 + 562,500 + (300,000)(0.10453)}$$

$$= \sqrt{602,500 + 31,359}$$

$$= \sqrt{633,859}$$

$$= 796.153 \text{ lb}$$

$$\tan A = \frac{(F_1 \sin B)}{(F_1 \cos B + F_2)}$$

$$= \frac{(200)(\sin 84°)}{[(200)(\cos 84°) + 750]}$$

$$= \frac{(200)(0.99452)}{(200)(0.10453) + 750}$$

$$= \frac{198.904}{770.906}$$

$$= 0.25801$$

Therefore: $A = 14°28'3''$

In the final type of problem involving concurrent forces there are more than three forces and their resultant must be found. A graphical presentation of this type of problem is presented in Figure 4-16. Note that coordinate axes x and y are used to define the plane in which the forces are found. In this way it will be easier to find the components of the sum of forces required (i.e., F_x and F_y). To find the sum of F_x and F_y, use a table similar to the following:

Force	F_x	F_y
F_1	$F_1 \cos A_1$	$F_1 \sin A_1$
F_2	$F_2 \cos A_2$	$F_2 \sin A_2$
F_3	$F_3 \cos A_3$	$F_3 \sin A_3$
F_4	$F_4 \cos A_4$	$F_4 \sin A_4$
	ΣF_x	ΣF_y

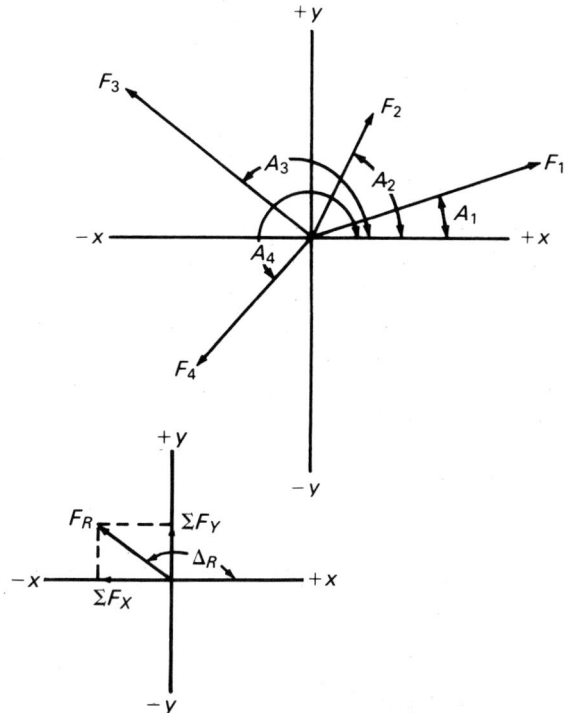

Fig. 4-16. Multiple concurrent forces.

With this table, and the resulting summations, it is possible to find the resultant of the concurrent forces by using the following formulas:

$$F_R = \sqrt{(\Sigma F_X)^2 + (\mathrm{E} F_y)^2}$$

$$\cos A_R = \frac{\Sigma F_X}{F_R}$$

$$\tan A_R = \frac{\Sigma F_Y}{\Sigma F_X}$$

FORCES AND A COUPLE

A couple is considered to be the resultant of two parallel forces of equal magnitude in opposite directions; the resultant will be equal to zero. In this situation, the couple will tend to produce a rotating

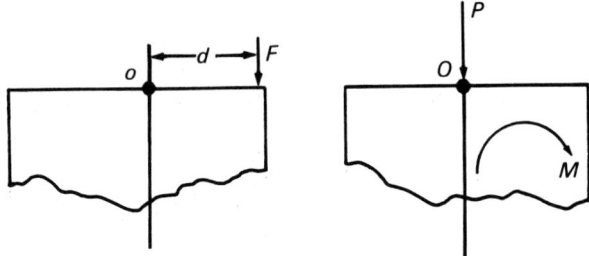

Fig. 4-17. Finding a force and a couple.

motion, which is called the moment of the couple. A typical problem here would be to find a couple and force that function as a single force. Here, the following formulas are used:

$$P = F$$
$$M = Fd$$

Where F = the single force, M = moment of the couple, P = the force passing through a point 0, and d = distance from 0 to F. These elements are illustrated in Figure 4-17. Here, M will generate a rotating force about point 0 and in the same direction as F.

Another problem would be to find the resultant, F_R, of a single force and a couple (Figure 4-18). The formulas used here are:

$$F_R = F$$
$$d = \frac{M}{F_R}$$

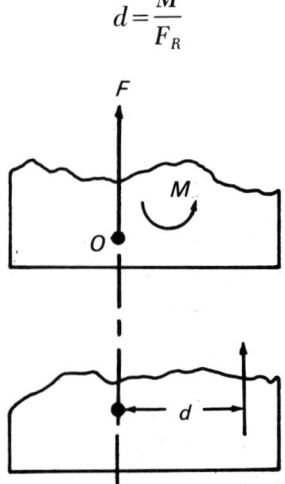

Fig. 4-18. Finding resultant of a force and couple.

A typical problem would be to find the couple of a moment with 45 kg of force passing through a point that is 340 m from a parallel and equal force. To solve this problem, the following procedure is used:

$$M = Fd$$
$$= (45)(340)$$
$$= 15,300 \text{ kgm or}$$
$$= 150,041.745 \text{ Nm (converted to Newton meter)}$$

PARALLEL FORCES

To calculate the resultant of coplanar parallel forces (Figure 4-19), select any point P in the plane so that a perpendicular distance d can be measured to the parallel forces (i.e., F_1, F_2, F_3, F_4, and F_5. The resultant is then calculated as a sum of the forces, where:

$$R = \Sigma F \text{ or}$$
$$R = F_1 + F_2 + F_3 + F_4 + \ldots$$

Thus, in our example, $R = 20 + 12 + 42 + 66 + 11 = 151$ kg.

The algebraic sum of the moments of forces about P is calculated as the sum of the products of each force times their distance from P (clockwise moments will have negative values, while counterclockwise moments have positive values). This is expressed as

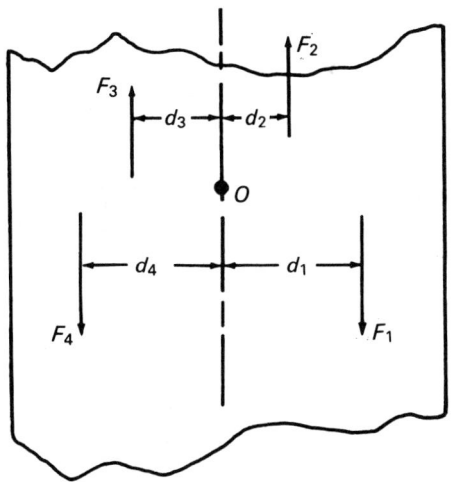

Fig. 4-19. Parallel forces.

100 • BASIC CONCEPTS OF MECHANICS

$$M_p = F_1 d_1 + F_2 d_2 + \cdots$$

The distance of the resultant from P is determined by the following relationship:

$$d = \frac{\Sigma M_p}{R}$$

It should be noted that if the resultant is equal to zero, then the resultant of the system of parallel forces will be a couple, ΣM_p. If the algebraic sum of the moments of forces about P is also zero, then the resultant will be a single force, R. The system will be in equilibrium if both R and M_p are equal to zero.

NONINTERSECTING FORCES

The last type of problem to be considered here deals with situations that involve nonintersecting forces. Shown in Figure 4-20 is a coplanar, nonconcurrent, nonparallel force system. To solve this problem, a point O is randomly placed in the plane. Two perpendicular axes are drawn through point O and labeled as x and y. Once this is accomplished, the following steps should be followed:

1. Locate a point on each force. This point can be placed at any location along the lines of action of the forces.

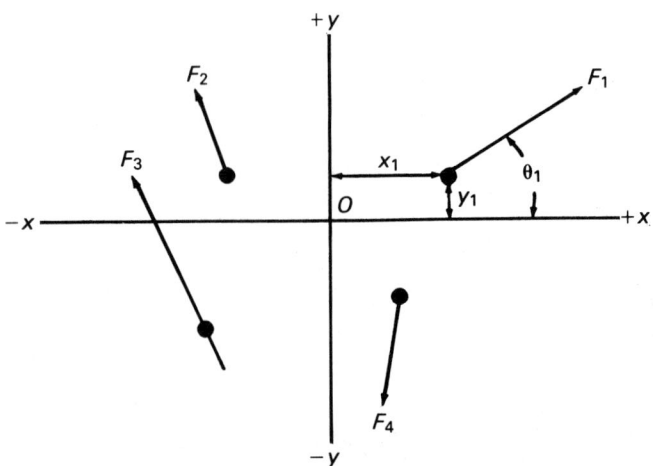

Fig. 4-20.   Nonintersecting forces.

2. Calculate the coordinates of each force by determining its point's location vis-à-vis the x and y axes. Each point, then, will have two coordinate values, which may have either positive or negative values. Note that all y-axis values (i.e., y_1, y_2, ...) that are below the x axis will be negative, and all above the x axis will be positive. All x-axis values (i.e., x_1, x_2, ...) that are to the left of the y axis will be negative, while those to the right of the y axis will be positive.

3. Measure the *counterclockwise*-direction angle (θ) that each line of action has with the x axis and label each (e.g., θ_1, θ_2, ...). The components of each force is then calculated for each axis. The x-axis component is the product of the force and the cosine of its measured angle, and the y-axis component is the product of the force and the sine of the measured angle.

4. The moment of force about point O for each force is the difference of two products. The first is a product of the force's x coordinate, magnitude of force, and sine of the measured angle. The second is a product of the force's y coordinate, magnitude of force, and cosine of the measured angle.

These four steps provide the basic data for calculating the system's resultant and moments of force. A quick reference for these steps is provided in Table 4-1.

To find the resultant of the system, make the computations with the following formulas:

$$R = \sqrt{(\Sigma F_X)^2 + (\Sigma F_Y)^2}$$

$$d = \frac{\Sigma M_o}{R}$$

$$\cos \theta_R = \frac{\Sigma F_X}{R}$$

$$\tan \theta_R = \frac{\Sigma F_Y}{F_X}$$

(*Note*: d is measured from O and is perpendicular to the resultant's line of action.)

For example, this procedure would be used in finding the resultant for a system of three coplanar, noncurrent forces with the following known data:

102 • BASIC CONCEPTS OF MECHANICS

$$F_1 = 100 \text{ N}, \quad x_1 = 5.0 \text{ m}, \quad y_1 = -1.0 \text{ m}, \quad \theta_1 = 270°$$
$$F_2 = 200 \text{ N}, \quad x_2 = 4.0 \text{ m}, \quad y_2 = 1.5 \text{ m}, \quad \theta_2 = 50°$$
$$F_3 = 300 \text{ N}, \quad x_3 = 2.0 \text{ m}, \quad y_3 = 2.0 \text{ m}, \quad \theta_3 = 60°$$

To solve this problem, the following calculations and procedures would be used:

$$F_{x1} = 100 \cos 270° = (100)(0) = 0 \text{ N}$$
$$F_{x2} = 200 \cos 50° = (200)(0.64279) = 128.558 \text{ N}$$
$$F_{x3} = 300 \cos 60° = (300)(0.50000) = 150.000 \text{ N}$$

$$F_{y1} = 100 \sin 270° = (100)(-1) = -100.000 \text{ N}$$
$$F_{y2} = 200 \sin 50° = (200)(0.76604) = 153.208 \text{ N}$$
$$F_{y3} = 300 \sin 60° = (300)(0.86603) = 259.809 \text{ N}$$

$$M_{o1} = (5)(-100) - (-1)(0) = -500 \text{ Nm}$$
$$M_{o2} = (4)(153.208) - (1.5)(128.588) = 419.995 \text{ Nm}$$
$$M_{o3} = (2)(259.809) - (2)(150.000) = 219.618 \text{ Nm}$$

Force	Coordinates of F			Components of F		Moment of F About O
F	x	y	θ	F_x	F_y	$M_o = xFy - yFx$
$F_1 = 100$	5	−1	270°	0	−100.00	−500.000
$F_2 = 200$	4	1.5	50°	128.558	153.208	419.995
$F_3 = 300$	2	2	60°	150.000	259.809	219.618
				278.558	313.017	139.613

$$R = \sqrt{(278.558)^2 + (313.017)^2}$$
$$= \sqrt{175{,}574.2017}$$
$$= 419.016 \text{ N}$$
$$d = \frac{139.613}{419.016}$$
$$= 0.333\ 193 \text{ m}$$
$$\tan \theta_R = \frac{313.017}{278.558}$$
$$= 1.12370$$
$$= 48°20'1''$$

Graphically, the resultant appears as shown in Figure 4-21.

Table 4-1. Nonintersecting Forces Formulas

	Force Coordinates of F			Components of F	
F	x	y	θ	F_X	F_Y
F_1	x_1	y_1	θ_1	$F_1 \cos \theta_1$	$F_1 \sin \theta_1$
F_2	x_2	y_2	θ_2	$F_2 \cos \theta_2$	$F_2 \sin \theta_2$
F_3	x_3	y_3	θ_3	$F_3 \cos \theta_3$	$F_3 \sin \theta_3$
F_4	x_4	y_4	θ_4	$F_4 \cos \theta_4$	$F_4 \sin \theta_4$
.	.	.	.	.	.
.	.	.	.	.	.
				ΣF_X	ΣF_Y

Moment of Force About O

Force	
	$M_O = xF_Y - yF_X$
F_1	$x_1 F_1 \sin \theta_1 - y_1 F_1 \cos \theta_1$
F_2	$x_2 F_2 \sin \theta_2 - y_2 F_2 \cos \theta_2$
F_3	$x_3 F_3 \sin \theta_3 - y_3 F_3 \cos \theta_3$
F_4	$x_4 F_4 \sin \theta_4 - y_4 F_4 \cos \theta_4$
.	.
.	.
	M_O

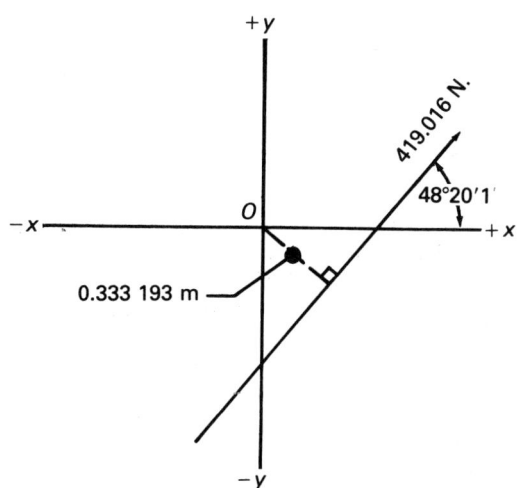

Fig. 4-21. Resultant presentation.

104 • BASIC CONCEPTS OF MECHANICS

Exercises:

6. Calculate the components of a force F, where $F = 240$ kips, the two component forces are $120°$ to each other, and the second component force is $45°$ to F.

7. Find the components F_1 and F_2 which are perpendicular to each other when $F = 475$ kg.

8. What is the resultant of two concurrent forces that are $145°$ to each other, where their magnitudes are $F_1 = 45$ lb and $F_2 = 32$ lb?

9. Calculate the resultant for the concurrent-force system shown in Figure 4-22.

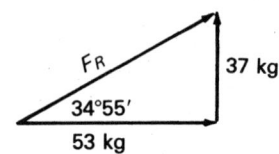

Fig. 4-22. Problem 9.

10. What is the moment of force about a point if the force exerted is 350 kg at a distance of 2.75 m from that point? What would be the moment of force about a point if the force exerted was 2.55 tons (t) at distance of 7.8 yd from a point?

11. Find the resultant for the single force and couple illustrated in Figure 4-23.

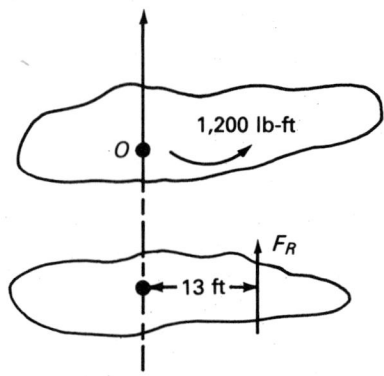

Fig. 4-23. Problem 11.

12. Calculate the resultant of three coplanar nonconcurrent forces for the following force system:

$F_1 = 15$ kg, $x_1 = 4$ cm, $y_1 = 3.5$ cm, $\theta_1 = 115°$.
$F_2 = 5$ kg, $x_2 = -2$ cm, $y_2 = -1.75$ cm, $\theta_2 = 10°$.
$F_3 = 24$ kg, $x_3 = 7$ cm, $y_3 = 8$ cm, $\theta_3 = 45°$.
$F_4 = 12$ kg, $x_4 = 6.5$ cm, $Y_4 = -2$, $\theta_4 = 15°$.

Mathematical Solutions: Noncoplanar Forces

There are many times when forces within a system are located in different planes. This arrangement is more typical of realistic, three-dimensional problems. This section presents mathematical procedures and formulas used in solving noncoplanar-force systems.

PARALLEL NONCOPLANAR SYSTEMS

An example of a system of parallel noncoplanar forces is illustrated in Figure 4-24. To find the resultant for this system, it is first necessary to draw in a three-axis coordinate system. This system is labeled as x, y, and z. Point O of the coordinate axes is the point of origin through which all axes pass. Another feature of the axes is that they are all perpendicular to each other.

The next step is to measure the distances that the forces are from the x and y axes. (*Note*: The distance from the z axis is not used here, since the forces shown are parallel to it, but are point

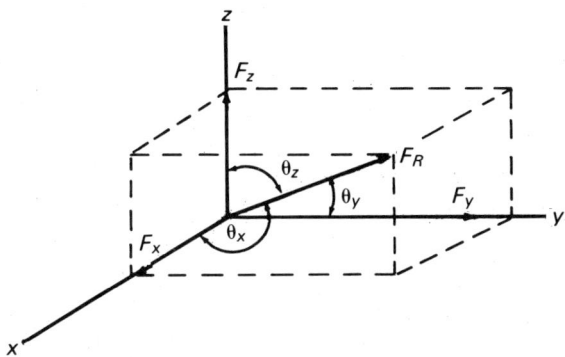

Fig. 4-24. Parallel noncoplanar force system.

locations on the x and y axes.) Set up the distance measures in an organized table such as:

Force	Coordinates of Force F	
F	x	y
F_1	x_1	y_1
F_2	x_2	y_2
F_3	x_3	y_3
F_4	x_4	y_4
.	.	.
.	.	.
.	.	.
ΣF		

The resulting magnitude of the system will be equal to the algebraic sum of all the forces. Hence:

$$F_r = \Sigma F = F_1 + F_2 + F_3 + F_4 + \cdots$$

The next procedure is to find the moment of each force about the x and y axes. When setting off moments about these axes, a positive ($+$) value will be assigned to counterclockwise moments, and negative ($-$) values to clockwise moments. To determine positive or negative values, carefully label the positive and negative ends of each axis. Set the results of the moments in a table as follows:

Moments M_X and N_Y Due to F

M_X	M_Y
$F_1 y_1$	$F_1 x_1$
$F_2 y_2$	$F_2 x_2$
$F_3 y_3$	$F_3 x_3$
$F_4 y_4$	$F_4 x_4$
.	.
.	.
.	.
ΣM_X	ΣM_Y

When dealing with the moment of a force, the perpendicular distance from a given point is called the *moment arm*. In our illustration, the perpendicular distance is measured from the point of origin, O. Thus, the distances are measures along the x and y axes, and noted as x_R and y_R, respectively. To calculate the moment arms of the resultant, the following formulas are used:

$$x_r = \frac{\Sigma M_Y}{F_R}$$

$$y_R = \frac{\Sigma M_X}{F_R}$$

If the sum of moments due to F are both equal to zero, then the resultant will be a force along the z axis; if the resulting force F_R is also zero, then the system will be in equilibrium. If the moments are not equal to zero, but F_R is, then the resultant will be a couple that lies in a plane that is parallel to the z axis and have an angle θ_R to be measured in a counterclockwise, or positive, direction from the positive side of the x axis. This angle is then calculated with the following formulas:

$$M_R = \sqrt{(\Sigma M_X)^2 + (\Sigma M_Y)^2}$$

$$\sin \theta_R = \frac{\Sigma M_X}{M_R}$$

To find the resultant for a parallel noncoplanar force system, we will use the following data:

$F_1 = 10$ kg, $x_1 = -2.0$ mm, $y_1 = 2.0$ mm
$F_2 = 20$ kg, $x_2 = 1.7$ mm, $y_2 = 1.8$ mm
$F_3 = 25$ kg, $x_3 = 1.5$ mm, $y_3 = -2.0$ mm

To solve this problem the following calculations are made:

$$F_R = 10 + 20 + 25 = 55 \text{ kg}$$

$F_1 y_1 = (10)(2) = 20$ kgmm
$F_2 y_2 = (20)(1.8) = 36$ kgmm
$F_3 y_3 = (25)(-2) = -50$ kgmm

$F_1 x_1 = (10)(-2) = -20$ kgmm
$F_2 x_2 = (20)(1.7) = 3.4$ kgmm
$F_3 x_3 = (25)(1.5) = 37.5$ kgmm

108 • BASIC CONCEPTS OF MECHANICS

$$\Sigma M_X = (20) + (36) + (-50) = 6 \text{ kgmm}$$
$$\Sigma M_Y = (-20) + (3.4) + (37.5) = 20.9 \text{ mm}$$
$$x_R = \frac{20.9}{55} = 0.38 \text{ mm}$$
$$y_R = \frac{6}{55} = 0.1091 \text{ mm}$$
$$M_R = \sqrt{(6)^2 + (20.9)^2} = 21.744 \text{ kg mm}$$
$$\sin \theta_R = \frac{6}{21.744} = 0.27594 \text{ therefore:}$$

Therefore:
$$\theta_R = 16°1'5''$$

NONPARALLEL NONCOPLANAR SYSTEMS

The configuration of most force systems will be nonparallel and noncoplanar and involve a great number of different forces. Figure 4-25 is an illustration of three nonparallel noncoplanar forces. In most cases, this type of system will have a resultant that is noncoplanar and a couple. Sometimes the couple is combined into two skewed forces. As might be surmised, all previous force systems were simpler and less complex than this type of system.

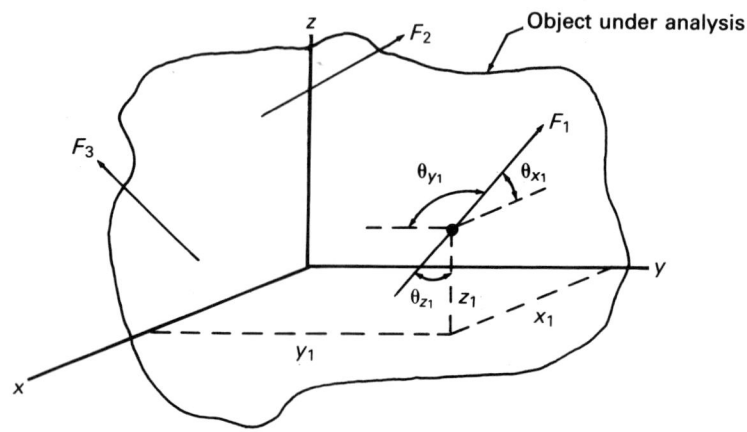

Fig. 4-25. Nonparallel noncoplanar force system.

For ease of understanding and application, the formulas and methods presented here are only for the three forces shown in Figure 4-25. For more complex and common systems, the same procedures and formulas should be used.

The first step in solving this type of problem is to place a three-axis coordinate system (x, y, and z) at any location within the object under analysis. Select a point on the line of action for each force and then measure its location on each axis so that F_1 will have coordinates x_1, y_1, and z_1. Then measure the angle of each line of action to each coordinate axis: θ_{x1}, θ_{y1}, and θ_{z1}. To keep track of these data, make use of a tabulation table such as:

Force	Coordinates of Force					
F	x	y	z	θ_X	θ_Y	θ_Z
F_1	x_1	y_1	z_1	θ_{X1}	θ_{Y1}	θ_{Z1}
F_2	x_2	y_2	z_2	θ_{X2}	θ_{Y2}	θ_{Z2}
F_3	x_3	y_3	z_3	θ_{X3}	θ_{Y3}	θ_{Z3}

Next, calculate the components of each force by use of the following table of formulas (again, a table is used to keep track of the procedure):

Force	Component of Force		
F	F_X	F_Y	F_Z
F_1	$F_1 \cos x_1$	$F_1 \cos y_1$	$F_1 \cos z_1$
F_2	$F_2 \cos x_2$	$F_2 \cos y_2$	$F_2 \cos z_2$
F_3	$F_3 \cos x_3$	$F_3 \cos y_3$	$F_3 \cos z_3$
	ΣF_X	ΣF_Y	ΣF_Z

The resultant of the force system, F_R, can now be calculated by the formula:

$$F_R = \sqrt{(\Sigma F_X)^2 + (\Sigma F_Y)^2 + (\Sigma F_Z)^2}$$

To find the angle that the resulting force is at in the coordinate system, use the following formulas:

$$\cos\theta_{XR} = \frac{\Sigma F_X}{F_R}$$

$$\cos\theta_{YR} = \frac{\Sigma F_Y}{F_R}$$

$$\cos\theta_{ZR} = \frac{\Sigma F_Z}{F_R}$$

The last series of calculations involved here deals with the moments of the forces. The moments M_X, M_Y, and M_Z about each axis are determined by the components of each force, namely, F_X, F_Y, and F_Z. Positive (+) moments will move in a counterclockwise direction, while negative (−) moments will move in a clockwise direction. To calculate the moments, use the formulas outlined in the following table:

Force	\multicolumn{3}{c}{Moments of Force About Axes x, y, and z}		
F	$M = yM_Z - zM_Y$	$M = zF_X - xF_Z$	$M = xF_Y - yF_X$
F_1	$M_{X_1} = y_1 F_{Z_1} - z_1 F_{Y_1}$	$M_{Y_1} = z_1 F_{X_1} - x_1 F_{Z_1}$	$M_{Z_1} = x_1 F_{Y_1} - y_1 F_{X_1}$
F_2	$M_{X_2} = y_2 F_{Z_2} - z_2 F_{Y_2}$	$M_{Y_2} = z_2 F_{X_2} - x_2 F_{Z_2}$	$M_{Z_2} = x_2 F_{Y_2} - y_2 F_{X_2}$
F_3	$M_{X_3} = y_3 F_{Z_3} - z_3 F_{Y_3}$	$M_{Y_3} = z_3 F_{X_3} - x_3 F_{Z_3}$	$M_{Z_3} = x_3 F_{Y_3} - y_3 F_{X_3}$
	ΣM_X	ΣM_Y	ΣM_Z

The components of the resultant couple M will be the algebraic sums ΣM_X, ΣM_Y, and ΣM_Z. To find M, use the following formula:

$$M = \sqrt{(\Sigma M_X)^2 + (\Sigma M_Y)^2 + (\Sigma M_Z)^2}$$

The resultant couple M will be positioned at specific angles to each of the three axes, which can be calculated by the formulas:

$$\cos\theta_X = \frac{\Sigma M_X}{M}$$

$$\cos\theta_Y = \frac{\Sigma M_Y}{M}$$

$$\cos\theta_Z = \frac{\Sigma M_Z}{M}$$

Exercises

13. Calculate the resultant for the parallel noncoplanar force system shown in Figure 4-26.

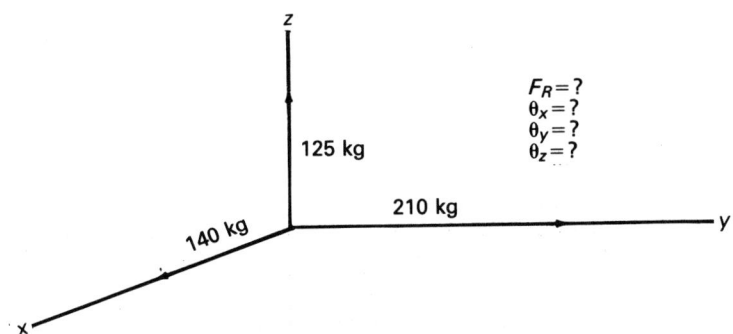

Fig. 4-26. Problem 13.

14. Calculate the resultant for the nonparallel noncoplanar force system illustrated in Figure 4-27.

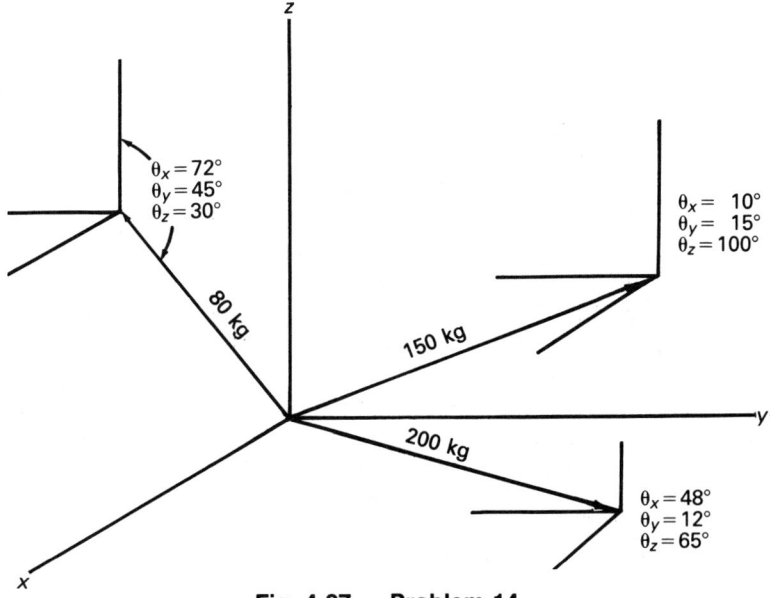

Fig. 4-27. Problem 14.

Friction

The application of mechanics to real-world situations is seldom accomplished with only force system diagrams. It incorporates the use of actual equipment designs, loads, and specifications. Problematic to many pieces of equipment is friction, a force that acts on machinery components.

Coefficient of Friction

Friction is created in the form of resistance as one body moves or slides across another. In mechanics, friction is divided into two broad categories. The first, *static friction*, is the friction that is built up as force is applied to a body up to the point where sliding does not occur. An example of this would be a stationary box that does not slide when force is applied to it. The second category of friction is *kinetic friction*, which is generated by a body as it actually slides across another body.

The coefficient of friction is a relationship between the normal force and friction force of a material. This coefficient is frequently represented as the Greek letter *mu* (μ), which has a mathematical relationship represented as:

$$\mu = \frac{Fr}{N}$$

where Fr is the friction force and N is the normal force.

The coefficient of friction is required in solving many mechanics problems. To determine this value, tables should be used that give the coefficient of friction for various materials. Table 4-2 presents the static and kinetic coefficient of friction for common materials.

The coefficient of friction can also be calculated by means of an inclined plane (Figure 4-28). The block of material is made up of one material and the plane of another for which the coefficient is required. The plane is raised slowly until the block just begins to slide; at this point the angle of inclination is noted. As the block's motion impends, or is just about to start, the weight (W) of the block parallel to the inclined plane is just enough to balance the maximum friction of force. The angle (ϕ) of the incline is then carefully measured. To find the coefficient of friction, the following formulas can be used:

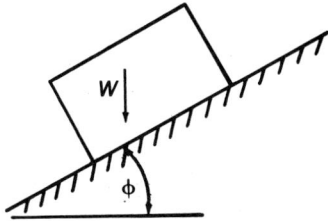

Fig. 4-28. Inclined plane.

$$\mu = \frac{W \sin \phi}{W \cos \phi} \text{ or}$$
$$= \frac{\sin \phi}{\cos \phi} \text{ or}$$
$$= \tan \phi$$

Thus, it is possible to conclude that the coefficient of friction is equal to the tangent of the angle where motion would impend. This angle is sometimes referred to as the *angle of repose*.

Inclined-Plane Wedge Problems

There are two basic types of inclined-plane wedge problems: those that take friction into consideration and those that do not. This section presents three basic inclined-plane wedge problems, along with formulas for both friction and nonfriction considerations. For ease of understanding, the following notations will be used for identifying quantities:

W = weight of the body
N = normal force
Fp = force to pull the body up
Fp_1 = force to pull the body down
Fp_2 = force to hold the body stationary

CASE NO. 1

The first inclined-plane problem is illustrated in Figure 4-29. Here, the pulling force is parallel to the inclined plane. When friction is to be neglected, the components of this problem can be solved for by using the following formulas:

BASIC CONCEPTS OF MECHANICS

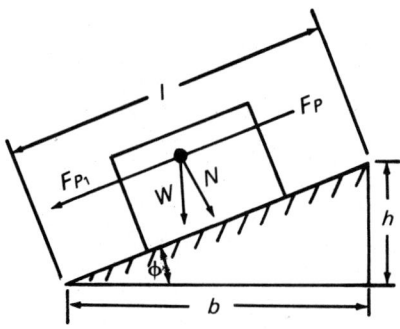

Fig. 4-29.   Case No. 1.

$$Fp = (W)\left(\frac{h}{l}\right) \text{ or}$$
$$Fp = (W)(\sin \phi)$$

$$W = (Fp)\left(\frac{l}{h}\right)$$
$$W = (Fp)(\csc \phi)$$

$$N = (W)\left(\frac{b}{l}\right) \text{ or}$$
$$N = (W)(\cos \phi)$$

A typical problem would be to solve for the amount of pulling force for a static teflon-coated body weighing 35 lb on a 24° inclined teflon-coated skid. The following procedure would be used:

$$\begin{aligned} Fp &= (W)(\sin \phi) \\ &= (35)(\sin 24°) \\ &= (35)(0.40674) \\ &= 14.2359 \text{ lb} \end{aligned}$$

When friction is to be taken into account, the problem becomes more involved, and the following formulas are used:

$$Fp = W(\mu \cos \phi + \sin \phi)$$
$$Fp_1 = W(\mu \cos \phi - \sin \phi)$$
$$Fp_2 = W(\sin \phi - \mu \cos \phi)$$

As can be seen, additional forces must be accounted for. With the same parameters as presented in the previous problem, the following calculations would be made for determining all forces acting upon the static body (note that the coefficient of friction was taken from Table 4-2, where the static coefficient for teflon-teflon is 0.04):

$$Fp = W(\mu \cos \phi + \sin \phi)$$
$$= 35 \,[(0.04)(\cos 24° + \sin 24°)]$$
$$= 35 \,[(0.04)(0.91355 + 0.40674)]$$
$$= 1.8484 \text{ lb}$$

$$Fp_1 = W(\mu \cos \phi - \sin \phi)$$
$$= 35[(0.04)(\cos 24° - \sin 24°)]$$
$$= 35[(0.04)(0.91355 - 0.40674)]$$
$$= 0.7095 \text{ lb}$$

$$Fp_2 = W(\sin \phi - \mu \cos \phi)$$
$$= 35 \,[\sin 24° - (0.04)(\cos 24°)]$$
$$= 35[0.40674 - (0.04)(0.91355)]$$
$$= 0.3702 \text{ lb}$$

(*Note*: If the value of Fp_2 were negative $(-)$, this would have meant that just to maintain equilibrium, the computed weight must push down on the body so that it will be perpendicular to the plane.)

CASE NO. 2

In the second inclined-plane problem, the pulling force acting upon the body is not parallel to the inclined surface (see Figure 4-30a). To account for this variation in force, the following formulas should be used when friction can be neglected:

$$Fp = W\left(\frac{\sin \phi}{\cos \phi}\right)$$

$$W = Fp\left(\frac{\cos \phi}{\sin \phi}\right)$$

$$N = W\left[\frac{\cos (\alpha + \phi)}{\cos \phi}\right]$$

When the pulling force (Fp) is at a right angle to the vertical weight of the body (Figure 5-30b), then the following formulas are used:

Table 4-2. Coefficient of Friction(μ)

Material	Coefficient of Friction	
	Clean	Lubricated
Static Coefficient		
Steel	0.8	0.16
Copper-lead alloy	0.22	—
Phosphor bronze	0.35	—
Aluminum bronze	0.45	—
Brass	0.35	0.19
Cast iron	0.4	0.21
Bronze	—	0.16
Hard carbon	0.14	0.11–0.14
Graphite	0.1	0.1
Tungsten carbide	0.4–0.6	0.1–0.2
Plexiglas	0.4–0.5	0.4–0.5
Polystyrene	0.3–0.35	0.3–0.35
Teflon	0.04	0.04
Aluminum-aluminum	1.35	0.30
Cadmium-cadmium	0.5	0.05
Chromium-chromium	0.41	0.34
Copper-copper	1.0	0.08
Iron-iron	1.0	1.15–0.20
Magnesium-magnesium	0.6	0.08
Nickel-nickel	0.7	0.28
Zinc-zinc	0.6	0.04
Glass-glass	0.9–1.0	0.1–0.6
Glass-metal	0.5–0.7	0.2–0.3
Hard carbon-carbon	0.16	0.12–0.14
Graphite-graphite	0.1	0.1
Tungsten carbide–tungsten carbide	0.2–0.25	0.12
Plexiglas-plexiglas	0.8	0.8
Polystyrene-polystyrene	0.5	0.5
Teflon-teflon	0.04	0.04
Wood-wood	0.2–0.5	—
Wood-metal	0.2–0.6	—
Brick-wood	0.6	—
Leather-wood	0.3–0.4	—
Leather-metal	0.6–0.4	—
Kinetic Coefficient		
Hard steel–hard steel	0.42	0.03–0.12
Mild steel–mild steel	0.57	0.09–0.19

(*continued*)

Table 4-2. (continued)

Material	Coefficient of Friction	
	Clean	Lubricated
Mild steel–cadmium silver	—	0.10
Mild steel–phosphor bronze	0.34	0.17
Nickel–mild steel	0.64	0.18
Aluminum–mild steel	0.47	—
Magnesium–mild steel	0.42	—
Teflon-teflon	—	0.04
Copper–mild steel	0.36	0.18
Nickel-nickel	0.53	0.12
Aluminum-aluminum	1.4	—
Copper–cast iron	0.29	—
Tin–cast iron	0.32	—
Glass-glass	0.40	0.09–0.12
Glass-nickel	0.56	—
Copper-glass	0.53	—
Bronze–cast iron	0.22	0.78
Cast iron–cast iron	0.15	0.06–0.70

$$Fp = W\left(\frac{h}{b}\right) \text{ or } \quad W = Fp(\cot \phi)$$

$$Fp = W(\tan \phi)$$

$$N = \frac{W}{\cos \phi}$$

$$W = Fp\left(\frac{b}{h}\right) \text{ or } \quad N = W(\sec \phi)$$

When friction is taken into account, the angle of repose (γ) must be used. This can be calculated by using the coefficient of friction as the tangent of the angle of repose. Thus, if the coefficient of friction is 0.2, then the angle will be computed to 11°18′36″. The two formulas used here are:

$$Fp = W\left[\frac{\sin(\alpha + \gamma)}{\cos(\beta - \gamma)}\right]$$

(The second formula is used when Fp is perpendicular to W.)

$$Fp = W[\tan(\alpha + \gamma)]$$

An example of a problem that might be encountered concerns a graphite body that is being pulled up a graphite ramp that lies at

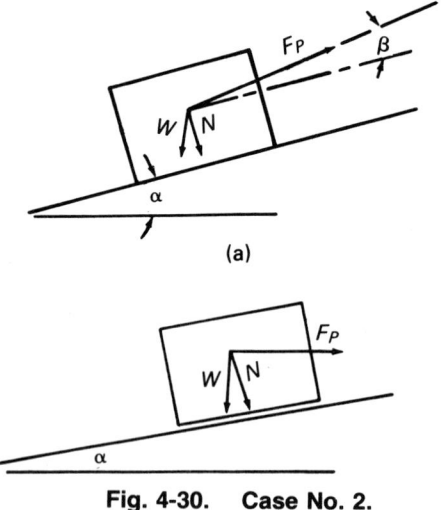

Fig. 4-30. Case No. 2.

an angle of 18°. The pulling force is in an upward 10° direction from the ramp's surface, and the weight of the block is 350 kg. To find the force required to pull the block up the ramp, the following procedure would be used:

$$\mu = 0.1 = \tan \gamma$$
Therefore: $\gamma = 5°42'38''$

$$\begin{aligned}
Fp &= W \frac{\sin (\alpha + \gamma)}{\cos (\beta - \gamma)} \\
&= 350 \frac{\sin (18° + 5°42'38'')}{\cos (10° - 5°42'38'')} \\
&= 350 \left(\frac{\sin 23°42'38''}{\cos 4°17'22''} \right) \\
&= 350 \left(\frac{0.40211}{0.99720} \right) \\
&= 141.1337 \text{ kg}
\end{aligned}$$

CASE NO. 3

The final type of wedge problem involves a downward force on a wedge. Presented in Figure 4-31 are two types of situations, each with a different set of data. In the first case, normal force is applied

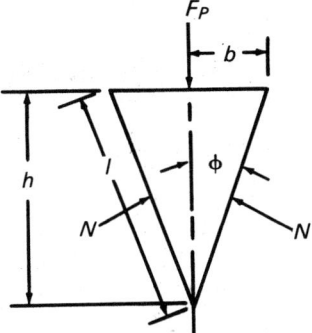

Fig. 4-31. Case No. 3.

perpendicular to the sides of the wedge while the pulling force is downward and perpendicular. The set of formulas used here are:

$$Fp = 2N\left(\frac{b}{l}\right) \text{ or}$$
$$Fp = 2N(\sin \phi)$$

$$N = Fp\left(\frac{l}{2b}\right) \text{ or}$$
$$N = Fp(\csc \phi)$$

With friction:

$$Fp = 2N(\mu \cos \phi + \sin \phi)$$

In the second situation, normal force is not applied perpendicular to the sides, but is perpendicular to the downward-pulling force. Here, the following formulas are employed:

$$Fp = 2N\left(\frac{b}{h}\right) \text{ or}$$
$$Fp = 2N(\tan \gamma)$$

$$N = Fp\left(\frac{h}{2b}\right) \text{ or}$$
$$N = Fp\frac{(\cot \phi)}{2}$$

When friction is accounted for:

$$Fp = 2N[\tan (\phi + \gamma)]$$

Exercises

15. Calculate the pulling force that is parallel to a 15° incline for a 250-kg body that is on that incline. What would be the pulling force if the body's weight increased by 50 kg?

16. Calculate the pulling force for the same two weights as in the preceding problem if the body is identified as brick and the incline is wood.

17. What is the pulling force and normal force for a 2-t block being pulled up a 5° incline by a force that is 15° to the incline?

18. What would be the pulling and normal forces for the body in problem 17 if it were perpendicular to the downward weight of the block?

19. Calculate the same forces identified in problem 17, but consider friction in your procedures. The body is made out of aluminum and the incline is mild steel (the coefficient of friction here is 0.61).

20. Using Figure 4-31 as a guide, calculate the pulling force when the normal force of 125 lb is perpendicular to the downward-pulling force of a wedge that has a coefficient of friction of 0.06. The total angle of this wedge is 30°.

Levers

The use of levers is commonly found in industrial equipment and processing procedures. To assure that the proper size equipment is selected to a particular job, it is often necessary to determine weight and force capacities. The methods used to specify this data here can be either SI metric or U.S. customary units. As previously discussed, the weight of a mass W in kg will be the same as a force of Wg in newtons, where g is the gravitational force of 9.81 m/s².

There are three general types of lever problems encountered in mechanics. The first is shown in Figure 4-32, with a given weight (W) at one end, a downward force (F) at the other, and a fulcrum in between. The formulas used in this problem are:

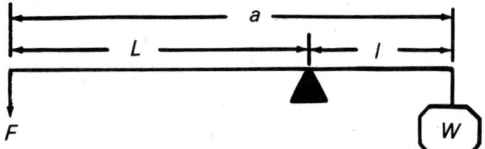

Fig. 4-32.   First lever arrangement.

$$F : W = l : L$$

$$FL = Wl$$

$$F = \frac{Wl}{L}$$

$$W = \frac{FL}{l}$$

$$L = \frac{Wa}{W + F} \text{ or}$$

$$L = \frac{Wl}{F}$$

$$l = \frac{Fa}{W + F}$$

$$l = \frac{FL}{W}$$

As an example, let's find the amount of force that would be required to balance a lever if a pulling weight of 2400 lb was exerted at one end of the lever, $l = 14$ ft, and $L = 38$ ft. The solution to this problem is:

$$\begin{aligned}F &= \frac{Wl}{L} \\ &= \frac{(2400)(14)}{38} \\ &= 884.211 \text{ lb}\end{aligned}$$

In another typical problem $F = 245$ kg, $W = 758$ kg, $l = 2.5$ cm, and the length of L must be calculated to maintain lever balance (i.e., equilibrium). The solution would be:

$$L = \frac{Wl}{F}$$
$$= \frac{(758)(2.5)}{245}$$
$$= 7.735 \text{ cm}$$

The second type of lever problem is illustrated in Figure 4-33, where the fulcrum is at the end of the lever, the force is an upward-acting force at the opposite end, and the weight is in the middle. The formulas used here are:

$$F : W = l : L$$
$$FL = Wl$$
$$F = \frac{Wl}{L}$$
$$W = \frac{FL}{l}$$
$$L = \frac{Wa}{W - F} \text{ or}$$
$$L = \frac{Wl}{F}$$
$$l = \frac{Fa}{W - F}$$
$$l = \frac{FL}{W}$$

In an example of a situation dealing with this problem, $l = 1.5$ m, $L = 3.0$ m, and $F = 1345$ lb. To find the amount of weight required to keep the lever in equilibrium, the following procedure is used:

$$W = \frac{FL}{l} = \frac{(1345)(1.5)}{3} = 672.5 \text{ lb}$$

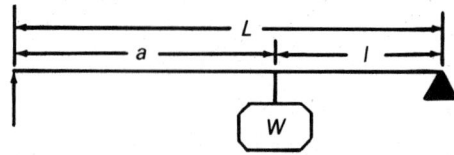

Fig. 4-33. Second lever arrangement.

In another problem $F = 2.1$ tonnes, $W = 13.4$ tonnes, and $a = 10$ yards. To find the length of L to secure equilibrium, the following calculations are made:

$$L = \frac{Wa}{W - F}$$
$$= \frac{(13.4)(10)}{(13.4 - 2.1)}$$
$$= 11.8584 \text{ yd}$$

The last type of lever problem deals with more complex and realistic forces. The lever problem illustrated in Figure 4-34 assumes two additional forces or weights in the lever. The formulas here are:

$$F_x = Wa + Pb + Qc$$
$$x = \frac{Wa + Pb + Qc}{F}$$
$$F = \frac{Wa + Pb + Qc}{x}$$

In a typical problem $W = 12$ kg, $P = 22$ kg, and $Q = 7$ kg; $a = 14$ cm, $b = 17$ cm, and $c = 20$ cm. If $x = 16$ cm, what should F be to maintain equilibrium?

$$F = \frac{Wa + Pb + Qc}{x}$$
$$= \frac{(12)(14) + (22)(17) + (7)(20)}{16}$$
$$= 42.625 \text{ kg}$$

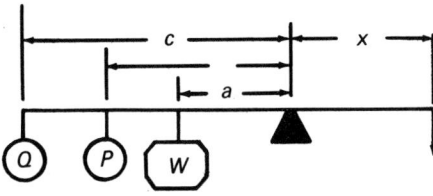

Fig. 4-34. Third lever arrangement.

Exercises

Using Figure 4-32 as a guide, find the missing quantities in question:

21. Find F when $L = 32$ in., $a = 46$ in., and $W = 142$ lb.
22. Find W when $F = 22$ kips, $L = 23$ cm, and $l = 21$ cm.
23. What should the length of l be to maintain equilibrium when $F = 45$ kg, $W = 133$ kg, and $a = 240$ mm?

Using Figure 4-33 as a guide, solve for the following problems:

24. To maintain balance, what should the upward force of a lever be if the central weight is 4400 g, $L = 300$ mm, and $a = 120$ mm?
25. Find L if $W = 2$ t, $F = 0.75$ t, and $l = 20$ yd.
26. If $l = 23$ m, and L is 46 m, what should the ratio between F and W be?

Using Figure 4-34 as a guide, solve the following problems:

27. Calculate for F if $a = 2.5$ in., $b = 5$ in., $c = 7.5$ in., $x = 4$ in., $Q = 25$ lb, $P = 48$ lb, and $W = 35$ lb.
28. Assuming that $F = 300$ lb in problem 27, how long must lever arm x be made to achieve equilibrium?

CHAPTER 5

Complex Mechanics

- Wheels and Pulleys
- Center of Gravity and Moments of Inertia
- Velocity and Acceleration
- Work and Power
- Centrifugal Force

Chapter 4 presented an overview of formulas and mathematical calculations used in the basic problems of mechanics that have applications to various fields of mechanical technology and engineering. In most cases, however, these basic concepts are difficult to apply to everyday situations encountered on the job. Most problems within the field of mechanics involve more complex statics as well as dynamics.

The variety of problems existing in industry are inexhaustible, even when dealing with common processes. Thus, it would be impossible to discuss the solution of all problems and their related formulas. It is therefore the intent of this chapter to present those formulas and problems that are frequently found in more complex statics systems and dynamics problems.

Wheels and Pulleys

Wheel-and-pulley formulas are also representative of many industrial problems. Presented here are two general categories of wheel-

and-pulley formulas for lifting or moving loads. The first category is for simple wheel-and-pulley systems, and the second includes differential pulleys.

Simple Pulley Systems

The first set of formulas to be presented deals with a drum on which a lifting rope is wound. Figure 5-1 illustrates this arrangement. The formulas associated with this configuration are:

$$F : W = r : W$$
$$FR = Wr$$
$$F = \frac{Wr}{R}$$
$$W = \frac{FR}{r}$$
$$R = \frac{Wr}{F}$$
$$r = \frac{FR}{W}$$

An example of a pulley arrangement is where the radius of the drum (around which a lifting rope is wound) has a diameter of 6 in. and a periphery gear has a diameter of 24 in. The periphery gear is mounted to the same shaft as the drum and is used to transmit

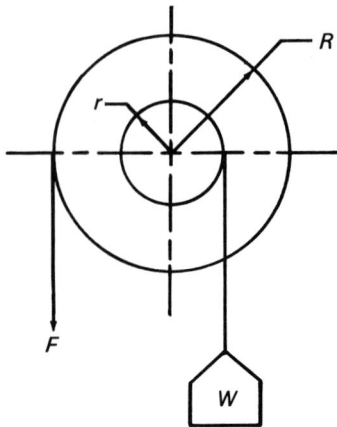

Fig. 5-1. Drum and rope wheel and pulley arrangement.

power to the drum. To calculate the force that must be exerted to raise 3400 lb, the following calculations are used:

$$F = \frac{Wr}{R}$$
$$= \frac{(3400)(6)}{24}$$
$$= 850 \text{ lb}$$

Two other wheel-and-pulley arrangements, which are related to the one just discussed, are illustrated in Figure 5-2. In Figure 5-2(a), the rate or velocity at which the weight (W) is raised will be one-half the rate or velocity of the force applied at F. Therefore:

$$F = \frac{1}{2W} = \frac{W}{2}$$

In the second situation, Figure 5-2(b), the following formulas apply:

$$F : W = \sec \theta : 2$$
$$F = W \frac{(\sec \theta)}{2}$$
$$W = 2F(\cos \theta)$$

As an example of a problem employing these formulas, let's determine the amount of force needed to lift a weight of 3500 kg when $\theta = 12°$.

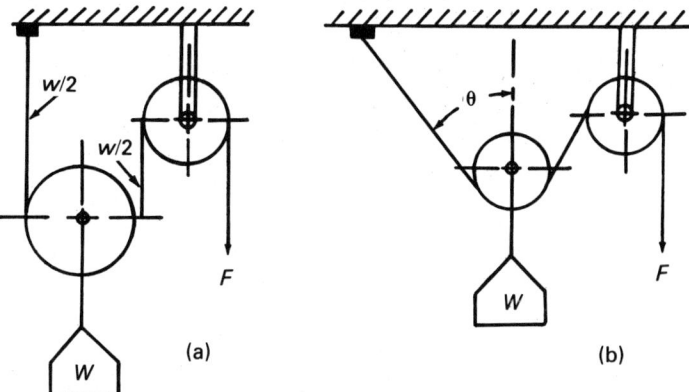

Fig. 5-2. Two-wheel-and-pulley arrangement.

$$F = \frac{W(\sec \theta)}{2}$$
$$= \frac{3500(\sec 12°)}{2}$$
$$= \frac{3500(1.02234)}{2}$$
$$= 1789.095 \text{ kg}$$

The third example of wheels and pulleys is illustrated in Figure 5-3 and addresses wheel-and-pulley problems with multiple blocks and strands. Here, n will equal the total number of strands or parts of rope used in the pulley system. In this configuration, the rate at which the weight is raised will be equal to $1/n$ of the velocity of the

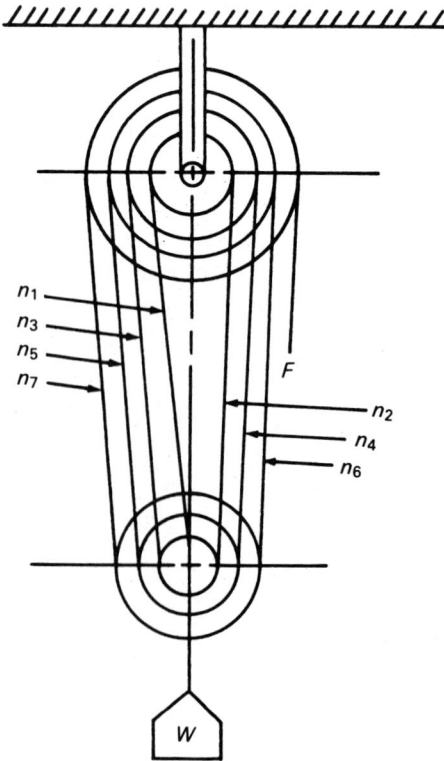

Fig. 5-3. Multiple block arrangement.

applied force (F). The formula used to calculate the applied force needed to lift a given weight is:

$$F = \frac{W}{n}$$

Figure 5-3 is an illustration of a triple and quadruple block with a total of seven strands of rope. To determine the amount of force needed to lift 540 lb of weight with this arrangement, the following calculation is made:

$$F = \frac{W}{n} = \frac{540}{7} = 77.143 \text{ lb}$$

The final example of a wheel-and-pulley system is a series of gears placed contiguous to one another (Figure 5-4). Here, letters A, B, C, and D represent the pitch circles (pitch diameters) of each gear. The two basic formulas used with this arrangement are:

$$F = \frac{Wrr_1r_2}{RR_1R_2}$$

$$W = \frac{FRR_1R_2}{rr_1r_2}$$

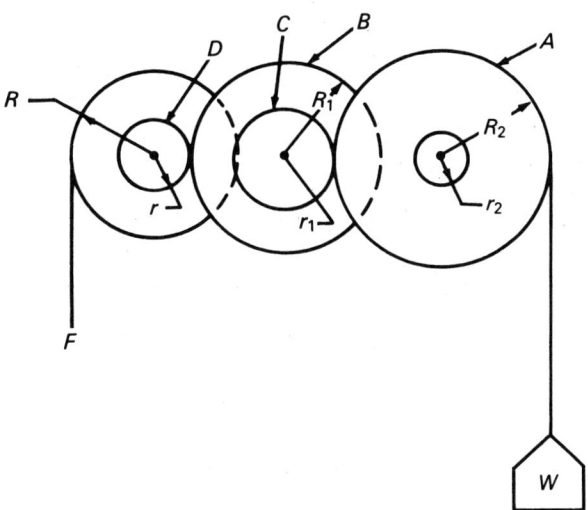

Fig. 5-4.   Series pulley system.

130 • COMPLEX MECHANICS

If the pitch diameters of the four gears shown in Figure 5-4 are 32, 30, 16, and 14 cm, respectively, with $R = 18$ cm, $R_1 = 22$ cm, $R_2 = 23$, $r = 9$ cm, $r_1 = 12$ cm, and $r_2 = 4$ cm, to find the amount of force needed to lift a weight of 750 kg, the following procedure would be used:

$$F = \frac{Wrr_1r_2}{RR_1R_2}$$
$$= \frac{(750)(9)(12)(4)}{(18)(22)(23)}$$
$$= 35.573 \text{kg}$$

Differential Pulley

Unlike simple pulley systems, a differential pulley requires the use of a chain-and-sprocket arrangement so that the chain will not slip as it moves across the face of the pulley. Shown in Figure 5-5 is an example of such an arrangement. The basic formulas related to this system are:

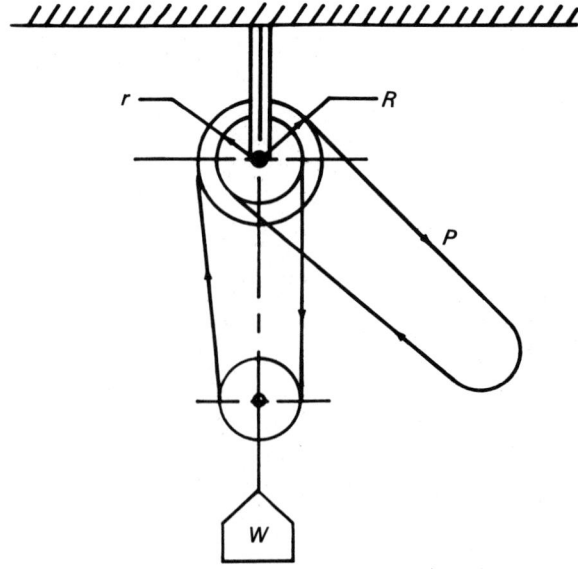

Fig. 5-5. Differential Pulley.

$$PR = \frac{W(R-r)}{2}$$

$$Fp = \frac{W(R-r)}{2}$$

$$W = 2\frac{(Fp)(R)}{R-r}$$

An example of a typical problem is to find the amount of pulling force required to lift a weight of 340 lb with a differential pulley when $R = 5$ in. and $r = 3$ in. This is solved as follows:

$$Fp = \frac{W(R-r)}{2R}$$
$$= \frac{340(5-3)}{(2)(5)}$$
$$= 68 \text{ lb}$$

Exercises

1. Referring to Figure 5-1, what is the maximum weight that can be lifted with a pulley system that has a drum diameter of 4 cm, periphery gear diameter of 24 cm, and a downward force of 333 kg?

2. What force would be needed to lift a 1000-kg weight on the pulley given in problem 1?

3. Referring to Figure 5-2(a), at what velocity will a 100-lb weight be lifted when a force of 25 lb is applied?

4. In a pulley arrangement similar to that shown in Figure 5-2(b), calculate the force needed to lift a 45-kip weight when $\theta = 30°$.

5. What weight can be lifted with the pulley given in problem 4 if a force of 50 lb is applied?

6. In a two-double-block pulley system, how much force is needed to lift a weight of 340 lb?

7. Given a pulley arrangement similar in appearance to that in Figure 5-4, pitch diameters of the gears for A, B, C, and D equal to 130, 128, 112, and 110 cm, respectively, and $R = 112$ cm, $R_1 = 114$ cm, $R_2 = 115$ cm, $r = 55$ cm, $r_1 = 56$ cm, and $r_2 = 54$

cm, how much force would be required to lift a weight of 560 kg?

8. For the same pulley given in problem 7, what is the maximum amount of weight that can be lifted with a force of 35 kg?

9. How much pulling force is required to lift a load of 250 lb with a differential pulley with a drum radius of 4 in. and a periphery radius of 8 in. applied?

Center of Gravity and Moments of Inertia

There are times when certain types of data are required to solve applied mechanics problems. This is especially true when the shape of an object will influence the way it behaves. Two important properties here are center of gravity and moments of inertia.

Center of Gravity

To locate the center of gravity for any moving machine part is important, since it will dictate the type of motion that a part will exhibit. As a case in point, the distance above ground level of the center of gravity of a forklift truck will determine whether or not it will overturn when negotiating certain curves with given loads.

The most basic type of problem is to find the center of gravity for a system of particles that vary in weight and are located in space. An example of this is shown in Figure 5-6, where a system is defined

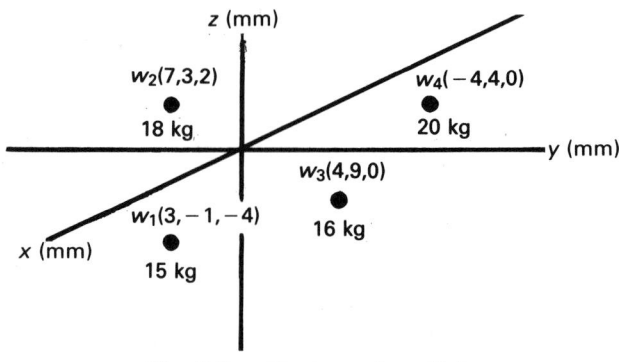

Fig. 5-6. System of particles.

by a three-axis coordinate system. To find the center of gravity for this system, we must first find the resultant, or total weight, by summing the weights of each individual particle (w_1). In the following formulas, the upper case sigma (Σ) is used to designate the sum of a series of value (e.g., $\Sigma = n_1 + n_2 + n_3 + \ldots$). Thus:

$$W = \Sigma(w_1)$$
$$= 15 + 18 + 16 + 20$$
$$= 69 \text{ kg}$$

To locate the x, y, and z coordinates of the center of gravity, the following formulas are used:

$$x = \frac{\Sigma(w_1 x_1)}{\Sigma(w_1)}$$
$$= \frac{(15)(3) + (18)(7) + (16)(4) + (20)(-4)}{69}$$
$$= 2.25 \text{ mm}$$

$$y = \frac{\Sigma(w_1 y_1)}{\Sigma(w_1)}$$
$$= \frac{(15)(-1) + (18)(3) + (16)(9) + (20)(4)}{69}$$
$$= 3.81 \text{ mm}$$

$$z = \frac{\Sigma w_1 z_1}{\Sigma(w_1)}$$
$$= \frac{(15)(-4) + (18)(2) + (16)(0) + (20)(0)}{69}$$
$$= -0.35 \text{ mm}$$

Therefore, the coordinates of the center of gravity for this system are $x = 2.25$ mm, $y = 3.81$ mm, $z = -0.35$ mm.

Before we progress to the center of gravity for more complex figures, it is necessary to understand the property known as *centroids*. A centroid is a geometric property that represents the central point of a line, surface, or solid. To find the centroid, the surface is divided into small areas noted as ΔA. Other factors considered are density (ρ) and thickness (t). Therefore, the three formulas used to find the coordinates of the centroid are:

$$x = \frac{\Sigma(\Delta A_l t \rho x_l)}{\Sigma(\Delta A_l t \rho)}$$

$$y = \frac{\Sigma(\Delta A_l t \rho x_l)}{\Sigma(\Delta A_l t) \rho}$$

$$z = \frac{\Sigma(\Delta A_l t \rho x_l)}{\Sigma(\Delta A_l t \rho)}$$

The term *centroid* is frequently used as the location of the center of gravity when geometry is the only concern. If the densities to objects are nonuniform, the centroid and center of gravity will differ. Presented here are formulas for the center of gravity for geometric objects.

LINES

The center of gravity for a line is its midpoint, and is determined by halving its length (l). The coordinate of this point will be solved by the formula $g_o = \frac{1}{2}$. The coordinates for the center of a circular arc are calculated by the following formulas (see Figure 5-7):

$$x_o = \frac{r \sin c}{\text{rad } c}$$

$$y_o = \frac{(2r \sin^2 1/2c)}{\text{rad } c}$$

Note that rad c is angle c measured in radians.

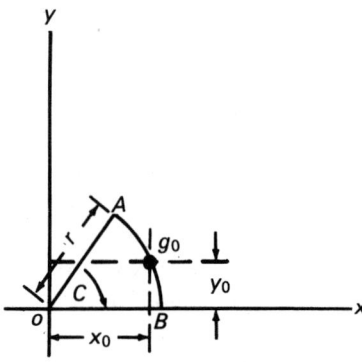

Fig. 5-7. Center of gravity of a circular arc.

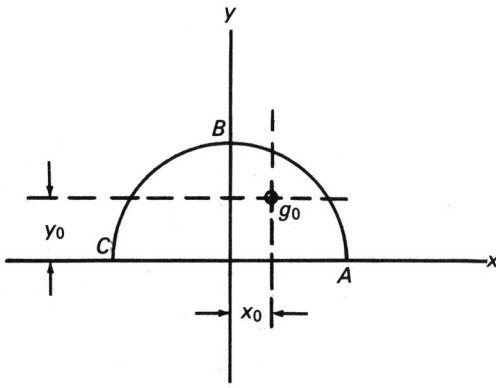

Fig. 5-8. Center of gravity of quadrant *AB* and Semicircle *ABC*.

Shown in Figure 5-8 are two types of lines: quadrant and semi-circumference. To calculate the center coordinates of the quadrant, the following formulas are used:

$$x_o = y_o = \frac{2r}{\pi}$$

The center coordinates formulas for the semicircumference are:

$$x_o = 0$$
$$y_o = \frac{2r}{\pi}$$

PLANE AREAS

The basic plane figure is the triangle, whose center is determined by the intersection of lines drawn from the vertices to the midpoint of their opposite side. The center of gravity for parallelograms is found by the intersection of their diagonals.

To find the center of gravity for a trapezoid (Figure 5-9), the following three formulas have been derived:

$$c = \frac{h(a+2b)}{3(a+b)}$$
$$d = \frac{h(2a+b)}{3(a+b)}$$
$$e = \frac{a^2 + ab + b^2}{3(a+b)}$$

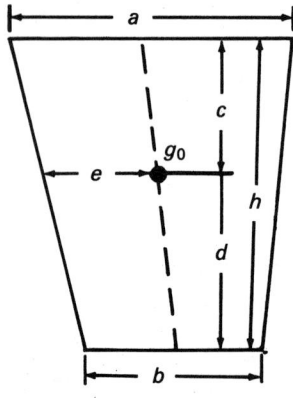
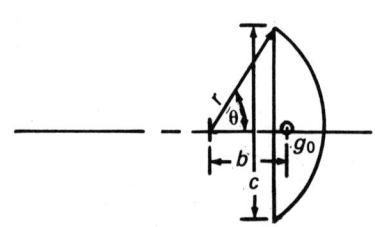

Fig. 5-9. Center of gravity of a trapezoid.

Fig. 5-10. Center of gravity of a circle segment.

The center of a circle segment (Figure 5-10) is found with either of the following formulas, where A = the area of the segment:

$$b = \frac{c^3}{12A} \text{ or}$$
$$b = \frac{2r^3 \sin^3 \theta}{3A}$$

Similarly, the center of gravity for a circle sector (Figure 5-11), where A = the area of the sector, is calculated by using the following formulas:

$$b = \frac{2rc}{3l} \text{ or}$$
$$b = \frac{r^2 c}{3A} \text{ or}$$
$$b = \frac{(38.197)(r \sin \theta)}{\theta}$$

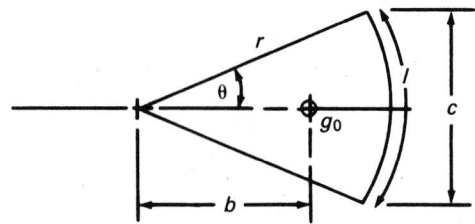

Fig. 5-11. Center of gravity of a circular sector.

Center of Gravity and Moments of Inertia • 137

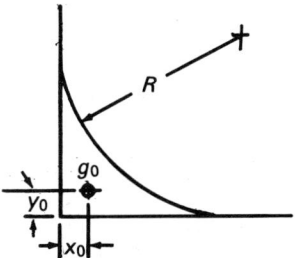

Fig. 5-12. Center of gravity of a spandrel.

The center of gravity for the area of spandrel or fillet (Figure 5-12) will make use of the following formula:

$$x_o = y_o = 0.2234R$$

The center of gravity for the area of a complete parabola [Figure 5-13(a)] is calculated with the following formula:

$$a = \frac{3h}{5}$$

To find the center of gravity for a half of a parabola [Figure 5-13(b)], use the following formulas:

$$a = \frac{3h}{5}$$

$$b = \frac{3w}{8}$$

SOLIDS

Center-of-gravity problems are frequently associated with solids or volumes. Here, the center of gravity must be calculated to maintain

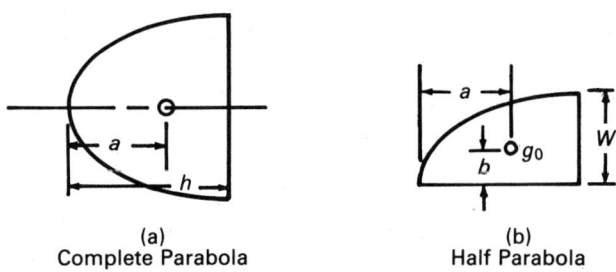

(a)
Complete Parabola

(b)
Half Parabola

Fig. 5-13. Center of gravity of a parabola.

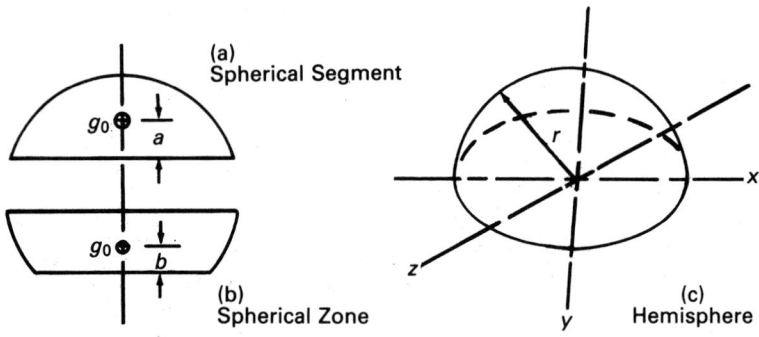

Fig. 5-14. Spherical solids.

balance in product handling and for proper equipment installation. Several solid figures are considered basic to most problems. One of these involves finding the center of gravity for the spherical surface of segments and zones of spheres [Figure 5-14(a) and 5-14(b)]. To find the center of gravity, distances a and b must be determined, as follows:

$$a = \frac{h}{2}$$

$$b = \frac{H}{2}$$

The center of gravity for a hemisphere, as illustrated in Figure 5-14(c), only requires that you know what the radius is. From this it is possible to calculate the distance from the flat base along the centerline. The distance a is determined by:

$$a = \frac{3r}{8}$$

The center of gravity for a solid cylinder is located along the centerline that cuts through its center and halfway through the solid. To find the same location in a hollow cylinder [Figure 5-15(a)] that has only one end (e.g., base), the following formulas are applied:

$$a = \frac{2h^2}{4h + d}$$

If the cylinder is cut at an inclined angle [Figure 5-15(b)], the following formulas are used:

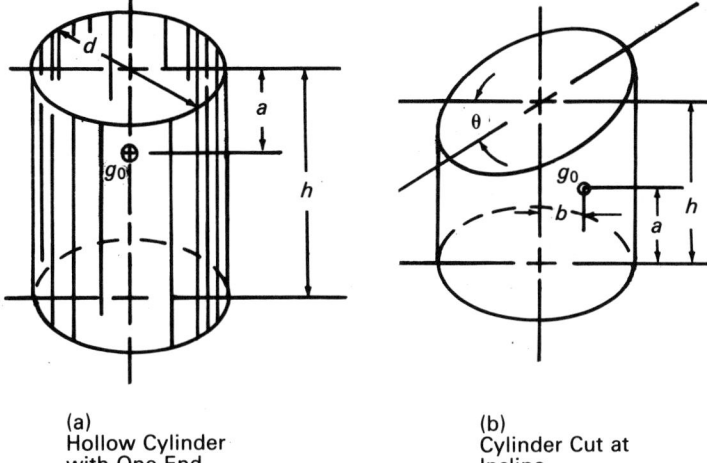

(a)
Hollow Cylinder
with One End

(b)
Cylinder Cut at
Incline

Fig. 5-15. Cylinders.

$$a = \frac{h}{2} + \frac{r^2 \tan^2 \theta}{8h}$$
$$b = \frac{r^2 \tan \theta}{4h}$$

The center of gravity for a solid pyramid or cone is found at a location that will be one-fourth the distance from the base and along the centerline running from the apex to the center of the base (Figure 5-16). This location is made with the following formula:

$$a = \frac{h}{4}$$

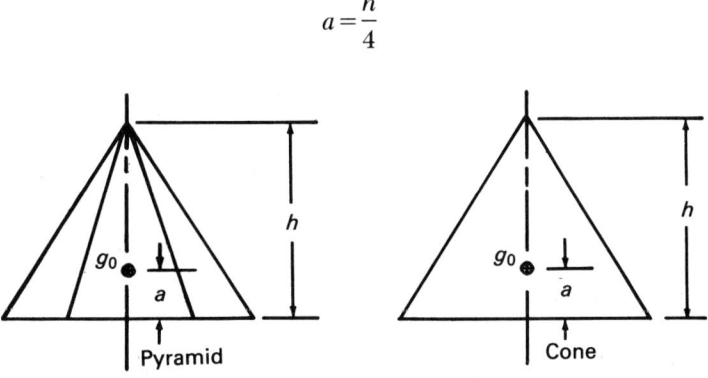

Pyramid

Cone

Fig. 5-16. Pyramid and cone.

140 • COMPLEX MECHANICS

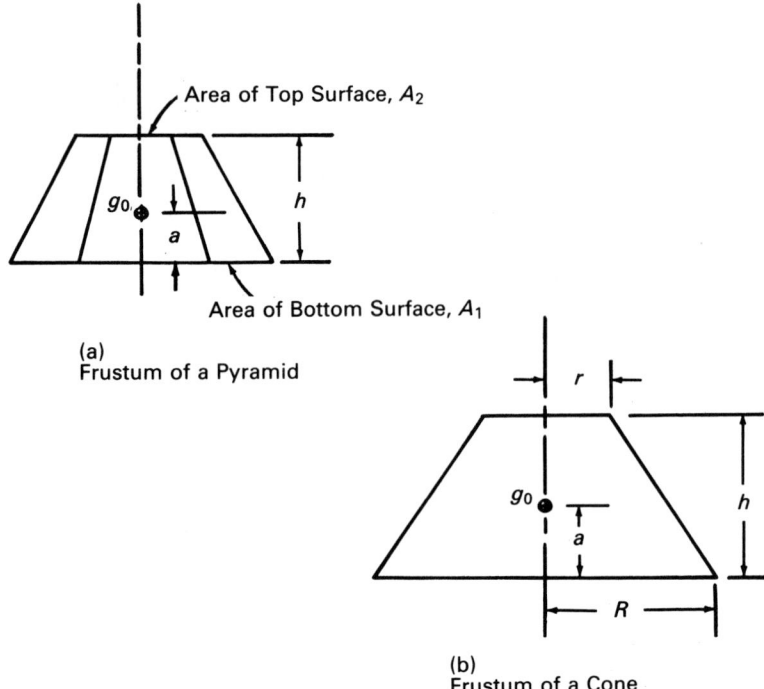

(a) Frustum of a Pyramid

(b) Frustum of a Cone

Fig. 5-17. Frustrum of pyramid and cone.

The center of gravity for a solid frustum of a pyramid will be found along the centerline that joins the center of each base [Figure 5-17(a)]. Its location on this centerline is measured from the bottom base and makes use of the surface area of the base (A_1) and top (A_2). The formula used here is:

$$a = \frac{h(A_1 + 2\sqrt{A_1 A_2} + 3A_2)}{4(A_1 + \sqrt{A_1 A_2} + A_2)}$$

The same concept applies to finding the center of gravity for the frustum of a cone [Figure 5-17(b)], except that the related formula uses the radius for the top surface (r) and bottom surface (R). Thus:

$$a = \frac{h(R + 2r)}{3(R + r)}$$

Exercises

10. Find the center of gravity for four points whose weight and location are: $A = 125$ kg $(5, -8, 7)$, $B = 90$ kg $(0, -5, 6)$, and $C = 290$ kg $(7, 6, -7)$.

11. Determine the location of the center of gravity for the trapezoid shown in Figure 5-18.

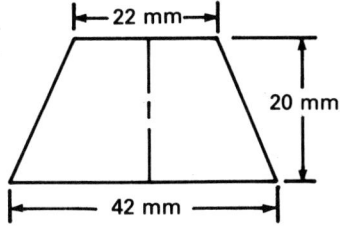

Note: The center of gravity will always be on the line joining the centers of the top and bottom lines.

Fig. 5-18. Problem 11.

12. What is the center of gravity for a straight line whose length is 54.74 in.?

13. Determine the x and y coordinates for the center of gravity of a spandrel whose sides are 5 m.

14. What will be the location of the center of gravity for half of a sphere whose radius is 72 cm?

15. Calculate the center of gravity for the two cylindrical solids illustrated in Figure 5-19.

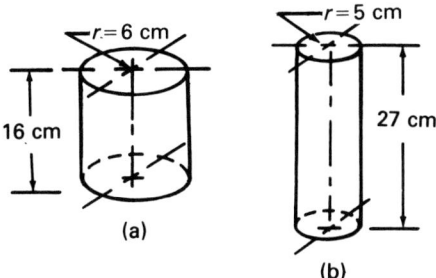

Fig. 5-19. Problem 15.

16. At what height above the base is the center of gravity for a solid cone whose base diameter is 3.2 m and total height is 12 m?

142 • **COMPLEX MECHANICS**

17. Calculate the center of gravity for the two frustums shown in Figure 5-20.

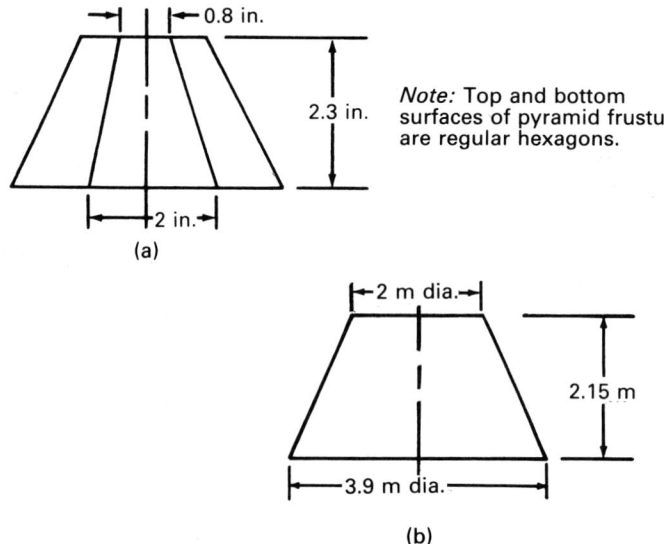

Fig. 5-20. Problem 17.

Moments of Inertia

A property of areas and solids in the field of engineering is the moment of inertia. Considered here are *area* and *polar* moments of inertia. The area moment of inertia is sometimes referred to as the *second moment of area* and is used in the design and analysis of beams and shafts. A moment of inertia about an axis that is perpendicular to the plane of the area is called the polar moment of inertia and is used for shafts and beams that are subjected to torque.

MOMENT OF INERTIA OF AN AREA

Using Figure 5-21 as an example, we note that the area shown is divided into numerous smaller areas equal to a. Their distances are measured from the x and y axes. The moment of inertia of the area in relationship to the x axis will be equal to the sum of the quantities

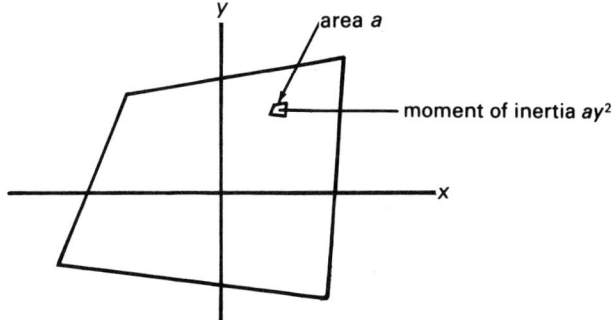

Fig. 5-21. Area made up of numerous small areas.

obtained by multiplying each a area by the square of its distance from the x axis along the y axis. Each small area, then would be equal to ay^2, and the whole area would be the sum of all ay^2, and be represented as Σay^2.

The symbol used to designate moment of inertia is I, with a subscript, such as I_x, designating the axis to which it refers. Hence:

$$I_x = \Sigma ay^2$$

The units expressing moments of inertia are somewhat different than those found in other expressions. Because we multiply the area (measured in square units) by y^2, the units are raised to the fourth power. Thus, the units are expressed in terms such as in.^{4} or mm^4.

The moments of simple areas are the basis for calculating more complex areas. Shown in Figure 5-22 are illustrations of each figure to be covered here, namely, rectangular, triangular, and circular areas.

The moment of inertia for a rectangular area will be equal to $\frac{1}{12}$ times the width times the depth cubed, where $b =$ width and $h =$ depth. This is represented in the formula:

$$I_x = \frac{1}{12} bh^3$$

To find the moment of inertia of a rectangular area 5 cm wide and 10 cm deep, with respect to a central axis parallel to the base [noted as G in Figure 5-22(a)], the following calculation is made:

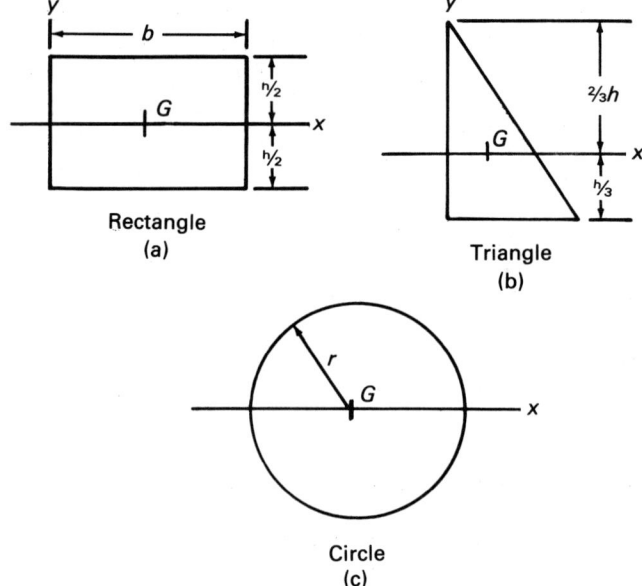

Fig. 5-22. Moments of inertia for simple areas.

$$I_x = \left(\frac{1}{12}\right)bh^3$$
$$= \left(\frac{1}{12}\right)(5)(10^3)$$
$$= 416.667 \text{ cm}^4$$

To find the moment of inertia for the same rectangle, but relative to the y axis, the following calculation is made:

$$I_y = \frac{1}{12}bh^3$$
$$= \left(\frac{1}{12}\right)(10)(5^3)$$
$$= 104.167 \text{ cm}^4$$

The moment of inertia for a triangular area is calculated by use of the following formula:

$$I_x = \frac{1}{36}bh^3$$

To determine the moment of inertia for the area of a right triangle that is 13 in. wide and 16 in. deep, with respect to the x axis, the following procedure is used:

$$I_x = \frac{1}{36}bh^3$$
$$= \frac{1}{36}(13)(16^3)$$
$$= 1479.111 \text{ in.}^4$$

The moment of inertia for a circular area with respect to the axis along its diameter, where r is the circle's radius, is:

$$I_x = \frac{\pi}{4}r^4$$

To calculate the moment of inertia for a circle with a diameter of 6 m, with respect to the x and y axes, the following manipulations are made:

$$I_x = I_y = \frac{\pi}{4}r^4$$
$$= \frac{3.14159}{4}6^4$$
$$= 1017.875 \text{ m}^4$$

To calculate the moment of inertia for a semicircle, simply halve the formula:

$$I_x = \frac{\pi}{8}r^4$$

The formula for the moment of inertia of a quadrant is therefore expressed as:

$$I_x = \frac{\pi}{16}r^4$$

MOMENT OF INERTIA OF COMPOSITE AREAS

Composite areas are areas that are made up, or divided into, simple areas such as rectangles, triangles, and circles. Here, the moments of inertia will be either added to or subtracted from each other. One

146 • COMPLEX MECHANICS

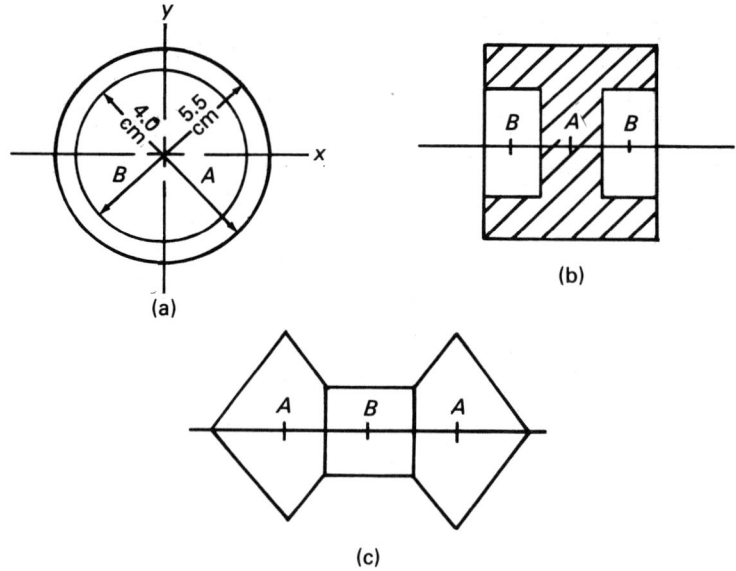

Fig. 5-23. Moment of inertia for composite areas.

such example is the ring illustrated in Figure 5-23(a). The moment of inertia is calculated for the 5.5-cm-radius circle (A); then the moment of inertia of the smaller circle, with radius of 4.0 cm (B), is subtracted from it. Thus:

$$I_X = I_{XA} - I_{XB}$$

The moment of inertia for the composite area in Figure 5-23(b) is calculated by using the following formula:

$$I_X = I_{XA} - 2I_{XB}$$

The moment of inertia for the composite area in Figure 5-23(c) can be found by the following formula:

$$I_X = 2I_{XA} + I_{XB}$$

POLAR MOMENTS OF INERTIA

The polar moments of inertia can be specified in terms of the x, y, and z axes, or I_x, I_y, and I_z. The procedures used here are the same as for calculating moment of inertia of areas. These formulas are summarized in Table 5-1.

Center of Gravity and Moments of Inertia • 147

Table 5-1. Moment of Inertia Formulas for Solids and Volumes

Solid		Moments of Inertia*
Rectangular parallelepiped		$I_X = \left(\dfrac{m}{12}\right)(a^2 + c^2)$ $I_Y = \left(\dfrac{m}{12}\right)(b^2 + c^2)$ $I_Z = \left(\dfrac{m}{12}\right)(b^2 + c^2)$
Circular cylinder		$I_X = \dfrac{mr^2}{4} + \dfrac{ml^2}{12}$ $I_Y = \dfrac{mr^2}{2}$ $I_Z = I_X$
Right-circular cone		$I_X = \dfrac{3mr^2}{20} + \dfrac{mh^2}{10}$ $I_Y = \dfrac{3mr^2}{10}$ $I_Z = I_X$
Sphere		$I_X = \dfrac{2mr^2}{5}$ $I_Y = I_X$ $I_Z = I_X$
Hemisphere		$I_X = \dfrac{2mr^2}{5}$ $I_Y = I_X$ $I_Z = I_X$

*Note: m = mass of volume

Exercises

Solve for the following moment-of-inertia area problems illustrated in Figure 5-24. Find both I_x and I_y.

18. Rectangular area in Figure 5-24(a).

19. Triangular area in Figure 5-24(b).

20. Triangular area in Figure 5-24(c).

21. Semicircle area in Figure 5-24(d).

22. Quadrant in Figure 5-24(e).

148 • COMPLEX MECHANICS

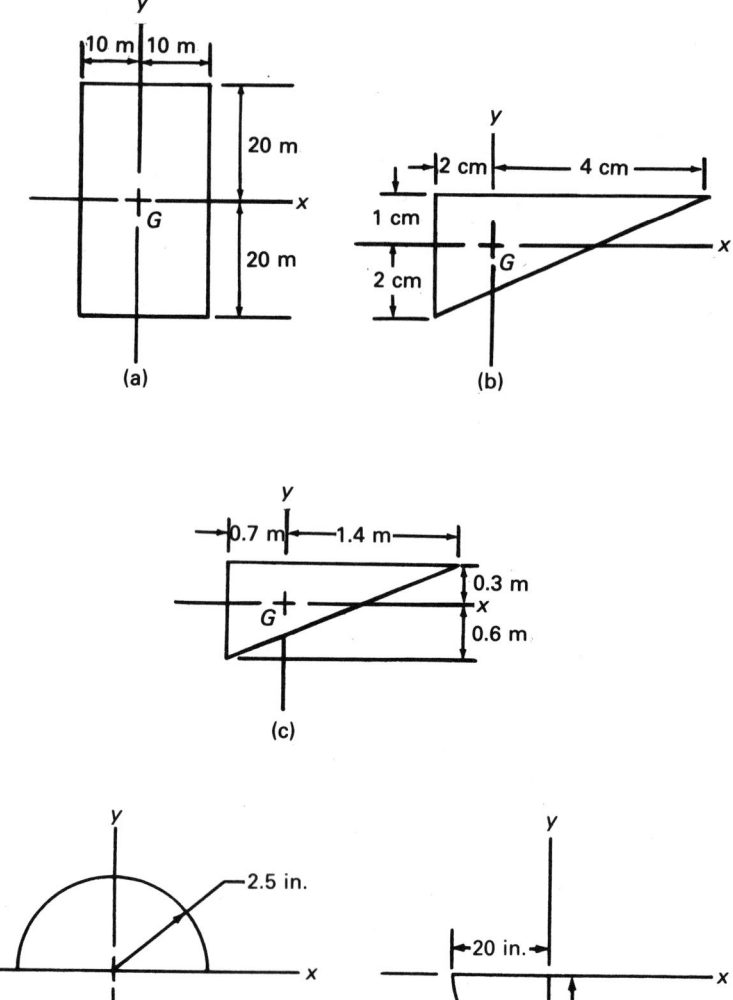

Fig. 5-24. Moment-of-inertia problems.

Find the moment of inertia of the composite areas illustrated in Figure 5-25.

Center of Gravity and Moments of Inertia • 149

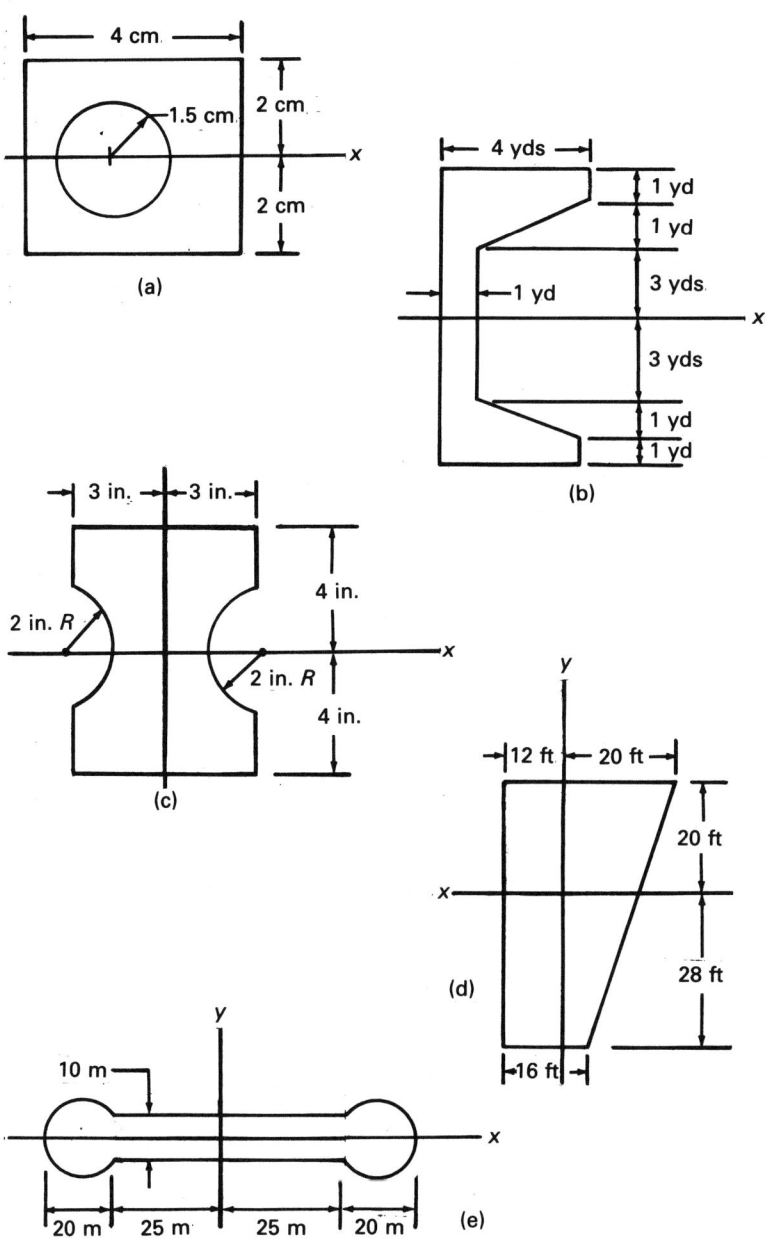

Fig. 5-25. Moment-of-inertia composite area problems.

23. Find I_x for area in Figure 5-25(a).
24. Find I_x for area in Figure 5-25(b).
25. Find both I_x and I_y for composite area in Figure 5-25(c).
26. Find both I_x and I_y for composite area in Figure 5-25(d).
27. Find I_x for composite area in Figure 5-25(e).

Find the polar moment of inertia for the masses illustrated in Figure 5-26.

28. Find the moment of inertia along all three axes for the rectangular parallelepiped in Figure 5-26(a).
29. Determine I_x, I_y, and I_z for the cone in Figure 5-26(b).
30. Find the moment of inertia for the sphere in Figure 5-26(c).

Velocity and Acceleration

Motion is the process of moving a body from one location or position to another. It is a process whereby a body progressively changes its position. There are two terms that are usually associated with motion. The first is *velocity* (v), which is the rate of change of distance (d) over time (t) of a body, and is expressed as:

$$v = \frac{d}{t}$$

The second term is *acceleration* (a), which is the rate of change of velocity relative to time, expressed as:

$$a = \frac{v}{t}$$

It should be noted that when velocity and acceleration are given per unit of time (e.g., second, minute, or hour), this will be known as *instantaneous speed*. In some cases, it will be necessary to determine *average speed* ($\bar{v}$). Average speed is calculated by dividing the total distance that it covers divided by the period of time (t) in which the distance was covered. Thus:

$$\bar{v} = \frac{d}{t}$$

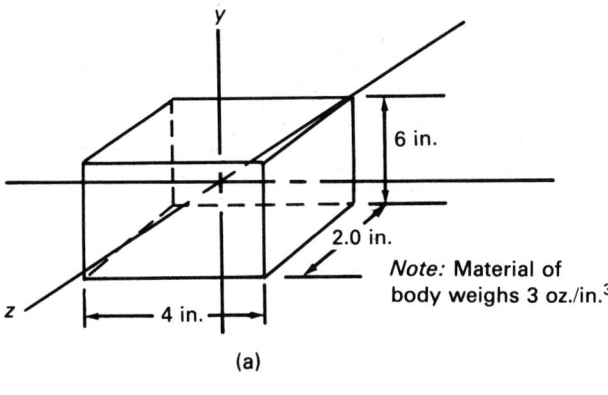

(a)

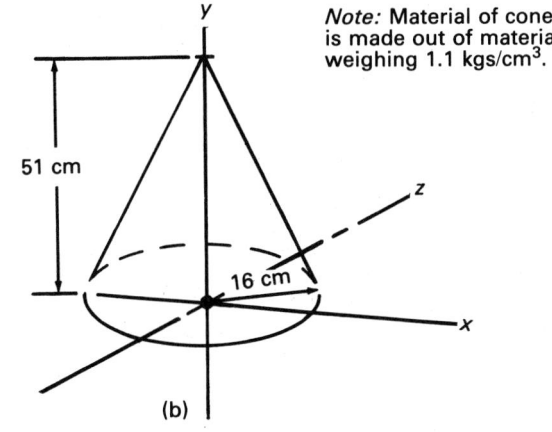

(b)

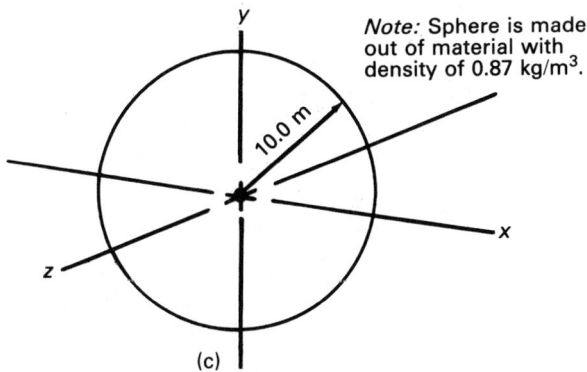

(c)

Fig. 5-26. Polar-moments-of-inertia problems.

An example of such a problem is to determine the average speed for a vehicle that travels a total distance of 175 m in 27 second(s). To find the average speed, the following calculation is made:

$$\bar{v} = \frac{d}{t}$$
$$= \frac{175}{27}$$
$$= 6.4815 \text{ m/s}$$

Motion will be *uniform* when the body's velocity is constant over the entire time of movement. As the velocity of a body increases over time, it is said to *accelerate*; *deceleration* occurs as velocity decreases over time. When acceleration is uniform, then the motion is known as *uniformly accelerated motion*.

Constant Velocity Motion

There are several definitions that one should be aware of when working with kinematic problems. These deal with the direction in which a body is moving. The first is *linear motion*, which pertains to bodies that travel in a "linear," or straight-line, path. *Angular motion* is movement of a rotating body; it is measured in radians and expressed by the Greek letter omega (ω). An example of angular motion is the sweeping of a second hand on a clock. A common unit of angular velocity is the revolution per minute (rpm), where 1 rpm = 0.105 rad/s.

When working with constant velocity motion, there are two basic sets of formulas that one would use. In these formulas, the following notations are employed:

d = distance traveled or moved
t = time or duration of movement
θ = angle of rotation (expressed in radians)
ω = angular velocity

The formulas used for constant linear velocity are:

$$d = vt$$
$$v = \frac{d}{t}$$
$$t = \frac{d}{v}$$

Constant angular velocity formulas are:

$$\theta = \omega t$$
$$\omega = \frac{\theta}{t}$$
$$t = \frac{\theta}{\omega}$$

To calculate the constant angular velocity for a body that has moved 43° over a period of 2.75 s, the following calculations would be made (note: rad = radian).

$$\omega = \frac{\theta}{t}$$
$$= \frac{0.75049}{2.75}$$
$$= 0.2729 \text{ rad/s}$$

The relationship that exists between linear and angular motion is precise. This can be expressed in the following relationship of a point at a distance r (radius of the circular motion) from the center of rotation:

$$v = r\omega$$

An example of this relationship in measured units is: m/s = mrad/s. Furthermore, the distance traveled by the body during rotation through a given angle can be expressed as:

$$d = r\theta$$

Using the same units of measure, this relationship can be given as: m = mrad.

Exercises

31. Calculate the distance covered by a body that is traveling at a constant linear velocity of 34 m/s over a period of 15 min.
32. What is the length of time that it would take a vehicle to travel a distance of 12 km if its constant linear velocity was 52,000 m/hr?
33. What is the constant angular velocity of a second hand on a clock over a 15-s period?

34. If the constant angular velocity of a rotating ball was 5.0 rad/s, how many degrees would it travel in 5 s?

35. What would be the distance traveled by a rotating ball that is 12 ft from the center of rotation and traveling at a constant velocity so that it covered a total of 450°?

Constant Acceleration Linear Motion

When a body moves along a straight path, it is known as linear motion. Here, two types of situations can exist. The first is where the body begins to move from rest and accelerates to a given or final speed. The second case is where a body is already moving at an initial or original velocity and accelerates or decelerates to a final velocity. In this book, the original velocity will be noted as v_o and the final velocity is v.

There are a number of different formulas available for both linear motion situations, depending on what data are available. Table 5-2 is a presentation of linear motion with constant acceleration formulas that are used for uniform acceleration from rest and initial velocity.

A typical problem concerns a vehicle whose initial speed was 88 m/s and acceleration (deceleration) was -2m/s^2. It must now be determined what speed was attained by the vehicle after 30 s. To solve this problem, the following calculations are made:

$$\begin{aligned} v &= v_o + at \\ &= 88 + (-2)(30) \\ &= 88 - 60 \\ &= 28 \text{ m/s} \end{aligned}$$

Another example problem involves a car moving at a rate of 80 km per hour (km/hr) when the driver hits the brakes to avoid an accident. If it takes the car 3 s to slow down to 40 km/hr, what is the rate of deceleration, the length of time that it will take the car to come to a complete halt, and the total distance it will have traveled? To answer these questions, the following calculations are made:

$$a = \frac{v - v_o}{t}$$
$$= \frac{40 - 80}{3}$$
$$= -13.333 \text{ m/s}$$
$$t = \frac{v - v_o}{a}$$
$$= \frac{0 - 80}{-13.333}$$
$$= 6.000 \text{ s}$$
$$d = \frac{t(v + v_o)}{2}$$
$$= \frac{6(40 + 80)}{2}$$
$$= 360 \text{ m}$$

Exercises

36. A man notes that a lightning bolt struck a tree 2.1 mi away from him. Assuming that sound travels at a speed of 1130 ft/s, how long will he have to wait until he hears the thunder?

37. What is the ground speed of an airplane whose air speed is 2000 m/hr when it is flying into a head wind of 56 m/hr? What distance can it cover, relative to the ground, in 10 h?

38. A driver slams on the brakes of a vehicle that is moving at 50 mi/hr, and it takes 4 s to come to a complete halt. What was the vehicle's acceleration?

39. How long would the vehicle in problem 38 take to come to a complete halt if it were traveling at a speed of 80 mi/hr?

40. How long would it take the vehicles in problems 38 and 39 to decelerate to a speed of 20 mi/hr?

41. An object starts from rest and moves in a linear path with a constant acceleration. If it covers a distance of 64 m in 4 s, what would be its final velocity? How much time would be required for it to cover half the distance?

42. If the initial speed of the object in problem 41 were 10 m/s, how far would it travel in 1 min?

COMPLEX MECHANICS

Table 5-2. Formulas for Linear Motion with Constant Acceleration

Unknown Quantity	Formula to Find
Uniform Acceleration from Rest	
Distance	$d = \dfrac{at^2}{2}$
	$d = \dfrac{vt}{2}$
	$d = \dfrac{v^2}{2a}$
Final velocity	$v = at$
	$v = \dfrac{2d}{t}$
	$v = \sqrt{2ad}$
Time	$t = \dfrac{2d}{v}$
	$t = \sqrt{2d/a}$
	$t = \dfrac{v}{a}$
Acceleration	$a = \dfrac{2d}{t^2}$
	$a = \dfrac{v^2}{2d}$
	$a = \dfrac{v}{t}$
Uniform Acceleration from Original Velocity (v_o):	
Distance	$d = v_o t + \dfrac{at^2}{2}$
	$d = \dfrac{(v + v_o)t}{2}$
	$d = (v^2 - v_o^2)2a$
	$d = vt - \frac{1}{2}at^2$
Final velocity	$v = v_o + at$
	$v = \dfrac{2d}{t} - v_o$
	$v = \sqrt{v_o^2 + 2ad}$

(continued)

Table 5-2. (*continued*)

Unknown Quantity	Formula to Find
	$v = \dfrac{d}{t} - \dfrac{at}{2}$
Original velocity	$v_o = \sqrt{v^2 - 2ad}$
	$v_o = \dfrac{2d}{t} - v$
	$v_o = v - at$
	$v_o = \dfrac{d}{t} - \dfrac{at}{2}$
Time	$t = \dfrac{v - v_o}{a}$
	$t = \dfrac{2d}{v + v_o}$
Acceleration	$a = \dfrac{v^2 - v_o^2}{2d}$
	$a = (v - v_o)t$
	$a = \dfrac{2(d - v_o t)}{t^2}$
	$a = \dfrac{2(vt - d)}{t^2}$

Constant Acceleration and Rotational Motion

Similar to the two types of linear motion problems, there are also two types of rotational motion problems: from rest and from initial velocity. When working with rotational motion, rotation is measured in terms of angular distance (θ) in radians, angular acceleration (α) is given in rad/s/s, final angular velocity is noted as ω, and initial angular velocity is ω_o.

The various formulas used for solving problems of constant acceleration and rotational motion are presented in Table 5-3.

A common application of angular velocity problems is the calculation of revolutions per minute (rpm). An example is a wheel rotating at a rate of 185 rpm that is slowed to 120 rpm in 0.250 s. To calculate the angular deceleration, it is necessary to employ con-

158 • COMPLEX MECHANICS

Table 5-3. Formulas for Constant Acceleration and Rotational Motion

Unknown Quantity	Formula to Find

Uniform Acceleration from Rest

Angular distance of rotation	$\theta = \dfrac{\alpha t^2}{2}$
	$\theta = \dfrac{\omega t}{2}$
	$\theta = \dfrac{\omega^2}{2}$
Final angular velocity	$\omega = \alpha t$
	$\omega = \dfrac{2\theta}{t}$
	$\omega = \sqrt{2\alpha\theta}$
Time	$t = \dfrac{2\theta}{\omega}$
	$t = \sqrt{2\theta/\alpha}$
	$t = \dfrac{\omega}{\alpha}$
Angular acceleration	$\alpha = \dfrac{2\theta}{t^2}$
	$\theta = \dfrac{\omega^2}{2}$
	$\alpha = \dfrac{\omega}{t}$

Uniform Acceleration from Original Velocity (ω_o)

Angular distance of rotation	$\theta = \omega_o t + \dfrac{\alpha t^2}{2}$
	$\theta = \dfrac{(\omega + \omega_o)t}{2}$
	$\theta = \dfrac{\omega^2 - \omega_o^2}{2}$
	$\theta = \dfrac{\omega t - \alpha t^2}{2}$
Final angular velocity	$\omega = \omega_o + \alpha t$
	$\omega = \dfrac{2\theta}{t} - \omega_o$

(continued)

Table 5-3. (*continued*)

Unknown Quantity	Formula to Find
Original angular velocity	$\omega = \sqrt{\omega_0^2 + 2\alpha\theta}$ $\omega = \dfrac{\theta}{t} + \dfrac{t\alpha}{2}$ $\omega_0 = \sqrt{\omega^2 - 2\alpha\theta}$ $\omega_0 = \dfrac{2\theta}{t} - \omega$ $\omega_0 = \omega - \alpha t$ $\omega_0 = \dfrac{\theta}{t} - \dfrac{\alpha t}{2}$
Time	$t = \dfrac{\omega - \omega_0}{\alpha}$ $t = \dfrac{2\theta}{\omega + \omega_0}$
Angular acceleration	$\alpha = \dfrac{\omega - \omega_0^2}{2}$ $\alpha = \dfrac{\omega - \omega_0 \theta}{t}$ $\alpha = \dfrac{2(\theta - \omega_0 t)}{t^2}$ $\alpha = \dfrac{2(\omega t - \theta)}{t^2}$

version Tables 5-4 or 5-5. With this, the following calculations are made:

$$\alpha = \dfrac{\omega - \omega_o}{t}$$
$$= \dfrac{12.57 - 19.37}{0.250}$$
$$= -272 \text{ rad/s/s}$$

Exercises

43. What will be the angular distance of rotation for a point on the outside of a flywheel over a period of 0.185 s that has a angular velocity of 94.5 rad/s from rest?

Table 5-4. Conversion of Angular Velocity in RPM to Radians per Second (rad/s)

rpm	Angular Velocity (rad/s)									
	0	1	2	3	4	5	6	7	8	9
0	0.00	0.10	0.21	0.31	0.42	0.52	0.63	0.73	0.84	0.94
10	1.05	1.15	1.26	1.36	1.47	1.57	1.67	1.78	1.88	1.99
20	2.09	2.20	2.30	2.41	2.51	2.62	2.72	2.83	2.93	3.04
30	3.14	3.25	3.35	3.46	3.56	3.66	3.77	3.87	3.98	4.08
40	4.19	4.29	4.40	4.50	4.61	4.71	4.82	4.92	5.03	5.13
50	5.24	5.34	5.44	5.55	5.65	5.76	5.86	5.97	6.07	6.18
60	6.28	6.39	6.49	6.60	6.70	6.81	6.91	7.02	7.12	7.23
70	7.33	7.43	7.54	7.64	7.75	7.85	7.96	8.06	8.17	8.27
80	8.38	8.48	8.59	8.69	8.80	8.90	9.01	9.11	9.21	9.32
90	9.42	9.53	9.63	9.74	9.84	9.95	10.05	10.16	10.26	10.37
100	10.47	10.58	10.68	10.79	10.89	11.00	11.10	11.20	11.31	11.41
110	11.52	11.62	11.73	11.83	11.94	12.04	12.15	12.25	12.36	12.46
120	12.57	12.67	12.78	12.88	12.98	13.09	13.19	13.20	13.40	13.51
130	13.61	13.72	13.82	13.93	14.03	14.14	14.24	14.35	14.45	14.56
140	14.66	14.76	14.87	14.97	15.08	15.18	15.29	15.39	15.50	15.60
150	15.71	15.81	15.92	16.02	16.13	16.23	16.34	16.44	16.55	16.65
160	16.75	16.86	16.96	17.07	17.17	17.28	17.38	17.49	17.59	17.70
170	17.80	17.91	18.01	18.12	18.22	18.33	18.43	18.53	18.64	18.74
180	18.85	18.95	19.06	19.16	19.27	19.37	19.48	19.58	19.69	19.79
190	19.90	20.00	20.11	20.21	20.32	20.42	20.52	20.63	20.73	20.84
200	20.94	21.05	21.15	21.26	21.36	21.47	21.57	21.68	21.71	21.89

44. Calculate the angular acceleration of a pulley wheel that is at rest and then rotates 15° over a period of 0.01 s.

45. What is the angular distance of rotation for a rotating gear whose initial angular velocity is 5.45 rad/s and reaches its final angular velocity of 6.50 rad/s in 1.250 s?

46. How long would it take a flywheel to make one complete revolution (360°) when its angular velocity increases from 1.18 rad/s to 2.50 rad/s?

47. A drive wheel rotating at 280 rpm is slowed to 210 rpm over a period of ⅝ s when a load is applied. What angular acceleration is experienced by the drive wheel?

Table 5-5. Conversion of Angular Velocities

Angular Velocities	Conversion Multipliers	
	rad/s to rpm	rpm to rad/s
1	9.55	0.1047
2	19.10	0.2094
3	28.65	0.3142
4	38.20	0.4189
5	47.75	0.5236
6	57.30	0.6283
7	66.84	0.7330
8	76.39	0.8378
9	85.94	0.9425

48. Given the same drive wheel in problem 47, suppose its rpm rate was reduced from 180 to 98 over period of 1.2 s. Calculate its angular deceleration.

Work and Power

Two important concepts frequently encountered in industrial situations are work and power. The calculation of these factors is extremely important in determining motor and drive design and specifications, plus maximum loads that can be handled by industrial equipment. This section reviews work and power concepts and common formulas employing these factors.

Basic Definitions

Whenever a force is used to produce motion in a body, the force undergoes displacement. The term used to describe this occurrence is known as *work* (K), and it is defined as force times the distance through which the force acts, or:

$$K = Fd$$

Examples of units of measure used to express work are the foot-pound (ft-lb) and joule (j).

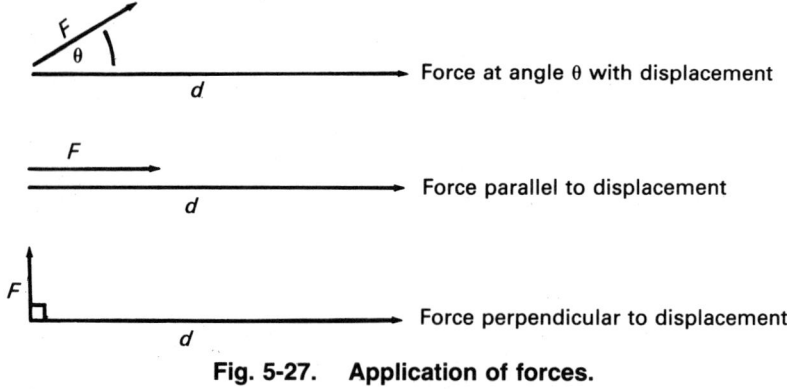

Fig. 5-27. Application of forces.

Figure 5-27 illustrates three examples of how force can be applied. In the first case, the force is at angle θ with displacement, so that:

$$K = Fd \cos \theta$$

In the second case, the force is parallel to displacement. Hence:

$$K = Fd \cos 0° = Fd$$

In the last case, the force is perpendicular to displacement, where:

$$K = Fd \cos 90° = 0$$

The rate at which work is being done, or at which energy is being dissipated, is known as *power*. Power (P) is defined as the amount of work accomplished over a given period of time, or:

$$P = \frac{K}{t}$$

The units of measure used for expressing power are generally the watt and joule (1 watt = 1 j/s = 1 Nm/s).

Work and Power Formulas

Presented in Table 5-6 are the formulas that can be used to solve many work and power problems. Again, these formulas can be used with either customary U.S. or metric units. The formulas are often used to calculate work and power when applied forces and velocities

Table 5-6. Work and Power Formulas

Unknown Quantity	Formula to Find*
Distance	$d = \dfrac{Pt}{F}$ $d = \dfrac{K}{F}$ $d = \dfrac{(550t)(hp)}{F}$
Velocity (constant or average)	$v = \dfrac{P}{F}$ $v = \dfrac{K}{Ft}$ $v = \dfrac{550hp}{F}$
Time	$t = \dfrac{Fd}{P}$ $t = \dfrac{K}{Fv}$ $t = \dfrac{FS}{550hp}$
Force (constant or average)	$F = \dfrac{P}{v}$ $F = \dfrac{K}{d}$ $F = \dfrac{K}{vt}$ $F = \dfrac{550hp}{v}$
Power	$P = Fv$ $P = \dfrac{Fd}{t}$ $P = \dfrac{K}{t}$ $P = 550hp$
Work	$K = Fd$ $K = Pt$ $K = Fvt$ $K = (550t)(hp)$

(*continued*)

COMPLEX MECHANICS

Table 5-6. (*continued*)

Unknown Quantity	Formula to Find*
Horsepower	$hp = \dfrac{Fd}{550t}$
	$hp = \dfrac{P}{550}$
	$hp = \dfrac{Fv}{550}$
	$hp = \dfrac{K}{550t}$

*Formulas that employ the factor of 550 are not to be used with SI metric units.

are accounted for at a particular point. As discussed earlier, horsepower (hp) is a traditional unit of power used in engineering and is equal to 745.6999 watts, or 550 ft-lb/s.

As an example of how these formulas can be used, let's determine the minimum horsepower rating for a motor that must be able to lift 3500-lb loads 30 ft in 45 s. This can be determined by using the following procedures:

$$hp = \frac{Fd}{500t}$$
$$= \frac{(3500)(30)}{(550)(45)}$$
$$= \frac{105,000}{24,750}$$
$$= 4.24.$$

In another case a casting block weighing 300 lb is lifted 6 ft in 5 s by a crane. To find the amount of power used, the following calculations are employed:

$$P = \frac{Fd}{t}$$
$$= \frac{(300)(6)}{5}$$
$$= 360 \text{ ft-lb/s or}$$
$$= 360/550$$
$$= 0.655 \text{ hp}$$

Exercises

49. A force of 130 N lifts a 12-kg load to a height of 8 m. How much work is accomplished by the force?

50. If the same force used in problem 49 were used to push that load the same distance horizontally, how much work would be accomplished?

51. A towing line that makes an angle of 20° with a track exerts 80 N of force to pull a load. How much work does it accomplish over a distance of 1 km?

52. What is the minimum amount of horsepower required to lift a weight of 2500 lb a total of 20 ft in 30 s?

53. How much work would be accomplished in problem 52?

54. Calculate the amount of work required to move 750 lb a distance of 100 ft using 20 ft-lb/s of power.

Centrifugal Force

Centrifugal force is the outward force pulling on a rotating body. By comparison, *centripetal force* is the inward force that provides a body uniform circular motion. When an object rotates about a central point or axis that differs from the center of the object (e.g., a ball being whirled at the end of a string versus the spinning of the ball), it will exert an outward centrifugal force on the axis and thereby keep it from moving along a linear path.

The factors used in centrifugal force problems are:

F_c = centrifugal force
m = mass (kg) of the revolving body
w = weight (lb) of the revolving body
v = velocity at the r radius on the body
n = revolutions per minute (rpm)
r = radius distance measured from the center of the mass to the center of gravity of the revolving body
g = acceleration due to gravity (32.16 ft/s^2 or 9.81 m/s^2)

The basic formulas used with customary U.S. units are:

166 • COMPLEX MECHANICS

$$F_c = wv^2/gr \text{ or}$$
$$F_c = \frac{4wr\pi^2 n^2}{3600\,g} \text{ or}$$
$$F_c = \frac{wrn^2}{2933} \text{ or}$$
$$F_c = 0.000341 wrn^2$$

$$w = \frac{F_c rg}{v^2} \text{ or}$$
$$w = \frac{2933 F_c}{rn^2}$$

$$r = \frac{wv^2}{F_c g} \text{ or}$$
$$r = \frac{2933 F_c}{wn^2}$$

$$v = \sqrt{F_c rg/w}$$
$$n = \sqrt{2933 F_c/wr}$$

When SI metric units are used, weight per acceleration due to gravity (w/g) will be replaced by the mass (m) of the body. Hence, the following formulas are employed:

$$F_c = \frac{mv^2}{r} \text{ or}$$
$$F_c = \frac{mn^2(2\pi r)^2}{3600 r} \text{ or}$$
$$F_c = 0.01097 mrn^2$$

An example of a centrifugal force problem is to determine the centrifugal force applied to a wheel with a radius of 10 in., rotating at 500 rpm, with a total weight of 400 pounds. The centrifugal force would be calculated as follows:

$$F_c = \frac{wrn^2}{2933}$$
$$= \frac{(400)(0.833)(500)^2}{2933}$$
$$= 28{,}400.95 \text{ lb}$$

Note that in this problem the radius, which is given in inches, was converted to the foot measure, since all standard centrifugal force measures are given in relationship to the foot measure.

Exercises

55. Calculate the centrifugal force acting on a 10-kg body that is rotating 240 rpm at a distance of 2.3 m from the center of the axis.

56. Find the centrifugal force acting on a 10-lb ball rotating at 340 rpm at a distance of 28 in. from the center of rotation.

57. What is the velocity of the ball in problem 56?

CHAPTER 6

Strength of Materials

- **Fundamentals of Strength of Materials**
- **Simple Shear and Stress Formulas**
- **Torsion and Torque**
- **Strength of Compression Members**

An area of study closely associated with statics and dynamics is strength of materials. This field deals with the study of the effects of applied external forces on materials. This is essential for the correct selection and specification of production, construction, and/or fabrication materials when considering the design and construction of machinery, machine components, and structures.

The design of equipment and structures requires that the response of their materials to forces can be predicted with a high degree of confidence and reliability. Otherwise, performance, safety, and work-life expectancy could not be assured. This requires that both the designer and user know to what extent the product can perform under static and dynamic forces.

There are four major categories of material properties: mechanical, thermal, electrical, and chemical. Though all four are important, only mechanical and (to a more limited degree) thermal properties will be addressed in this chapter. Engineers, technologists, and technicians are frequently concerned with three basic types of mechanical properties that influence performance: *strength*, *elastic deformation*, and *elastic stability*.

There are also other, more complex elements that affect the performance of materials, such as environmental conditions and tribological (friction) factors. These, however, are beyond the scope and intent of this chapter.

Fundamentals of Strength of Materials

Force and the length of time that a part is subjected to a force are factors that must be considered when calculating strength of material properties. To analyze these properties, magnitude, location(s), and direction of force must be known; hence, the importance of understanding force systems. In addition, the duration, cycle, and/or repetitiveness of the forces are given in terms of the time element.

To understand more clearly the applications of formulas used in strength-of-material calculations, it is first necessary to gain an understanding of basic definitions and properties.

Basic Definitions

Most mechanical properties are determined by laboratory or in-line (testing during production) tests. Based upon specific measures, calculations are then made to present a property within the context of standardized measures. Before this can be accomplished, one must have an understanding of those terms commonly used in industrial testing and strength of materials.

Stress (S), or *mechanical stress*, is caused by the application of external forces and is expressed in terms of force per unit areas: force divided by area. Measures used here are lb/in.^{2} (psi), N/m^2, and Pa. (*Note*: When pascal units are used, it is often expressed in terms of megapascals or MPa). Thermal stress is caused by the difference in expansion due to temperature change, and chemical stress occurs as a result of corrosion or the beginning of a chemical reaction. In addition, there are three subcategories of mechanical stress:

- *Compressive stress* results from the compression or shortening of the material due to a squeezing force.
- *Tensile stress* results from the stretching of material due to a pulling force.
- *Shearing stress* is due to the shearing or cutting-off action where one part of the material will slide over another part as a result of an applied force.

170 • STRENGTH OF MATERIALS

Strain (ϵ), or *unit strain*, is the degree of dimensional change in a body when it is subjected to a load. Strain, then, is deformity that is calculated in terms of deformation (e) divided by length (L), and is normally expressed in terms of percent of deformity.

Stress-strain diagrams are used to illustrate various mechanical properties of materials in relation to the amount of stress and strain exerted. Presented in Figure 6-1 is an illustration of a stress-strain diagram for annealed steel.

The *proportional limit* is that point on a stress-strain diagram where the line changes from a straight-line relationship to a curved relationship.

The *elastic limit* is the maximum point at which the material can be deformed and still return to its original length when the load is removed. Material that is worked below its elastic limit is said to be stressed in its *elastic region*, while material exceeding the elastic limit is said to be stressed in its *plastic region*.

The *yield point* is that location on the stress-strain curve where the strain suddenly increases at a much greater rate than stress. It should be noted that not all materials will have a yield point.

Property Measures

In addition to properties that can be observed in tests and illustrated in a stress-strain diagram, there are a number of other mechanical properties that can be expressed in terms of measures or quantities.

Yield strength (S_Y) is the maximum amount of stress that can

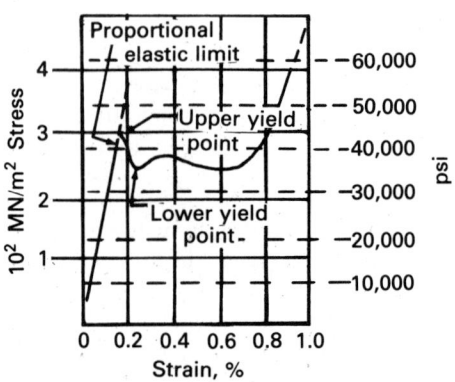

Fig. 6-1.   Stress-strain diagram.

be applied to a part without causing it to permanently deform. This property can only be specified for materials that have an elastic limit.

Tensile strength (S_T) is also referred to as the *ultimate strength* of material. In relation to the stress-strain curve, it is the maximum stress level obtained.

The *modulus of elasticity* (*E*), also known as *Young's modulus*, is a proportion of stress to strain as used to find either stress or strain when the other is known (the modulus of elasticity will be the same for both compression and tension). This property is usually expressed in terms of psi or Pa. When pascals are used, the quantity is often measured in GPa (gigapascals).

The *modulus of elasticity in shear* (E_S) is expressed in radians, but is still the proportion of stress to strain. The proportional limit, however, is relative to material shear.

Poisson's ratio (μ) is the proportion of lateral deformation to longitudinal deformation.

Simple Shear and Stress Formulas

Central to all strength-of-material calculations are simple shear and stress problems. Formulas used in these cases are crucial to most other problems encountered in the workplace.

Tensile and Compressive Stress

A part is placed under tension when forces are used to stretch or elongate it, while those under compression are squeezed or made smaller. The basic formula used to calculate stress in both cases is:

$$S = \frac{F}{A}$$

In this formula, *F* is the amount of tension or compressive force acting on the body and *A* is the cross-sectional area of the part for which stress is being calculated. To use this formula, the force must be perpendicular to the cross-sectional area and it must be applied along the central axis of the part. If any of these vary, this formula cannot be used.

For an example of how to use this formula, a test specimen cross section is round and has a diameter of 150 mm. If a load of

57 N is used to produce a tension load, the amount of tensile stress exerted on the piece is calculated as follows:

$$S = \frac{F}{A}$$
$$= \frac{57}{17671.44}$$
$$= 0.003\ 225\ 544\ \text{N/mm}^2$$
$$= 32,255.544\ \text{MPa}$$

Note: 1 N/mm² = 1 Pa
1,000,000 N/mm² = 1 MPa

The compression stress for a square rod, whose dimensions are 1.5 in. by 1.5 in., that is subjected to a squeezing force of 3550 lbs is calculated as follows:

$$S = \frac{F}{A}$$
$$= \frac{3550}{2.25}$$
$$= 1577.778\ \text{psi}$$

The most difficult aspect of solving for shearing stress is to determine the correct cross-sectional area. In Figure 6-2, we have two bodies that are joined and are exposed to a shearing force. Here, the cross-sectional area is simple (section A–A). Note that when force is exerted from one direction, a reactionary force must be applied

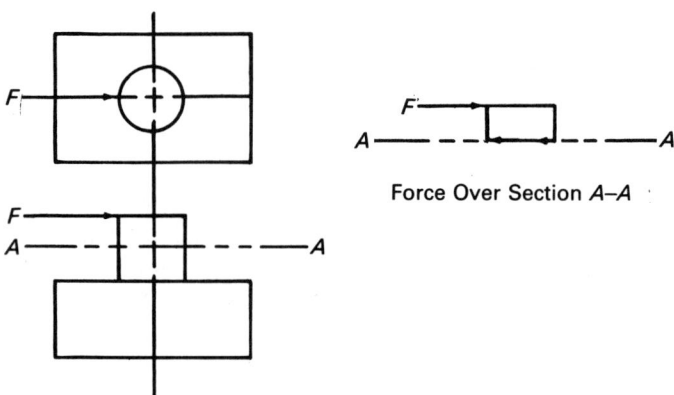

Fig. 6-2. Two bodies exposed to a shearing force.

in the opposite direction to maintain equilibrium. When there is deformity, there will be more shearing force than reacting force.

The basic formula used for calculating shearing stress employs a subscript *s*, which stands for shear. This formula is:

$$S_S = \frac{F}{A_S}$$

In Figure 6-3, section A–A cuts through the center of the pins, which act as tension members. When a load (force) is applied to the pins, shear stress will develop in the shaded area. Thus, A_S is calculated only for that section. The areas in shear are shown in section A–A, which are 10 mm by 20 mm. When a load of 550 N is applied, the shearing stress developed is determined as follows:

$$S_S = \frac{F}{A_S}$$
$$= \frac{550}{400}$$
$$= 1.375 \text{ Pa}$$

Note: There are two areas in shear; thus, the area is doubled.

Exercises

1. Calculate the stress that is exerted on a steel wire that has a diameter of 4 mm and is subjected to a tensile force of 1436 (give your answer in units of MPa).

2. A 1-in. square rod is subjected to a pulling force of 568 lb. What is the tensile stress encountered?

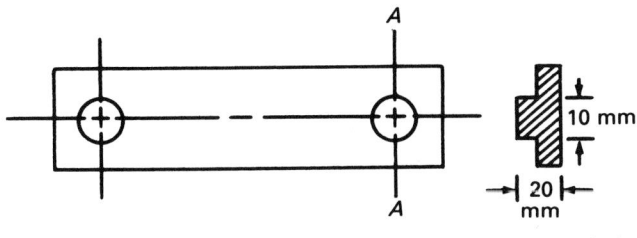

Fig. 6-3. Shear problem.

3. Using the illustration in Figure 6-4(a), calculate the stress exerted on the test specimen's cross section when a compressive force of 5400 lb is applied.

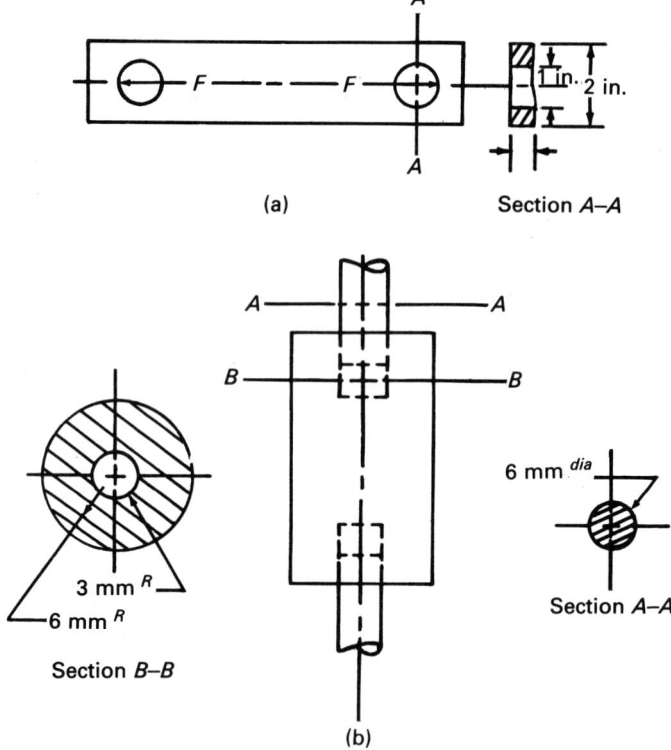

Fig. 6-4. Problems 3 and 4.

4. Two small rods are force-fitted into a center rod [Figure 6-4(b)]. Calculate the tensile stress for cross sections A–A and B–B when a total pulling force of 1232 N is applied (give your answer in units of Pa).

5. What is the maximum amount of force that can be applied to a 2 cm by 4 cm bar that has a compression strength of 60 Pa?

6. Using Figure 6-5, calculate the shearing stress at the root diameter of threads when the nut and bolt are placed under a load of 6400 lb. (*Hint*: The area under stress is cylindrical.)

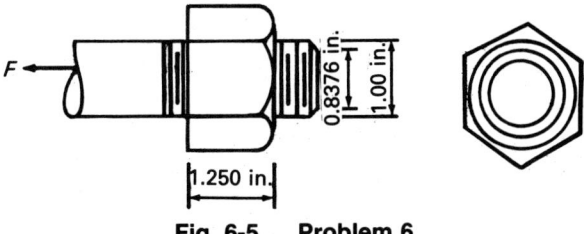

Fig. 6-5.   Problem 6.

7. Refering to Figure 6-6, calculate the shear stress exerted on the pin when a load of 62,800 N is mounted onto part B.

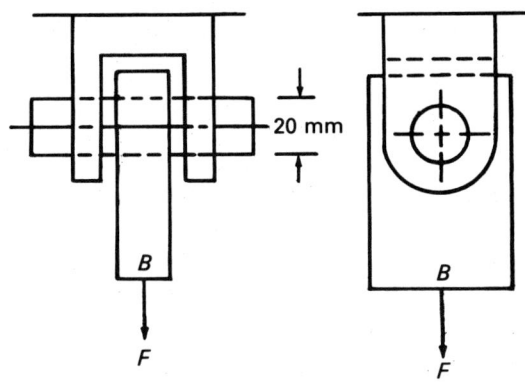

Fig. 6-6.   Problem 7.

Strain

All materials will deform to some degree when subjected to stress. By definition, deformation is any change of dimension or shape. In tension and compression, this means a change in length. If the original length of a bar is L, and it is subjected to tension, the deformity (e) will be the amount of stretch, so that the final length (L_F) can be expressed as:

$$L_F = L + e$$

where:

$$e = L_F - L$$

176 • STRENGTH OF MATERIALS

When the bar is subjected to compression, the final length will be:

$$L_F = L - e$$

where:

$$e = L - L_F$$

It should therefore be noted that:

$$e = L_F - L$$

where L is the original length of the stock. Both tension and compression strain are illustrated in Figure 6-7.

Strain is usually represented by the Greek letter epsilon (ϵ), where the basic formula for calculating simple strain is:

$$\epsilon = \frac{e}{L}$$

Since strain is the proportion of length to length, or one number to another in the same unit of measure, it will be expressed as a number with no dimension.

An example of how the strain formula is used is in the case of a rod whose original length is 300 mm and which is subjected to a load where its final, compressed length is 280 mm. The compressive strain is then calculated as:

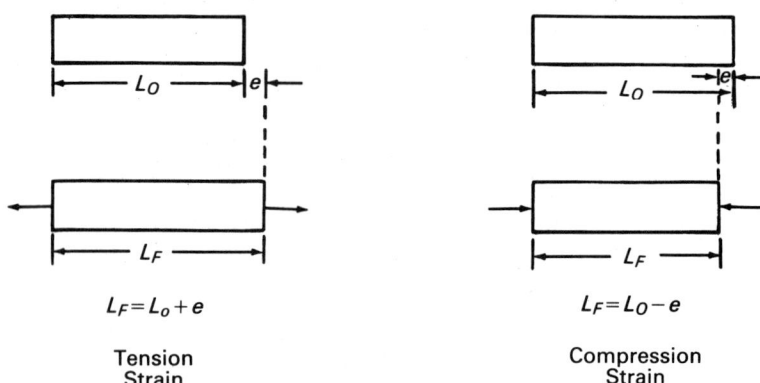

Fig. 6-7. Illustration of strain.

$$e = L - L_F$$
$$= 300 - 280$$
$$= 20$$

$$\epsilon = \frac{e}{L}$$
$$= \frac{20}{300}$$
$$= 0.6667$$

The characteristics of deformation caused by shear are quite different from those of tension and compression and are illustrated in Figure 6-8. Similar to shear stress, shear strain uses the same basic formula as found in tension and compression shear. Hence:

$$\epsilon_S = \frac{e}{L}$$

An example of calculating shearing strain involves a machine mounting that is 720 mm in length and which is deformed 0.125 mm after the workpiece is fastened. To calculate the amount of strain, the following procedure is used:

$$\epsilon_S = \frac{e}{L}$$
$$= \frac{0.125}{720}$$
$$= 1.736 \times 10^{-4}$$

Exercises

8. A bar is stretched a total of 0.185 in. from its original length of 72 in. Calculate its strain.

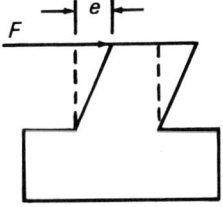

Fig. 6-8. Deformation in shear.

178 • STRENGTH OF MATERIALS

9. Find the strain of a steel plate whose length was 250 mm before it was compressed to 239 mm.

10. A round metal bar is twisted so that one end rotates 10°. If the bar has a diameter of 100 mm, what is the shearing strain at the surface of the bar?

Modulus of Elasticity

As already noted, modulus of elasticity is a proportion between stress and strain that is given in units of either GPa or psi. Table 6-1 is a listing of values of modulus of elasticity for tension and compression (Young's modulus), and in shear.

With a known modulus of elasticity, it is possible to solve a number of problems with unknown shear or stress quantities. For example, a piece of Tobin bronze is subjected to a shearing strain

Table 6-1. Modulus of Elasticity for Common Materials

Material	Young's Modulus (E) psi × 10^6	GPa	Modulus in Shear (E_S) psi × 10^6	GPa
Cast steel	28.5	196.5	11.3	77.91
Cold-rolled steel	29.5	203.40	11.5	79.29
Stainless steel 18-8	27.6	190.30	10.6	73.08
Other steels	28.6–30.0	197.2–206.8	11.0–11.9	75.8–82.1
Cast iron	13.5–21.0	93.1–144.8	5.2–8.2	35.9–56.5
Malleable iron	23.6	162.71	9.3	64.12
Copper	15.6	107.56	5.8	39.99
70-30 Brass	15.9	109.63	6.0	41.37
Cast brass	14.5	99.97	5.3	36.54
Tobin bronze	13.8	95.15	5.1	35.16
Phosphor bronze	15.9	109.63	5.9	40.68
Aluminum alloys	9.9–10.3	68.3–71.0	3.7–3.9	25.5–26.9
Monel metal	25.0	172.37	9.5	65.50
Inconel	31	213.74	11	75.84
Z-nickel	30	206.84	11	75.84
Beryllium copper	17	117.21	7	48.26
Elektron magnesium	6.3	43.44	2.5	17.23

of 0.00287. To find the shearing stress, the following procedures are used:

$$S_S = E_S \epsilon_S$$
$$= (35.16)(0.00287)$$
$$= 0.100\,9092 \text{ GPa}$$
$$= 1.009\,092 \text{ MPa}$$

A second example is to find the strain exerted on a malleable iron bar that is subjected to a tension stress of 11,250 psi:

$$\epsilon = \frac{S}{E}$$
$$= \frac{11,250}{23.6}$$
$$= 476.69$$

Most problems encountered in stress-strain relationships are more encompassing than these two situations. The basic formula used here is:

$$e = \frac{FL}{AE}$$

Using this particular formula, the amount of deformation will be in units of millimeters or inches. Other derivations of this formula are presented in Table 6-2.

Exercises

11. A cast-steel plate is stretched to a strain of 0.0009. Calculate its tensile stress in MPa and psi.

Table 6-2. Formulas for Comprehensive Problems

Unknown Quantity	Formula to Find
Deformation	$e = \dfrac{FL}{AE}$
Magnitude of force	$F = \dfrac{AEe}{L}$
Length prior to deformation	$L = \dfrac{AEe}{F}$

180 • STRENGTH OF MATERIALS

12. A piece of copper is subjected to a shearing strain of 0.000112. Find the shearing stress in MPa and psi.

13. What is the strain in a Monel metal bar that is subjected to a compressive stress of 22,000 psi?

14. Calculate the strain in a malleable iron bar that is subjected to a tension stress of 4.75 MPa.

15. Calculate the magnitude of the compressive force that is applied to an Inconel block $15 \times 20 \times 40$ cm so that it will be shortened a total of 0.0095 mm.

16. How much does an Elektron magnesium alloy bar stretch when it is subjected to a force of 0.060 MN, if it is $0.72 \times 0.06 \times 0.02$ m in size?

Factor of Safety

One fundamental question asked in strength-of-material problems is how much stress can a particular material be subjected to and still provide service? This level of stress is used in design calculations and is referred to as the *working stress* (S_W). When calculations are made, the design must be such that the material remains below the proportional limit as well as significantly below the ultimate strength. This is accomplished by using a *factor of safety* (f), which is also referred to as the *reserve factor*.

This factor is used to give the designer information about the maximum stresses that the material should be subjected to within limits of safety. This is accomplished by dividing the ultimate strength of the material by the factor of safety. Hence:

$$S_W = \frac{S_T}{f}$$

Thus:

$$f = \frac{S_T}{S_W}$$

(*Note*: This factor of safety is for static loads only. When cyclic loads are applied, the formula is more complex and incorporates the factors of *elastic limit* (S_E) and *design stress* (S_D), where $f = S_E/S_D$ and $S_E <$ the yield stress.)

Table 6-3. Factor of Safety for Various Materials

Material	Steady Stress	Cyclic Stress
Brick and stone	15	25
Timber	8	10
Cast iron	6	10
Wrought iron	4	6
Structural steel	4	6

The factor of safety will generally fall between values of 1.5 and 10, where the higher the number, the lower the working stress. Because of the variety of materials on the market and the changes made in their properties by processing and manufacturing procedures, it is difficult to give an accurate listing. A representative sample, however, is presented in Table 6-3. It is therefore recommended that manufacturers be contacted directly for handbooks containing this information.

Thermal Deformation

The introduction of temperature (increases or decreases) to materials will result in stress— expansion when heated and contraction when cooled. The amount of expansion and contraction per material is expressed in terms of a *coefficient of expansion*, which is noted by the Greek letter alpha (α). Hence, the coefficient of expansion is the unit of deformation per change in degree of temperature. Table 6-4 lists the coefficient of expansion of common materials.

The amount of deformation occurring as a result of temperature change is calculated as the product of the temperature change (ΔT) and the coefficient of expansion. Thus:

$$\Delta L = \alpha \Delta T L$$

A typical thermal deformation problem involves a copper rod 36 in. long and subjected to a temperature increase of 214°F. To find the amount of deformation (i.e., change in length), the following calculations are made:

$$\begin{aligned} \Delta L &= \alpha \Delta T L \\ &= (9.2 \times 10^{-6})(214)(36) \\ &= 0.07088 \text{ in.} \end{aligned}$$

Table 6-4. Coefficients of Expansion

Material	Coefficients (α)	
	in./in./°F × 10⁻⁶	mm/mm/°C × 10⁻⁶
Aluminum		
2024-T3	12.6	22.7
6061-T6	13.5	24.3
7079-T6	13.7	24.7
Beryllium, QMV	6.4–10.2	11.5–18.4
Cast iron, gray	6.0	10.8
Copper (pure)	9.2	16.6
Copper alloys	10.0	18.0
Lead (pure)	29.3	52.8
Magnesium		
AZ31B-H24 (sheet)	14.5	26.1
HK31A-H24	14.0	25.2
Molybdenum, wrought	3.0	5.4
Nickel (pure)	7.2	13.0
Platinum	5.0	9.0
Silver (pure)	11.0	19.8
Stainless steel	6.1	11.0
Steel		
AISI C1020 (hot worked)	8.4	15.1
AISI 304 (sheet)	9.9	17.8
Tantalum	3.6	6.5
Titanium, B 120VCA (aged)	5.2	9.4
Tungsten	2.5	4.5

In another case, a stainless-steel wire that is 2.5 m long is subjected to a decrease in temperature of 50°C. To find the change in length, the following procedures are used:

$$\Delta L = \alpha \Delta T L$$
$$= (6.1 \times 10^{-5})(-60)(2500)$$
$$= -9.15 \text{ mm}$$

Exercises

17. Calculate the change in length of a tungsten rod 1.12 m long that is subjected to a temperature increase of 250°C.

18. What will be the overall length of a pure nickel bar whose

original length was 12 in. before its temperature was decreased by 150°F?

19. Determine the amount of deformation resulting from a heat-treating process that increases the temperature of a 48-in.-long forged bar of AISI C1020 steel a total of 1100°F.

20. Find the amount of force exerted upon a 15 × 37.5 × 320-mm wrought molybdenum bar that is pinned between two brackets (Figure 6-9) when its temperature is increased by a total of 224°C. (*Note:* The modulus of elasticity is 7.2×10^{-6}.)

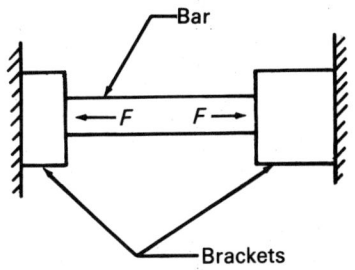

Fig. 6-9.   Problem 20.

Torsion And Torque

Torsion is the twisting of one end of an object with respect to the other end. This twisting action is caused by a phenomenon known as *torque*, which is the product of the magnitude of the force and its moment arm. Torsion is commonly found in circular shafts that are used to transmit power and/or motion between machine parts. Torque is a force or combination of forces that produces or tends to produce a rotating or twisting movement called torsion.

Torque

Torque is a *twisting moment* that is noted by the Greek letter tau (τ). It is the product of the magnitude of the force and the perpendicular distance from the force's line of action to the point of twist. Using Figure 6-10(a) as a simple illustration, to calculate the amount of torque at point O, we use the following formula:

$$\tau = FL$$

184 • STRENGTH OF MATERIALS

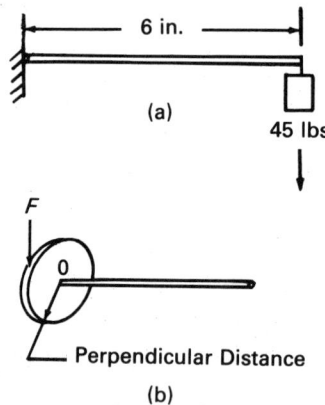

Fig. 6-10. Simple torque diagrams.

where L is the distance from the weight (w). Thus, if a 45-lb force is 6 in. from the point, the amount of torque generated is:

$$\tau = FL$$
$$= (45)(6)$$
$$= 270 \text{ lb-in.}$$

Note that torque is measured in units such as lb-in. and N-mm. In Figure 6-10(b), the perpendicular distance from the line of action to the point of twist is the radius of the wheel. The distance in this illustration is not measured along the shaft, since it rotates—though the same formula is used. Here again, you always multiply a force by a distance.

Torsional Shearing Stress

When a part is twisted, the maximum stress that is exerted is called its torsional shearing stress. The formula used to calculate the torsional shearing stress depends on the shape of the part. In all, there are five basic shapes that will be considered.

CIRCULAR SHAFTS

The formulas used to illustrate the relationships existing in the torsional shearing stress of a circular shaft are:

$$S_S = \frac{16\tau}{\pi d^3}$$

$$\tau = \frac{\pi d^3 S_S}{16}$$

To find the maximum torsional shearing stress when a torque of 3200 lb-in. is applied to a circular shaft 0.875 in. in diameter, the calculations used are:

$$S_S = \frac{16\tau}{\pi d^3}$$
$$= \frac{(16)(3200)}{(3.14159)(0.875^3)}$$
$$= \frac{51,200}{2.1046}$$
$$= 24,327.66 \text{ psi.}$$

In another example, it must be determined how much torque must be applied to a solid cylinder shaft 12 mm in diameter to cause a maximum torsional shearing stress of 65.4 MPa. This problem is solved as follows:

$$\tau = \frac{\pi d^3 S_S}{16}$$
$$= \frac{(3.14159)(12^3)(65.4)}{16}$$
$$= 22,189.67849 \text{ N-mm}$$

HOLLOW CIRCULAR BAR

The formulas used when working with a hollow circular bar [Figure 6-11(a)], are as follows:

$$S_S = \left[\frac{16}{\pi}\right]\left[\frac{d_2}{\pi(d_2^4 - d_1^4)}\right]$$

$$\tau = \frac{\pi(d_2^4 - d_1^4)S_S}{16 d_2}$$

To find the torque in a hollow circular shaft with a 10-mm outer diameter and 5-mm inner diameter when a maximum torsional

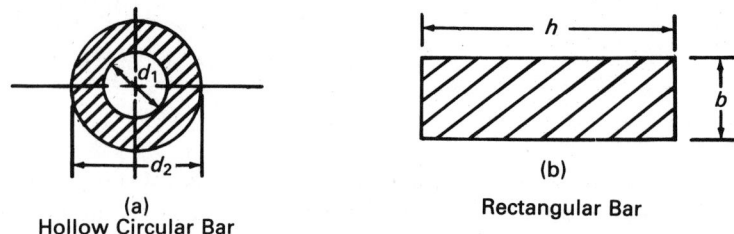

(a)
Hollow Circular Bar

(b)
Rectangular Bar

Fig. 6-11. Cross-sections for torsional shearing stress formulas.

shearing stress of 112 MPa is applied, the following procedure is used:

$$\tau = \pi(d_2^4 - d_1^4)\frac{S_S}{16d_2}$$

$$= (3.14159)(10^4 - 5^4)\frac{(112)}{(16)(10)}$$

$$= 20{,}616.684 \text{ N-mm}$$

RECTANGULAR BAR

In some situations, it may be necessary to determine the torsional shearing stress of a rectangular bar. Here, the bar will be a structural part of a machine or structure, rather that an integral drive component. In this case, torque would be calculated by the product of the load and the distance from the load. Torsional shearing stress calculations would employ the following formula [see Figure 6-11(b)]:

$$S_S = \frac{\tau(3h + 2b)}{b^2 h^2}$$

To find the maximum torsional shearing stress when a torque of 8500 lb-in. is applied to a 4×1.5-in. bar, the following procedure is used:

$$S_S = \frac{(8500)[(3)(1.5) + (2)(4)]}{(4^2)(1.5^2)}$$

$$= \frac{106{,}250}{36}$$

$$= 2951.389 \text{ psi}$$

Torsion and Torque • 187

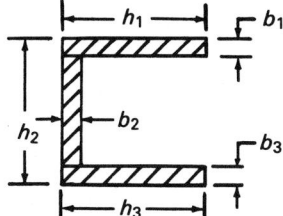

Fig. 6-12. Thin-walled open section.

THIN-WALLED OPEN SECTIONS

Figure 6-12 is an example of a thin-walled open section. Here, the channel is divided into narrow rectangles, where the maximum shear stress will occur at the center of the longest side of the thickest rectangular section. The formula used to find shear stress is:

$$S_S = \frac{3\tau b_{max}}{\Sigma h b^3}$$

In this formula, τ is the torque, b_{max} is the thickness of the thickest section, and $\Sigma h b^3$ is the sum of all sections' hb^3. Thus, for the channel in Figure 6-12:

$$\Sigma h b^3 = h_1 b_1^3 + h_2 b_2^3 + h_3 b_3^3$$

A typical problem is illustrated in Figure 6-13. To find the maximum shear stress when it is subjected to a torque of 1,500,000 N-mm, the following procedures are used:

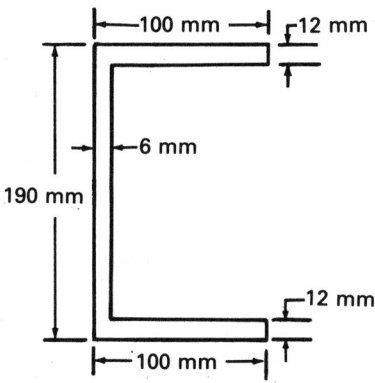

Fig. 6-13. Thin-walled open-section problem.

$$\Sigma hb^3 = 94(12^3) + 190(6)^3 + 94(12)^3$$
$$= 365{,}904 \text{ mm}^4$$

$$S_S = \frac{3\tau b_{max}}{\Sigma hb^3}$$
$$= \frac{(3)(1{,}500{,}000)(12)}{365{,}904}$$
$$= 147.58 \text{ MPa}$$

THIN-WALLED CLOSED SECTIONS

The formula used to calculate the torsional shear stress in thin-walled closed section, as illustrated in Figure 6-14, is:

$$S_S = \frac{\tau}{2At}$$

Here, τ = torque, A = the enclosed area at the midsection of the wall, and t = the thickness of the thinnest section of the wall.

If a torque of 12,000 lb-in. is applied to the bar shown in Figure 6-15, the maximum shearing stress would be calculated by the following procedures:

$$A = (3.05)(2.13) = 6.4965 \text{ in.}^2$$
$$S_S = \frac{\tau}{2At} = \frac{12{,}000}{(2)(6.4965)(0.12)} = 7696.45 \text{ psi}$$

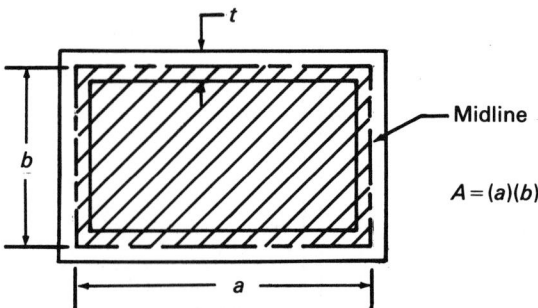

Fig. 6-14. Factors for torsional shear-stress problems.

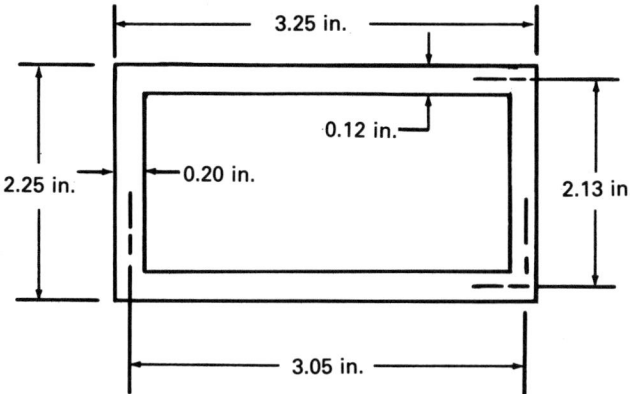

Fig. 6-15. Thin-walled closed-section problem.

Exercises

21. What is the maximum shearing stress in a solid shaft where the torque applied is 2640 lb-in. and its circular diameter is 2 in.?

22. Find the torque developed on a 150-mm-diameter solid spindle that has a maximum torsional shearing stress of 72.4 MPa.

23. If a solid circular shaft is subjected to a torque of 27,000 lb-in. and has a working stress of 9800 psi, what size diameter should be specified?

24. A pipe with an outside diameter of 160 mm and an inside diameter of 120 mm is subjected to a torque of 30,000,000 N-mm. Determine its maximum torsional shearing stress.

25. A rectangular 0.5 × 2-in. bar is used as a support bracket. If a load of 375 lb. is applied to it, find the maximum torsional shearing stress.

26. Calculate the shearing stress in the T-section illustrated in Figure 6-16(a).

27. What is the torsional shear stress in the thin-walled closed section in Figure 6-16(b)?

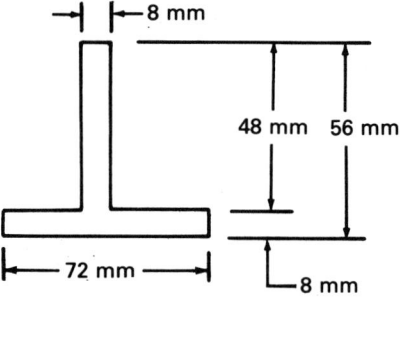

(a)

Note: Applied torque of 150,000 N-mm.

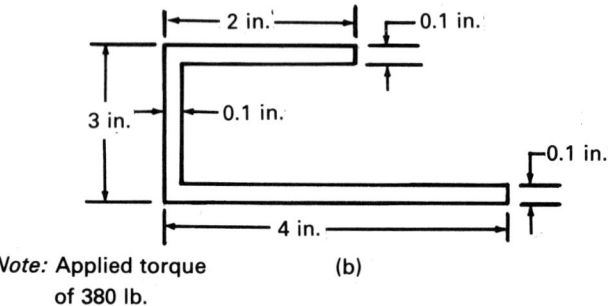

Note: Applied torque of 380 lb.

(b)

Fig. 6-16. Torque problems 25 and 26.

Angle of Twist

When a shaft is subjected to a force, including a twisting force, it will deform. The amount of deformation is important in the design of machine and equipment. Generally, there are two formulas used here: solid and hollow circular shafts. The formulas for finding the angle of twist (θ) are:

Solid circular bar:

$$\theta = \frac{32\tau L}{\pi E_s d^4}$$

Hollow circular bar:

$$\theta = \frac{32\tau L}{\pi E_s (d_2^4 - d_1^4)}$$

In these formulas, L = length of the bar, E_S = shearing modulus of elasticity (see Table 6-1), d = diameter, d_2 = outside diameter, and d_1 = inside diameter.

As an illustrative example, let's determine the angle of twist for a circular copper shaft that is 2.2 in. in diameter and 46 in. long and is subjected to a torque of 15,000 lb-in:

$$\theta = \frac{32\tau L}{\pi E_S d^4}$$
$$= \frac{(32)(15,000)(46)}{(3.14159)(5,800,000)(2.2^4)}$$
$$= \frac{22,080,000}{426,843,058}$$
$$= 0.05173 \text{ radians}$$
$$= 2°57'50''$$

Exercises

28. A solid circular shaft 6 in. in diameter and 12 ft long is made out of phosphor bronze. What is the angle of twist due to a torque of 200,000 lb-in.?

29. A hollow circular cast-steel shaft has an outside diameter of 300 mm, an inside diameter of 225 mm, and is 2.5 m long. Determine the angle of twist caused by a torque of 60×10^6 N-mm.

30. A malleable iron tube has a 3-in. outside diameter, a 2-in. inside diameter, and is 48 in. long. If the tube carries a torque of 54,000 lb-in., what is the angle of twist?

Strength of Compression Members

The study of the strength of compression members is based on four major concepts. The first two have already been covered and discussed in some detail: centroid of an area and moment of inertia. The other two are known as the *radius of gyration* and the *slenderness ratio*.

Radius of Gyration and Slenderness Ratio

The radius of gyration, also referred to as the *radius of inertia*, is denoted by the symbol r. The basic formula used here is:

$$r = \sqrt{I/A}$$

where I is the moment of inertia of the area and A is the size of the area.

Since the moment of inertia is used to solve problems along axes (i.e., I_X, I_Y, and I_Z), radius-of-gyration calculations are also made along axes. Consider, for example, a rectangle that has an area of 8 in.^{2} and a moment of inertia of 0.6678 in.^{4} with respect to the x axis. To find the radius of gyration along the x axis, the following calculation is made:

$$\begin{aligned} r_X &= \sqrt{I_X/A} \\ &= \sqrt{0.6678/8} \\ &= \sqrt{0.083475} \\ &= 0.2889 \text{ in.} \end{aligned}$$

Because compression problems are found in a wide variety of parts and products, specific formulas are used to calculate I and r for different shapes. Because of the breadth of shapes encountered, it is beyond the scope of this book to provide discussion for each. Table 6-5 is a summation of formulas used in the calculation of compression and related problems.

In the formulas in Table 6-5, the *section modulus* is used in the design of beams that are placed under loads. Considerations such as the bending moment ($M = SI/y$) and the cross-sectional area of the beam must be taken into consideration.

The slenderness ratio is the proportion of the unsupported length of the compressed member divided by the least radius of gyration. There is no symbol for the ratio, so it is simply identified as l/r. An example of how this is used is with a rectangular member whose cross-sectional area of 3.0×2.0 in. is bisected by the x and y axes. If $I_X = 2$ in.^{4} and $I_Y = 4.5$ in.^{4}, the least radius of gyration will be:

$$\begin{aligned} r &= \sqrt{I_X/A} \\ &= \sqrt{2/6} \\ &= 0.57735 \text{ in.} \end{aligned}$$

If the length of the compressed member is 8 ft, the slenderness ratio will be:

$$\frac{l}{r} = \frac{96}{0.57735}$$
$$= 166.277$$

Exercises

Calculate the radius of gyration for the following sections:

31. Circular area with a radius of 3 in., where the x and y axes serve as its centerlines (compute r_X and r_Y).

32. Circular tube with an inside radius of 0.50 in., outside radius of 0.60 in., and the x and y axes are its centerlines (find r_X).

Find the slenderness ratio for the following compressed sections:

33. A solid circular section 4 in. in length and with a radius of 0.50 in.

34. The support column illustrated in Figure 6-17, which is 96 in. long.

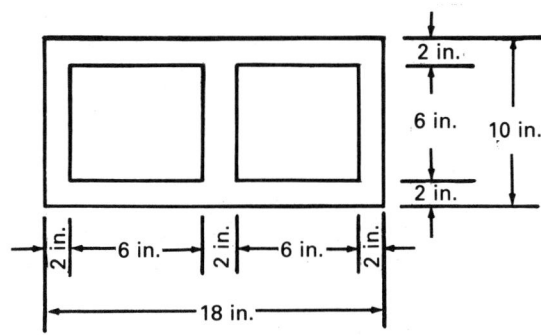

Fig. 6-17.   Problem 33.

Member Formulas

There are four major categories of compression member formulas, each being applicable to a certain type or design of member. These are: short-compression members, intermediate column members,

Table 6-5. Properties of Section

y = distance from neutral axis
A = cross-sectional area
I = moment of inertia
Z = section modulus (I/y)
r = radius of gyration ($\sqrt{I/A}$)

Section	y	A	I	Z	r
rectangle ($b \times d$)	$d/2$	bd	$bd^3/12$	$bd^2/6$	$d/\sqrt{12}$
diamond (side b, diagonal d)	$b/\sqrt{2}$	b^2	$b^2/12$	$b^3/6\sqrt{2}$	$b/\sqrt{12}$
triangle ($b \times d$)	$2d/3$	$bd/2$	$bd^3/36$	$bd^2/24$	$d/\sqrt{48}$

Shape	$\bar{y}$	A	I	Z	r
Circle	$d/2$	$\pi d^2/4$	$\pi d^4/64$	$\pi d^3/32$	$d/4$
Ellipse	a	πab	$\pi d^3 b/2$	$\pi a^2 b/4$	$a/2$
H-section	$d/2$	$2bt_1 + (d-2t_1)t_2$	$\dfrac{bd^3 - (b-t_2)(d-2t_1)^3}{12}$	$bd^3 - (b-t_2)(d-2t_1)^3$	$\left[\dfrac{bd^3 - (b-t_2)(d-2t_1)^3}{24bt_1 + 12t_2(d-2t_1)}\right]^{1/2}$
T-section	$\dfrac{d - d^2 t_2 + t_1^2(b-t_2)}{2(bt_1 + ht_2)}$	$bt_1 + ht_2$	$\tfrac{1}{3}[t_2 y^3 + b(d-y)^3 - (b-t_2)(d-y-t_1)^3]$	I/y	$\sqrt{I/A}$

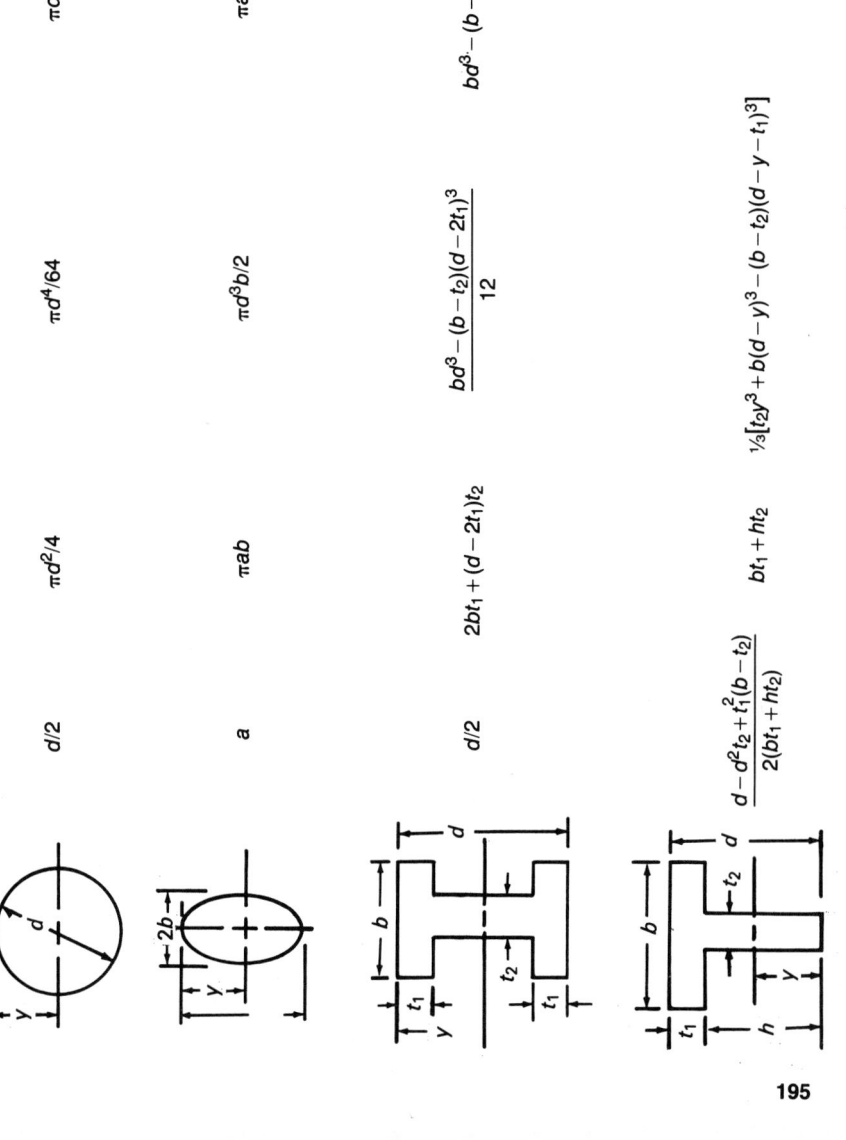

parabolic columns, Gordon-Rankin formula, and slender columns. The factors used here are:

F_P = compressive force
I = moment of inertia
$\dfrac{l}{r}$ = slenderness ratio
A = area of the cross section
C_C = constant
f = factor of safety
k = constant
S = stress
S_Y = yield point
E = modulus of elasticity

SHORT-COMPRESSION MEMBER.

Short-compression members are specified by the value of their slenderness ratio. As an example, short-compression steel members are those that have a slenderness ratio of less than 40, and aluminum and magnesium alloys with slenderness ratios below 10. The formula used for short-compression members is:

$$S = \frac{F_P}{A}$$

As you may remember, this formula is also used for calculating other simple stress problems.

INTERMEDIATE COLUMNS

The guideline used for identifying an intermediate column is one whose slenderness ratio is above its short-compression slenderness ratio (e.g., steel, between 40 and 150; and aluminum and magnesium alloys, between 10 and 70). For intermediate column problems the basic formula used is:

$$F_P = A\left[S - (k)\left(\frac{l}{r}\right) \right]$$

Here, k is a constant that varies according to stress, member design, and member material. This constant is available from manufacturers who supply column members. For examples, see the following formulas for intermediate columns made of low-carbon steel:

$$F_P = A\left(15,000 - 50\frac{l}{r}\right) \text{ U.S. customary units}$$

$$F_P = A\left(103 - 0.345\frac{l}{r}\right) \text{ SI metric units}$$

PARABOLIC COLUMNS

A second type of intermediate column is the parabolic column, which uses a variety of formulas according to the factors just outlined for intermediate columns. The American Institute of Steel Construction, Inc. (AISC), recommends the following:

$$\frac{F_P}{A} = \frac{S_Y}{f(1 - 9[k(l/r)]^2/C_C^2)}$$

The factor of safety is calculated as:

$$f = \frac{5}{3} + \left[\frac{3(k)(l/r)}{8C_C}\right] - \left[\frac{(k)(l/r)^3}{8C_C^3}\right]$$

The constant C_C is:

$$C_C = \sqrt{2\pi^2 E/S_Y}$$

GORDON-RANKIN COLUMN FORMULAS

This fourth formula generally appears as:

$$\frac{F_P}{A} = \frac{S}{1 + B(l/r)^2}$$

SLENDER COLUMNS

By standard calculations, a member is considered to be a slender column if its slenderness ratio is above a specific value such as 150 for steel and 70 for aluminum or magnesium alloy. The formula used for solving slender column problems is also known as the *Euler formula*, where:

$$F_P = \frac{\pi^2 EI}{fl^2}$$

As an example of how these formulas are used, let's take a rectangular steel bar with dimensions of $12 \times 3 \times 200$ mm and a modulus of elasticity of 207 GPa. To find the working load of the member where the moment of inertia is 27 mm^4 and the factor of safety is 2.5, the following calculations are made:

$$F_P = \frac{\pi^2 EI}{fl^2}$$
$$= \frac{(3.14159^2)(200 \times 10^3)(27)}{(2)(200^2)}$$
$$= 666.197 \ N$$

Exercises

35. Calculate the working load for a steel angle support bracket that is used on a piece of production equipment. Its angle dimensions are $150 \times 150 \times 10$ mm and it is 3 m long. The formula recommended by the AISC is:

$$\frac{F_P}{A} = 110 - \frac{0.483 \ l}{r}$$

36. If a steel tube has an inside radius of 2.5 in., an outside radius of 3.0 in., is 32.0 in. long, and has an average compression of 20,000 psi, calculate its working load using the formula:

$$\frac{F_P}{A} = \frac{18,000}{1 + (1/18,000)(l/r)^2}$$

37. What is the working load during compression for a 6-cm square magnesium alloy bay that is 2 m long? The factor of safety for this part is designated as 1.60.

CHAPTER 7

Fluidics

- **Basic Properties of Fluids**
- **Static Fluid Concepts and Calculation**
- **Dynamic Hydraulic Systems**
- **Pneumatic Systems**

The field of fluidics deals with the application of fluids, usually under pressure, to provide power for work. Both liquids and gases are considered to be fluids. Examples are hydraulic systems that use oils and water-base fluids, and pneumatic systems that use air. In these systems, confined liquids behave as solids, and air in pneumatic systems exhibit a "spongy" characteristic. This chapter is a presentation and discussion of common formulas and relationships used by technicians and technologists who work with fluidics.

Basic Properties of Fluids

Shear force, tensile force, and compressive force are three types of force that may act on any object. Fluids placed under shear action will move continuously. Compressive forces when used in conjunction with fluidics is known as *pressure* and is denoted as a force per given unit area. Generally, there are three kinds of pressure:

1. *Atmospheric pressure* (P_{ATM}) refers to the force acting on an area due to the weight of the atmosphere. The term *vacuum* is often used to describe pressures below atmospheric.
2. *Gauge pressure* (P_G) is a measurement referred to in pressure

instrumentation. Gauge pressure is calculated as the difference between the pressure of the measured fluid (P) and atmospheric pressure. Thus:

$$P_G = P - P_{ATM}$$

3. *Absolute pressure* (P_{AB}) is the sum of both gauge and atmospheric pressures. Hence:

$$P_{AB} = P_G P_{ATM}$$

Density, Specific Weight, and Specific Gravity

Density is denoted by the Greek letter rho (ρ), and is defined as mass per unit volume. In fluid mechanics, the units are slugs/ft^3, lbf-s/ft^3, kg/m^3, and g/cm^3. The density ρ is equal to 0.002378 slug/ft^3, and 1 slug/ft^3 = 515.3788 kg/m^3. This can be expressed in the following formula:

$$\rho = \frac{M}{V}$$

where M is the mass of the body and V is its volume.

For example, the density of a hydraulic fluid has an observed mass of 25 cubic centimeters (cc) that is balanced by a 20-g weight. To find its density, the following calculations are made:

$$\begin{aligned}\rho &= \frac{M}{V}\\ &= \frac{20}{25}\\ &= 0.80 \text{ g/cm}^3 \text{ or}\\ &= 800.0 \text{ kg/m}^3\end{aligned}$$

The specific weight of a substance is denoted by the Greek letter gamma (γ) and is defined as the ratio of the body's weight to its volume. This is expressed as:

$$\gamma = \frac{w}{v}$$

In this formula, weight can be determined by multiplying the body's mass and gravitational constant ($w = Mg$). This constant is 32.1740 ft/s/s or 9.80665 m/s/s. To find the weights of two substances with masses of 7.5 slugs and 12.55 kg, respectively, the following calculations are used (note that the slug is used to define a mass unit in the British system, and 1 slug = 32 lbs):

$$w = Mg$$
$$= (7.5)(32.1740)$$
$$= 241.305 \text{ lb}$$

$$w = Mg$$
$$= (12.55)(9.80665)$$
$$= 123.073\ 458 \text{ N}$$

Specific weight is also referred to as the *weight density* of a substance. This measure is important when liquid measures are used (e.g., gallon and liter), but must be converted to weight. This is particularly true in the transporting of bulk liquids. For example, to find the specific weight of a liquid weighing 4200 N and with a volume of 58 m^3, the following procedures are used:

$$\gamma = \frac{w}{v}$$
$$= \frac{4200}{58}$$
$$= 72.414 \text{ N/m}^3$$

Specific volume, another term encountered in fluidics, is represented by the letter u. By definition, specific volume is the reciprocal of specific weight. The standard units of measure here are ft^3/lb and m^3/m. Thus:

$$u = \frac{1}{\gamma}$$

Specific gravity (Sg) is a dimensionless number; therefore, it is expressed without units. Specific gravity is defined as the ratio of a substance's density (specific weight) to the density (specific weight) of a standard substance. The standard used for computing the specific gravity for most substances is water, which has an assigned value of 1.

Density and specific gravity have an important relationship, expressed in the following formulas:

202 • FLUIDICS

$$\gamma = \rho g$$
$$\rho = \frac{\gamma}{g}$$

Since density is used to solve many problems, it is often necessary to know the density of substances that one is working with. Presented in Table 7-1 are the densities of common liquids at atmospheric pressure.

Table 7-1. Density of Common Liquids at Atmospheric Pressure

Temp. °C	0	20	40	60	80	100
°F	32	68	104	140	176	212
Liquids	slugs/ft^3 (kg/m^3)					
Ethyl alcohol	1.564 (806.05)	1.532 (789.56)	1.498 (772.04)	1.463 (754.00)		
Benzene	1.746 (899.85)	1.705 (878.72)	1.663 (857.07)	1.621 (835.43)	1.579 (813.78)	
Carbon tetrachloride	3.168 (1632.72)	3.093 (1594.07)	3.017 (1554.90)	2.940 (1515.21)	2.857 (1472.44)	
Gasoline S_g 0.68	1.345 (693.18)	1.310 (675.15)	1.275 (675.11)	1.239 (638.55)		
Glycerin	2.472 (1274.02)	2.447 (1261.13)	2.423 (1248.76)	2.398 (1235.88)	2.372 (1222.48)	2.346 (1209.08)
Kerosene S_g 0.81	1.630 (840.07)	1.564 (806.05)	1.536 (791.62)	1.508 (777.19)	1.480 (762.76)	
Machine oil S_g 0.907	1.778 (916.34)	1.752 (902.94)	1.727 (890.06)	1.702 (877.17)	1.677 (864.29)	1.651 (850.89)
Fresh water	1.940 (999.83)	1.937 (998.29)	1.925 (992.10)	1.908 (983.34)	1.885 (971.49)	1.859 (958.09)
Salt water	1.995 (1028.18)	1.998 (1024.57)	1.975 (1017.87)			

Exercises

1. What are the weights of three compounds whose masses are 5 slugs, 10 kg, and 5 gm, respectively?
2. Find the density of a fluid that has a mass of 25 m and is balanced by a weight of 18 g. Express the densities in g/cm^3, kg/m^3, and slugs/ft^3.
3. What are the specific weights for three substances whose weights and volumes are: (a) 7200 lb, 120 ft^3, (b) 500 dyn, 7000 cm^3, and (c) 3600 N, 50 m^3.

Viscosity

Viscosity is the resistance of fluids to flow and is related to the internal friction of the fluid itself. Thicker fluids will flow at a slower rate than thinner ones, indicating a higher internal friction. All fluids are made up of layers. The farther a layer is from a surface, the less will be its internal friction and resistance to flow.

The shear stress (τ) between adjacent layers of fluids is proportional to the fluid's viscosity. This is noted in the formula:

$$\tau = \mu\left(\frac{x}{y}\right)$$

where x is the relative velocity of the fluid, y is the thickness of the fluid, and μ is the constant of proportionality. μ is also known as the *absolute viscosity* and is expressed in units of lb/s/in.^{2} in reyns, or dyn/s/cm^2 in poise (1 poise = 1 reyn = $2.248 \times 10^{-6}[(2.54)(12)]^2$ lb-s/ft^2 = 2.089×10^{-3} lb-s/ft^2 = 2.089×10^{-3} slug/ft-s). Typical units of reyns and poise are the microreyn (0.000 001 reyn) and centipoise (0.01 poise).

A related expression is known as *kinematic viscosity*, which is represented by the Greek letter nu (ν) and defined as the ratio of absolute viscosity (poise) to mass density, or:

$$\nu = \frac{\mu}{\rho}$$

Kinematic viscosity can be expressed in terms of in.^{2}/s in Newts or cm^2/s in stokes (St). Here, the Centistoke (cSt) is frequently used. When using the metric system, it is frequently necessary to convert absolute viscosity, which is also referred to as *dynamic viscosity*, to *kinematic viscosity* by dividing absolute viscosity by specific gravity. Thus:

$$cSt = \frac{cP}{Sg}$$

To convert kinematic viscosity (v) in Newts to centistokes, the following formula is used:

$$v(\text{Newts}) = 0.001552\,[v(cSt)]$$

Dynamic viscosity can be found by using the following formula:

$$\mu(cP) = \rho v$$

where ρ = the density in g/cm^3 at the same temperature as the kinematic viscosity, and v = the kinematic viscosity in cSt.

An example of a typical problem is to find the kinematic viscosity of a lubrication fluid with a density of 0.94, which is observed in the laboratory to have a calibration constant (k) of 0.1180 in 198 s. To solve this problem, we multiply the constant by time:

$$\begin{aligned} v &= tk \\ &= (198)(0.1180) \\ &= 23.364 \text{ cSt} \end{aligned}$$

Dynamic viscosity is calculated as:

$$\begin{aligned} \mu &= \rho v \\ &= (0.94)(23.364) \\ &= 21.962\,16 \text{ cP} \end{aligned}$$

Exercises

4. Find the kinematic viscosity in cSt units for a fluid that has a density of 0.82. When tested in the laboratory through a calibrated capillary tube, it is observed to have a calibration constant of 0.1200 in 225 s.

5. What is the dynamic viscosity for problem 4?

Static Fluid Concepts and Calculation

Perhaps the most important concept associated with fluidics is pressure. Simply, pressure (p) is a unit of force (force acting on a unit area) and is expressed by the following formula:

$$p = \frac{F}{A}$$

where F is the total force applied and A is the area of a surface.

The mathematical expression for the amount of force at the bottom of a container that is filled with a fluid is calculated by the product of its volume (v) and density (ρ), or:

$$F = v\rho$$

From this relationship, it is now possible to find pressure with the formula:

$$p = \frac{v\rho}{A}$$

This formula can be expressed in simpler form as:

$$p = h\rho$$

where h is the height of the container. Thus, the amount of pressure at the bottom of a container is equal to the product of the fluid's density and depth. For example, if a liquid has a density of 857.07 kg/m^3 and is at a depth of 12 cm, the amount of pressure exerted at the bottom of its container will be:

$$\begin{aligned} p &= h\rho \\ &= (0.12)(857.07) \\ &= 102.8484 \text{ kg/m}^2 \end{aligned}$$

Pascal's Law

The fundamental law of fluid power was derived by Blaise Pascal in 1653. Known as Pascal's law, it states that "pressure applied to an enclosed fluid is transmitted undiminished to every portion of the fluid and the walls of the container." This law is illustrated in Figure 7-1, and can be expressed in the following mathematical relationship:

$$p = \frac{F_1}{A_1} = \frac{F_2}{A_2} = \frac{F_N}{A_N}$$

This relationship states that the pressure is transmitted through

206 • FLUIDICS

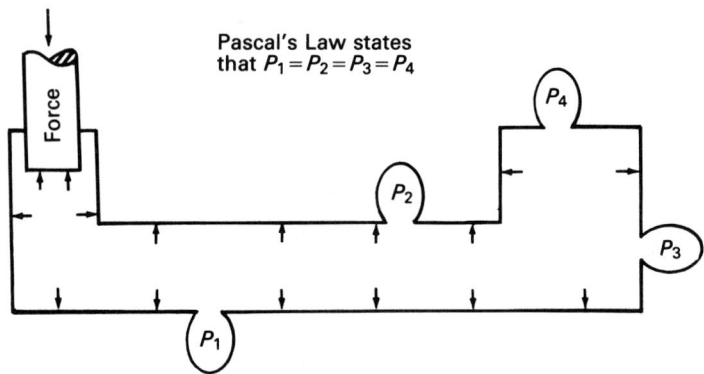

Fig. 7-1. Pascal's law.

all connecting sections at the same magnitude. It also follows that:

$$F_1 = \frac{A_1}{A_2} F_2$$

Products such as hydraulic lifts and brakes make use of the principle. Here, the hydraulic press is a force-multiplying device where the multiplication factor will be equal to the ratio of the areas of its pistons.

Fluid System Calculations

The fluid in a vessel or tube that creates pressure and increases the height of the liquid (e.g., standpipes and piezometer tubes) within the vessel or tube is known as the *pressure head*. The pressure head is most often used for reporting static fluid pressures within vessels at critical depths. The pressure of any pressure head may be obtained by the relationship $p = h$. When the liquid is water, the following equation may be used:

$$p = 0.43333h$$

where h is head in feet of fluid. If the fluid is not water, the following equation is used to give the pressure (in U.S. customary):

$$p = 0.43333Sgh$$

To solve problems where the depth of the head must be determined, the following formula is used for all fluids:

$$h = \frac{2.31p}{\text{Sg}} \text{ (U.S. customary) or}$$

$$h = \frac{p}{\gamma} \text{ (for any system)}$$

An example of this would be to determine at what depth one will encounter a pressure of 32 psi in a lake (*Note*: The Sg of water is 1). To solve this problem, the following procedure is used:

$$h = \frac{2.31p}{\text{Sg}}$$
$$= \frac{(2.31)(32)}{1}$$
$$= 68.16 \text{ ft}$$

An understanding of pressure head and the basic formulas covered thus far makes it possible to solve more complex problems. This section will present problems associated with measuring devices and other applied pressure situations.

MEASURING DEVICES

The simplest type of pressure measuring device is an open-tube manometer (Figure 7-2). This device consists of a U-shaped tube that contains a liquid. At one end, the liquid is exposed to the atmosphere, while the other is at the measured pressure. The pressure at the bottom of the first (left) column is found by using the formula:

$$p_1 = p_{ABS} + \rho g y_1$$

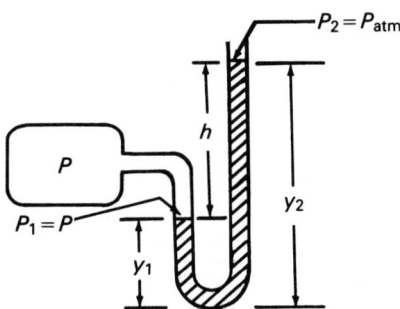

Fig. 7-2. Open-tube manometer.

while the pressure at the bottom of the second (right) column is found by:

$$p = p_{ABS} + \rho g y_2$$

Furthermore:

$$p_{ATM} = \rho g h$$

In all formulas, ρ = density and g = acceleration due to gravity.

Remember that p_{ABS} is the absolute pressure and p_{ATM} is the atmospheric pressure (since it is exposed to the atmosphere). The difference between the two ($p_{ABS} - p_{ATM}$) is known as the *gauge pressure*. Hence, the gauge pressure is proportional to the difference in liquid column heights.

An example of a typical problem is to compute the atmospheric pressure on a day when the height of a barometer is 84.2 cm. Here, we note that the density of mercury is 13.570 g/cm³ and its acceleration due to gravity is 980 cm/s. Thus:

$$p_{ATM} = \rho p h = (13.570)(980)(84.2)$$
$$= 1,119,742.12 \text{ dyn/cm} = 111,974.212 \text{ N/m}$$

Note: 100,000 N/m² = 1 bar, which is a common unit of measure for atmospheric pressures.

A second measuring device is the *differential manometer* (Figure 7-3). It is similar to the open-tube manometer except that both its ends are connected to different pressure sources. Since the pressure at one head is known, the measure, and unknown pressure, is found as the difference in pressure head in the two columns:

$$p = p_1 - p_2$$

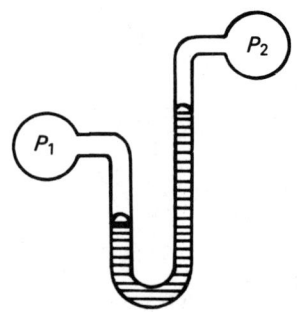

Fig. 7-3.   Differential monometer.

Exercises

6. An open-tube manometer, similar to that shown in Figure 7-2, is filled with mercury and exposed to an atmospheric pressure of 970 millibars. If $y_1 = 30.0$ mm and $y_2 = 80.0$ mm, what will be the absolute pressure at the bottom of the U-tube?

7. Calculate the absolute pressure in problem 6 at 50 mm below the free surface of the open end.

8. What is the absolute pressure of the gas in the tank in problem 6?

Dynamic Hydraulic Systems

A static hydraulic system has *potential energy* as a result of its elevations, high pressure, and capacity for movement. The system becomes dynamic with the addition of pumps and the reduction of resistance (friction) to movement. Loss of energy is a result of friction created by the flowing of fluids through pipes, tubes, and fittings and is expressed in terms of pressure drop and loss of fluid horsepower. Presented in this section are formulas used for determining the energy in hydraulic systems.

Hydraulic System Principles and Equations

Calculations made of hydraulic systems are based on several basic principles and equations that are central to the field of fluidics. Each is interrelated with the workings of all industrial hydraulic systems used today. In many cases, they serve as a basis for understanding how systems work and are therefore often used in combination with one another.

BERNOULLI'S EQUATION

The saving of energy in a fluid power system is described in Bernoulli's theorem, which states that the total energy of a fluid system will remain constant, since the total energy at one point in the system will be the same as the total energy at any other point in the system (Figure 7-4). Thus, Bernoulli's equation is written across two points (p_1 and p_2):

210 • FLUIDICS

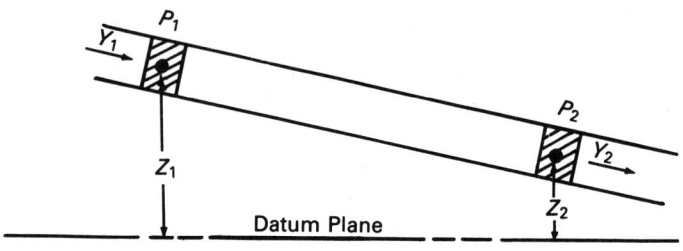

Fig. 7-4. Bernoulli's equation.

$$Z_1 + \frac{p_1}{\rho} + \frac{v_1^2}{2g} + W_I = Z_2 + \frac{p_2}{\rho} + \frac{v_2^2}{2g} + W_o + F$$

where Z = height above a datum plane, p = pressure, v = velocity, g = gravitational constant, ρ = density of fluid, F = energy losses due primarily to friction, W_I = power added by a pump, and W_o = power removed by a device such as motors. Presented in Table 7-2 are other formulas derived from Bernoulli's equation.

A typical problem using the principles of Bernoulli's equation is illustrated in Figure 7-5. To find the amount of pressure developed at the motor (p_2) if the flow rate is 20 gpm, the total friction loss is 10 ft. of fluid, and the pump has a 3-hp capacity (assume that the fluid has a specific gravity of 0.90 and density of 56.130 lb/ft³), the following procedures would be used:

$$W = \frac{3950P}{QSg} = \frac{(3950)(3)}{(20)(0.90)} = 658.333 \text{ ft}$$

$$v = \frac{0.321Q}{A} = \frac{(0.321)(20)}{0.7854} = 8.174 \text{ ft/s}$$

$$W = \frac{p_2}{\rho} + \frac{v_2^2}{2g} + F$$

$$p_2 = \rho \left[W_I - \left\{ \left(\frac{v_2^2}{2g} \right) + F \right\} \right]$$

$$= 56.130 \left[658.33 - \left\{ \frac{8.174^2}{(2)(32.1740)} + 10 \right\} \right]$$

$$= 36,332.650 \text{ lb/ft}^2$$

$$= 252.310 \text{ psi}$$

Dynamic Hydraulic Systems • 211

Table 7-2. Equations Related to Bernoulli's Theorem

Unknown	Factors	Equation
Power of fluid	P = power Q = flow volume Sg = specific gravity	$W = \dfrac{P}{QSg}$ $W = \dfrac{3950P}{QSg}$ (U.S. customary only)
Added power	p = pressure at some point v = velocity at some point F = energy loss to that point g = gravitational constant	$W_I = \left(\dfrac{p_2}{\rho}\right) + \left(\dfrac{v_2^2}{2g}\right) + F$
Removed power		$W_O = \dfrac{p_1}{\rho} + \dfrac{v_1^2}{2g} - F$
Pressure changes	if segment includes motor if segment include pump segment has no motor or pump	$p = W_O + F$ $p = W_I - F$ $p = F$
Fluid velocity	d = inside diameter of pipe or tube	$v = \dfrac{Q}{A}$ $v = \dfrac{Q}{d^2}$ $v = \dfrac{0.321Q}{A}$ (U.S. customary only) $v = \dfrac{0.408Q}{d^2}$ (U.S. customary only)

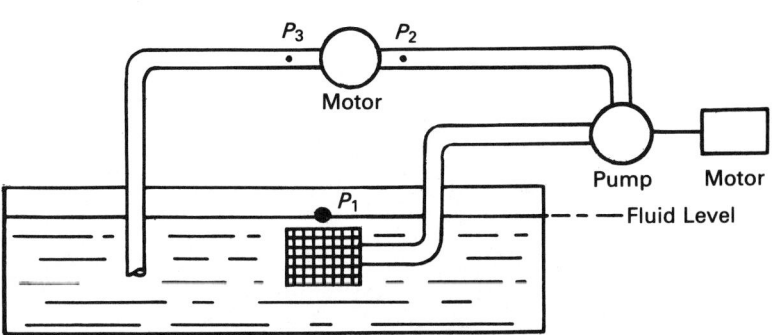

Fig. 7-5. Application of Bernoulli's equation.

CONTINUITY EQUATION

In most hydraulic system problems, the flow is considered to be constant or steady and velocity will therefore be a constant. As a result, the continuity equation is based upon the concept that during a steady flow the rate of fluid flow will be the same at any point in the conductor. The continuity equation and its derivatives are:

$$Q = Av$$
$$Q_1 = A_1 v_1 = Q_N = A_N v_N$$
$$\frac{v_1}{v_2} = \frac{A_2}{A_1} = \frac{d_2^2}{d_2^2}$$
$$v_2 = \frac{v_1 A_1}{A_2} = \frac{v_1 d_1^2}{d_2^2}$$

In these equations, Q = the constant flow rate, A = the cross-sectional area of the pipe or tube, and v = the velocity of the system's fluid. Q will normally be measured in terms of gallons per minute (gpm) or cubic meters per second (m³/s), A in square inches (in.²) or square meters (m²), and v in feet per second (fps) or meters per second (m/s).

TORRICELLI'S THEOREM

An illustration of Torricelli's theorem is shown in Figure 7-6. Here, the speed with which the fluid is discharged will be the same as the speed of a body falling from rest from the height h. This can be calculated as the square root of the product of twice the acceleration due to gravity, multiplied by the head generating the jet. Mathematically, this is expressed:

$$v = \sqrt{2gh}$$

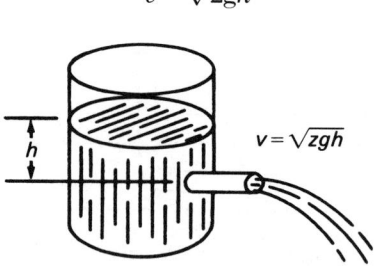

Fig. 7-6. Toricelli's theorem.

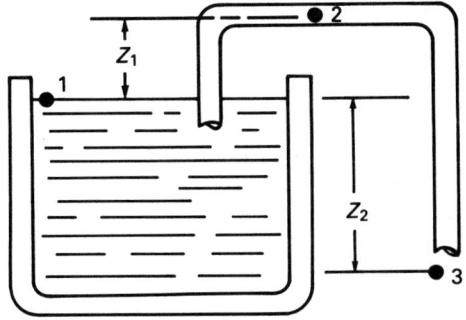

Fig. 7-7. The siphon.

This formula is based upon *ideal* conditions, when friction or other head losses do not occur. If the system is not ideal, then the following formula should be used:

$$v = \sqrt{2g(h - F)}$$

When a siphon is used (Figure 7-7), the following formulas apply:

$$v = \sqrt{2g(Z_1 - F)}$$

$$p_2 = -\rho \left[Z + \left(\frac{v^2}{2g} \right) + F \right]$$

REYNOLDS NUMBER

The critical velocity of a fluid (the span of velocities separating laminar and turbulent flow) is determined by the value of the Reynolds number. A value of less than 2000 indicates laminar or streamline flow, and values that exceed 4000 indicate turbulent flow. Reynolds numbers that range between 2000 and 4000 are indicative of transitional flow.

The Reynolds number, noted as N_R, may be calculated by using the following equation:

$$N_R = \frac{dv\rho}{\mu}$$

where d = inside diameter of the pipe or tube in feet, v = fluid velocity (ft/s), ρ = mass density (lb/ft), and μ = absolute viscosity (lb/ft = s).

When specific gravity is known, the following equation can be used:

$$N_R = \frac{124 dv\rho}{\mu} \text{ or}$$
$$N_2 = \frac{1140 dv\text{Sg}}{\mu}$$

Another formula that can be used when kinematic viscosity (cSt) is known is:

$$N_R = \frac{7740 vd}{v}$$

DARCY FORMULA

The Darcy formula is generally used by engineers who need to make crude-oil pipeline calculations. The basic formula is:

$$h_F = \frac{fLv^2}{d2g}$$

where f = a dimensionless friction factor, L = length of the pipe or tube, d = inside diameter of the pipe or tube, and v = the average velocity flow of the fluid. This formula can also be expressed in a form incorporating conventional pipeline units, as follows:

$$p = \frac{34.87 fB^2 \text{Sg}}{d^5}$$

where p = friction press drop in lb/in.^{2}-mi, f = friction factor, B = flow rate in bbl/h (42 gal/bbl), Sg = specific gravity of the oil, and d = inside diameter of the pipeline.

The friction factor in the Darcy formula is frequently determined by the formula:

$$f = \frac{64}{N_R}$$

Therefore, this formula can be substituted for f in the formula.

Exercises

9. Fluid flows from a 2-in.-diameter pipe at a rate of 18 gpm. Find the minimum diameter to which the pipe can be reduced so that the velocity of the fluid will not exceed 10 fps.

10. Fluid is flowing through a horizontal pipe that reduces from a 3-in. to a 1-in. diameter at a rate of 500 gpm. If the pressure in the 3-in.-diameter pipe is 1000 psi, what will be the pressure in the 1-in. pipe?

11. A water tank 40 ft in height has a valve located at its base. If the valve is opened, what will be the velocity of the flowing water?

12. A lubricant has a kinematic viscosity of 0.05 Newts and is flowing through a 1-in. pipe at a rate of 100 gpm. Calculate its Reynolds number.

13. What is the type of flow found in problem 12?

Hydraulic System Problems

Engineers, technologists, and technicians who work with hydraulic systems are often faced with calculations that address specific problems. Presented in this section are five common problem areas that are often encountered in the field of fluidics.

FRICTION AND WEAR

Both friction and wear occur in hydraulic systems as a result of fluids moving against the surfaces of pipes and tubing. Friction (F) is defined as any force that is parallel to the two surfaces in contact and that resists movement between the fluid and surface. A *coefficient of friction* (f) can be calculated by the basic formula:

$$f = \frac{F}{L}$$

where F = force required to balance opposition to movement and L = load forcing the two surfaces together. Wear, by contrast, is the

216 • FLUIDICS

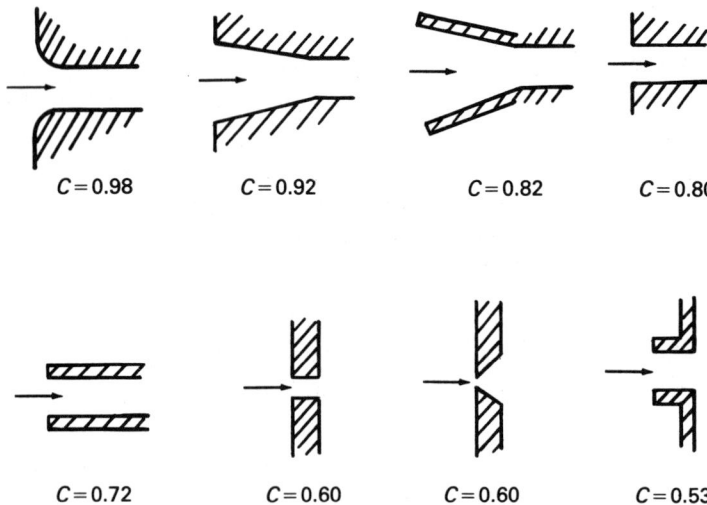

Fig. 7-8. Orifice designs with coefficient of friction.

permanent displacement of surface material that is caused by friction in the system.

The coefficient of friction is to be considered not as a constant but as a dimensionless expression. When there is streamline or laminar flow, $f = 64/N_R$. For transitional and turbulent flow, however, there is no true mathematical relationship. See Figure 7-8.

FRICTION LOSSES IN PIPES

Friction that is caused by the rubbing of fluids against pipes and tubes is always present in dynamic systems. Friction results in heat and is a loss of potential energy in the system. This loss is always calculated as a loss in head or pressure throughout the system; hence, *friction loss* or *friction drop*.

The basic formula used to calculate pressure drop due to friction is:

$$\Delta p = \frac{p}{Sw}$$

where Δ = change in pressure and Sw = specific weight. Other forms of this formula are:

$$\Delta h = \frac{fLv^2}{2gd_{FT}}$$

$$\Delta h = \frac{fLv^2}{5367d}$$

$$\Delta h = 0.1863 \frac{Lv^2}{d_{IN}}$$

Here, Δh = head loss (ft), f = friction factor, L = length of the pipe (ft), v = fluid velocity (ft/s), g = acceleration due to gravity (32.1740 ft/s/s), d_{FT} = inside diameter (ft), d_{IN} = inside diameter (inches). If the pressure drop is needed in psi, then the following equations can be used:

$$\Delta p = \frac{\rho L v^2}{288 g d_{FT}}$$

$$\Delta p = \frac{\rho L v^2}{772.8 d_{IN}}$$

$$\Delta p = \frac{fLv^2 Sg}{12.4 d_{IN}}$$

In these formulas, Δp = pressure drop and d = density (lb/ft^3).

Formulas that can be used for either turbulent or laminar flow, or when the type of flow is unknown, are:

$$\Delta h = \frac{\mu L v}{10.4 d_{IN}^2 \rho}$$

$$\Delta h = \frac{\mu L v}{649 d_{IN}^2 Sg}$$

$$\Delta p = \frac{\mu L v}{1497 d_{IN}^2}$$

Here, μ = absolute viscosity (cSt), L = length of pipe or tubing (ft), and all others as above.

A typical example of a problem involves a hydraulic oil having a specific gravity of 0.94 and a kinematic viscosity of 100 cSt and that flows through a pipe with an inside diameter of 1.250 in. at a rate of 50 gpm. To find the pressure drop and head loss per 10 ft of pipe, the following procedures are used:

To compute absolute viscosity:

$$\mu = \nu Sg$$
$$= (100)(0.94)$$
$$= 94 \text{ cp}$$

To compute fluid velocity:

$$v = \frac{Q(0.408)}{d^2}$$
$$= \frac{(50)(0.408)}{1.25^2}$$
$$= 13.056 \text{ ft/s}$$

To find head loss:

$$\Delta h = \frac{fLv}{649 d^2 Sg}$$
$$= \frac{(94)(10)(13.056)}{(649)(1.25^2)(0.94)}$$
$$= 12.875 \text{ ft per 10 ft}$$

To find pressure drop:

$$\Delta p = 0.433 Sgh$$
$$= (0.433)(0.94)(12.875)$$
$$= 5.240 \text{ psi}$$

FRICTION LOSS IN FITTINGS

Fluid-system fittings are any components other than straight pipes or tubing. Examples of fittings are elbow tees, valves, and bends. In most cases, fluid pressure drop will increase as it passes through fittings. Head loss resulting from flow through fittings will be approximately proportional to the square of the flow velocity. This is expressed as:

$$h_L = \frac{kv^2}{2g}$$

In this formula, k is a proportional constant of the length of straight pipe that has the same pressure drop as the fitting (Le) to the pipe diameter (d), and is mathematically found by:

$$k = \frac{Le}{d}$$

Table 7-3. Proportional Constant (*k*) for Selected Pipe Fittings

Fittings	k
Globe valve, perpendicular stem (fully opened)	340
Globe valve, y-pattern (fully opened)	160
Globe valve (fully opened)	13
Gate valve (3/4 opened)	35
Gate valve (1/2 opened)	160
Gate valve (1/4 opened)	900
Angle valve (fully opened)	145
Check valve, ball (fully opened)	150
Swing check valve (fully opened)	135
Elbow, 90° standard	30
Elbow, 90° long radius	20
Elbow, 90° square corner	57
Elbow, 45° standard	16
Tee, standard (through run)	20
Tee, standard (through branch)	60
Return bend, close pattern	50
Plug cock, full port (two-way, fully opened)	18
Plug cock, reduced port (three-way, straight through)	44
Plug cock, reduced port (three-way, through branch)	140

Presented in Table 7-3 is the proportional constant for selected pipe fittings.

Smooth pipe bends in pipes and tubing will cause less pressure drop than elbows. Their proportional constants, however, do not follow the same relationship as for other fittings. Here, k will vary according to the ratio of the curvature radius to pipe or tube diameter (r/d). Table 7-4 presents the proportional constant for smooth 90° bends in pipes and tubes.

The k values given are primarily used for turbulent fluid flow. In some cases, it will be necessary to make use of the k value for streamline flow. Though limited research has been conducted relative to resistance of valves and fittings in laminar flow, the proportional constant can be calculated by use of the following formula:

$$k_L = \frac{kN_R}{1000}$$

Table 7-4. Proportional Constant for Smooth 90° Bends

r/d	k
1	20
2	12
3	12
4	14
5	16
6	18
7	21
8	24
9	27
10	30
11	32
12	35
13	37
14	39
15	41
16	43
17	45
18	47

where k_L = constant for laminar flow, k = proportional constant, and N_R = Reynolds number.

A last factor to consider is the *flow coefficient*, which is also known as the *volume coefficient*. Noted as C_V, this coefficient is frequently used with control valves, which are rated according to their C_V value. This coefficient is found by the following formula:

$$C_V = \frac{29.9 d^2}{\sqrt{k}}$$

Other relationships are:

$$Q = 7.9 C_V \sqrt{\Delta p / \rho}$$

$$p = Sg \left(\frac{Q}{C_V} \right)^2$$

Exercises

14. Hydraulic fluid having a specific gravity of 0.89 and a kinematic viscosity of 150 cSt flows through a standard 1-in. pipe. If the rate of flow is 40 gpm, determine the type of flow found in the system.

15. Calculate the pressure drop and head loss per 100 ft of pipe for the system in problem 14. (*Note*: The inside diameter of a standard 1-in. pipe is 1.05 in.)

16. If the rate of flow in problem 14 increases to 200 gpm, the fluid velocity will be 74 ft/s and the Reynolds number will be 4020. If the friction factor is 0.04 calculate the pressure drop and head loss per 100 ft of pipe.

Pneumatic Systems

Pneumatic systems are used to provide or control power. Though pneumatic systems use compressed fluids, only air and gases will be considered here. All calculations made for pneumatic systems concerning volume and pressure changes incorporate the use of absolute pressure and temperature. As previously discussed, absolute pressure is the sum of the gauge and atmospheric pressures:

$$p_{ABS} = p_G + p_{ATM}$$

For example, if the atmospheric pressure is assumed to be 14.696 lb/in.^{2} (a common approximation used is 14.7 lb/in.^{2}) and the gauge pressure is 5 lb/in.^{2}, then the absolute pressure would be 19.7 lb/in.^{2}. A vacuum of 8 lb/in.^{2} would be an absolute pressure of 6.7 lb/in.^{2} (14.7 − 8 = 6.7).

Absolute temperature using the Fahrenheit scale would be:

$$T = t + 459.3 \text{ or (approximately)}$$
$$T = t + 460$$

Here, t is the thermometer reading in °F. The absolute temperature is often expressed in terms of degrees Rankine. If the Centigrade scale is used, the absolute temperature would be:

$$T = t + 273$$

If the temperature reading on a thermometer were 46°F, the absolute temperature would be 506°R.

Gas Laws

There are several gas laws that are frequently used to solve pneumatic system problems. The first is known as *Boyle's law*, and it states that when the temperature of a gas is held constant, its volume will vary inversely with its absolute pressure. Mathematically, this is expressed as follows:

$$\frac{p_1}{p_2} = \frac{V_2}{V_1}$$

Thus:

$$p_2 = \frac{p_1 V_1}{V_2} \text{ and}$$

$$V_2 = \frac{p_1 V_1}{p_2}$$

As an example of how Boyle's law is used, a water tank on a heating system has a capacity of 20 ft³ and is three-quarters full when the system heats up. To calculate the amount of pressure under which the system is operating, the following calculations are made (in this problem, V_2 is the unfilled portion of the tank):

$$p_2 = \frac{p_1 V_1}{V_2}$$
$$= \frac{(14.696)(20)}{5}$$
$$= 58.784 \text{ psia}$$
$$= 44.088 \text{ psig}$$

In this problem psia stands for pounds per square inch absolute and psig, for pounds per square inch gauge. Gas pressures should always be converted to psi gauge. This is accomplished by using the following relationships:

$$p_{ABS} = p_G + p_{ATM} \text{ or}$$
$$p_{ABS} = p_G + 14.696$$

Hence:

$$p_G = p_{ABS} - 14.696$$

The second gas law to be considered is *Charles's law*, which states that when confined gas is held at a constant pressure, its volume will be directly proportional to the absolute temperature. This is written as:

$$\frac{V_1}{V_2} = \frac{T_1}{T_2}$$

For example, if the total volume of a gas storage tank kept under a constant pressure is 28,000 ft³ when the temperature is 62°F, what will its volume be when the temperature rises to 89°F? To solve this problem, the temperature readings are converted to absolute temperature and the following calculations are made:

$$V_2 = \frac{V_1 T_2}{T_1}$$
$$= \frac{(28000)(549)}{522}$$
$$= 29{,}448.275 \text{ ft}^3$$

The third gas law is known as *Gay-Lussac's law*, and it states that when a volume of gas is confined and held constant, the resulting pressure will be directly proportional to the absolute temperature. Mathematically, this is expressed as:

$$\frac{p_1}{p_2} = \frac{T_1}{T_2}$$

An example of how this is used is in an air compressor that has a pressure of 250 psi when the midday temperature is 85°F. If the temperature decreases to 72°F, what will be the new pressure? This problem is solved by the following procedures:

$$p_2 = \frac{p_1 T_2}{T_1}$$
$$= \frac{(264.696)(545)}{532}$$
$$= 271.164 \text{ psig}$$

In this problem, temperature was converted to degrees Rankine and pressure to psia. Hence, the new pressure is read as psig.

The last gas law to be considered is the *combined gas law*. As its name implies, it is a combination of the previous laws. The combined gas law is stated mathematically as follows:

$$\frac{p_1 V_1}{T_1} = \frac{p_2 V_2}{T_2}$$

The three factors of pressure, volume, and temperature can be solved with the following variations:

$$p_2 = \frac{p_1 V_1 T_2}{V_2 T_1}$$

$$V_2 = \frac{p_1 V_1 T_2}{p_2 T_1}$$

$$T_2 = \frac{p_2 V_2 T_1}{p_1 V_1}$$

Exercises

17. A stainless-steel tank is placed on a water heating system that has a capacity of 6 ft³. If the tank is filled two-thirds full when the system heats up, what will be the operating pressure?

18. A gas tank has the capability of maintaining a constant pressure. If its total volume is 12,000 ft³ at a temperature of 55°F, what will its volume be when the temperature is raised 65°F?

19. If a compressed-air tank is brought to a pressure of 150 psi when the temperature is 95°F, what will its pressure be when the temperature cools to 65°? Give your answer in psia and psig.

20. One hundred cubic feet of air at p_{ATM} has a temperature of 70°. If it is compressed to 100 psi at a temperature of 110°F, what will its final volume be?

Flow Losses

The flow of air is similar to that of other fluids—when it flows through pipes and tubes it will lose energy as a result of friction. The basic formula used here is:

$$\Delta p = \frac{0.01 L C_1 C_2}{\rho}$$

where Δp = pressure drop, L = length of pipe or tubing, ρ = density of gas, C_1 = discharge factor, and C_2 = size factor. To find C_1 and C_2, see Tables 7-5 and 7-6.

Calculations made for gas flow losses in pipes and tubings follow the same procedures used in other fluid problems. The density of specific gases may be found by multiplying the tabular values (see Table 7-7) by the specific gravity of the gas itself.

Table 7-5. Discharge Factor

Total Pressure Change lb/hr	C_1
800	0.00064
900	0.00081
1,000	0.001
1,500	0.00225
2,000	0.004
2,500	0.062
3,000	0.009
3,500	0.0122
4,000	0.016
5,000	0.025
6,000	0.036
7,000	0.049
8,000	0.064
9,000	0.081
10,000	0.10
20,000	0.4
25,000	0.62
30,000	0.9
40,000	1.6
50,000	2.5
60,000	4.9
70,000	4.9
80,000	6.4
90,000	8.1
100,000	10.0
200,000	40
300,000	90
400,000	160
500,000	250
600,000	360
700,000	490
800,000	640
900,000	810
1,000,000	1,000

Table 7-6. Size Factor (C_z)

Pipe Size (nominal)	C_2
1/8	7,920,000
1/4	1,590,000
3/8	319,000
1/2	93,500
3/4	21,200
1	5,950
1 1/4	1,408
1 1/2	627
2	169
2 1/2	66.7
3	21.4
3 1/2	10
4	5.17
6	0.610

Table 7-7. Density of Air (lb/ft^3)

Temp. (°F)	Pressure (psi)							
	0	40	80	100	120	140	160	200
30	0.0811	0.302	0.522	0.633	0.743	0.853	0.964	1.185
40	0.0795	0.295	0.512	0.620	0.728	0.836	0.944	1.161
50	0.0782	0.291	0.504	0.610	0.717	0.823	0.929	1.142
60	0.0764	0.248	0.492	0.596	0.700	0.804	0.908	1.116
70	0.0750	0.279	0.483	0.585	0.687	0.789	0.891	1.095
80	0.0736	0.274	0.474	0.574	0.674	0.774	0.874	1.075
100	0.0709	0.264	0.457	0.554	0.650	0.747	0.843	1.036
120	0.0685	0.255	0.441	0.535	0.628	0.721	0.815	1.001
140	0.0662	0.246	0.427	0.517	0.607	0.697	0.787	0.967
150	0.0651	0.242	0.420	0.508	0.597	0.686	0.774	0.951
175	0.0626	0.233	0.403	0.488	0.573	0.659	0.744	0.914
200	0.0602	0.224	0.388	0.470	0.552	0.634	0.716	0.879
250	0.0559	0.208	0.361	0.437	0.513	0.589	0.665	0.817
300	0.0523	0.1945	0.337	0.408	0.479	0.550	0.622	0.764

CHAPTER 8

Cams and Gears

- Cams
- Gears

Cams and gears are used to transmit motion or power from one machine to another. This motion can be either rotating or reciprocating. Both devices are found throughout industry and require careful and precise calculations for their production and specification. This chapter provides formulas that are commonly used when working with cams and gears.

Cams

A cam can be defined as a mechanical device that is used to transmit and generate motion to a *follower* by direct physical contact. Cams are capable of delivering positive mechanical control with great accuracy and are usually mounted on rotating shafts—though some remain stationary while followers move about them. Because the cam is the result of a desired follower movement, its shape will be determined by the motion of the follower.

The two most common types of cam designs are the *plate*, or *O.D.*, *cam* and the *drum*, or *cylinder*, *cam*. Plate cams are usually shaped like a disk with the contour machined about its circumference. Here, the follower's line of action will generally be along the axis of rotation. A drum cam will have a machined "track" about its circumference, with the line of action parallel to the axis of rotation.

There are two other cam designs, variations of the plate and drum cams, that have some significant usage in industry. The first is known as the *conjugate cam* and is a variation of the plate cam where a cam track is machined into the face of the disk. The second is the *index cam*, which is a variation of the drum cam where the follower passes over the cam in an arclike motion.

Cam Nomenclature

To make effective use of cam formulas, it is first necessary to have an understanding of cam nomenclature. The basic terms used when working with cams are presented as follows:

The *follower displacement* is the position of the follower from a point of rest to one complete machine cycle (cam displacement) and is specified in either degrees or linear measurement.

Cam displacement is the cam motion from a point of rest relative to the follower mechanism. Graphically, this is illustrated in Figure 8-1 and is specified in terms of degrees or linear measurement.

The *cam profile* is the actual shape or contour of the cam's working surface.

The *base circle* is the smallest-radius circle, r_B, that can be drawn to the cam's profile.

The *trace point* is the centerline of the follower device that rides along the cam's profile. If the follower is a flat surface, then the trace point will be the outside edge that is in direct contact with the cam—producing an envelope.

The *pitch curve* is the continuous trace points that are generated as a result of the cam displacement.

The *prime circle* is a function of the roller radius and is related to the base circle. It is the smallest-radius circle within the pitch curve, and is drawn from the center of the cam.

The *pressure angle* is that angle between the direction where the follower wants to go and where the cam wants to direct it.

The *pitch point* is that location on the pitch circle where the pressure angle is the largest.

The *pitch circle* is that circle, r_P, that passes through the pitch point.

The *transition point*, which is sometimes referred to as the *crossover point*, is that position where the follower's acceleration changes from plus $(+)$ to minus $(-)$. That is, the force acting on

Cams • 229

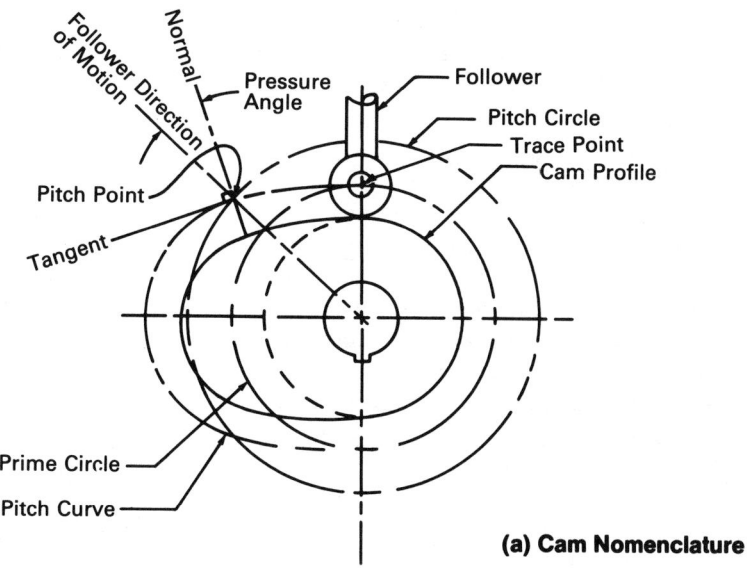

(a) Cam Nomenclature

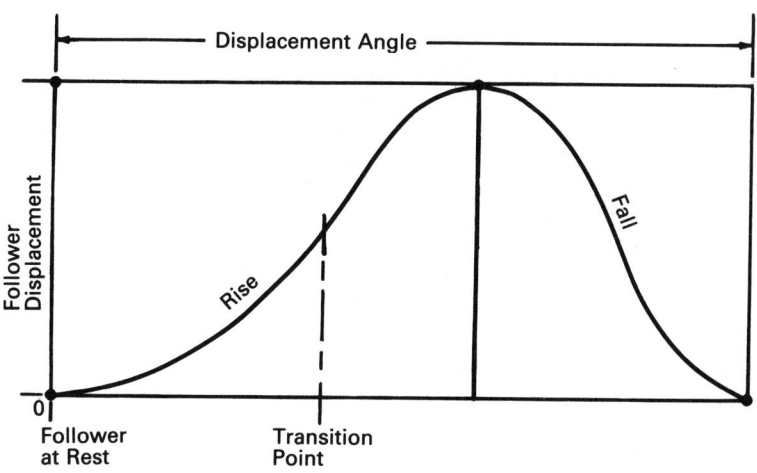

(b) Cam Displacement

Fig. 8-1. Cam nomenclature and displacement.

the follower changes its direction (e.g., from upward to downward or when the follower changes from one cam profile and crosses over to the opposite or conjugate one).

Displacement Diagrams

Most calculations made for cams are associated with design considerations. Because the motion of cam followers is of primary significance, its rate of speed and various positions are carefully planned and calculated. A tool used here is the *displacement diagram*. This diagram (Figure 8-2) is an accurately plotted curve that shows the displacement of the follower as ordinates drawn on a baseline that represents one complete cycle or revolution of the cam.

Displacement diagrams can be drawn to any convenient scale. The horizontal distances T_1, T_2, etc., are expressed in terms of degrees, radians, or time (seconds). The vertical distance h is used to show the rise and fall of the follower—also known as the *stroke* of the follower.

Because each cam is designed to accomplish a specific task, its displacement curve will be unique. However, all incorporate the same basic factors. These factors, and their symbols, are:

y = follower displacement
h = maximum follower displacement
t = time for cam to rotate through angle ϕ, in seconds
T = time for cam to rotate through angle β, in seconds
ϕ = a degree measurement of the cam angle rotation for follower displacement y

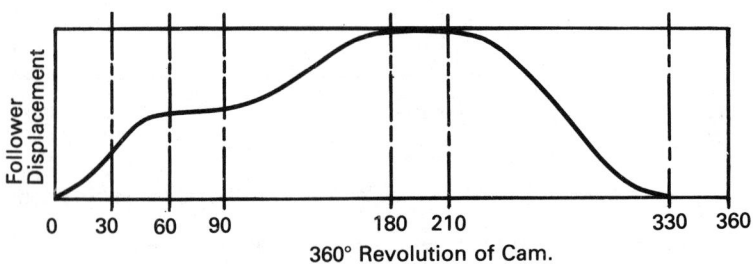

Fig. 8-2. Displacement diagram

β	= a degree measurement for cam angle rotation for total rise h
v	= follower velocity
a	= follower acceleration
N	= cam speed (usually in rpm)
ω	= angular velocity of the cam in degrees/s
ω_R	= angular velocity of the cam in rad/s
W	= effective weight
g	= gravitational constant (386 in./s^2)
R_{MIN}	= minimum radius to cam pitch curve
R_{MAX}	= maximum radius to cam pitch curve
r_F	= radius of cam follower (roller)
ρ	= radius of the curvature of the cam pitch curve within the path of the center of the follower's roller
R_C	= radius of the curvature of the actual cam surface contour
α_{MAX}	= specified maximum pressure angle in degrees
$R_{\alpha MAX}$	= radius from the cam center to a point on the pitch curve where α_{MAX} is located
ϕ_P	= rise angle in degrees
R_α	= the radius from the center of the cam to the pitch curve at α
ϕ	= a degree measure for rise angle

Several of the mathematical relationships that exists between these factors are:

$$t = \frac{\phi}{\omega} \text{ (through angle } \phi\text{)}$$

$$T = \frac{\beta}{\omega} \text{ (through angle } \beta\text{)}$$

$$\frac{t}{T} = \frac{\phi}{\beta}$$

$$\omega = \frac{\beta}{T} = \frac{\phi}{t} = 6N = \text{degrees/second}$$

$$\omega_R = \frac{\pi \omega}{180}$$

$$R_C = \rho - r_F \text{ (convex surfaces)} = \rho + r_F \text{ (concave surfaces)}$$

$$\phi_P = \alpha_{MAX} = R_{\alpha MAX}$$

$$\phi = \alpha = R_\alpha$$

Cam Formulas

As might be expected, there is a limitless number of displacement diagram configurations, representing the unique characteristics of a particular cam. Most diagrams, however, can be broken down into major groupings with specific characteristics pertaining to follower velocity and acceleration.

This section will describe the major types of displacement diagrams and their related formulas. A number of the formulas given in this section make use of *inverse trigonometric functions*, such as arcsin x. This notation can also be specified as the antisine of x, inverse sine of x, or $\sin^{-1} x$. The notation means "the principal angle whose sine is x." Thus, the principal angle is an angle between $-90°$ and $+90°$ in the case of arcsin and arctan, and between $0°$ and $180°$ in the case of arccos.

CONSTANT VELOCITY MOTION.

Presented in Figure 8-3 is an illustration of a constant velocity displacement diagram. As its name implies, the rate of motion is at a constant velocity and appears as a straight line. This type of motion is rarely used alone and is usually found at the start or finish of a cam rotation cycle.

The basic formulas used here assume that $0 < t < T$, so that $a = 0$. Therefore:

$$y = \frac{ht}{T} \text{ or}$$
$$y = \frac{h\phi}{\beta} \text{ and}$$
$$v = \frac{h}{T} \text{ or}$$
$$v = \frac{h\omega}{\beta}$$

To determine the displacement of the follower when $t = 2.3$ s, $T = 1.8$ s, and the maximum displacement of the follower is 3.54 in., the following procedure would be used:

$$y = \frac{ht}{T} = \frac{(3.54)(2.3)}{1.8} = 4.5 \text{ in.}$$

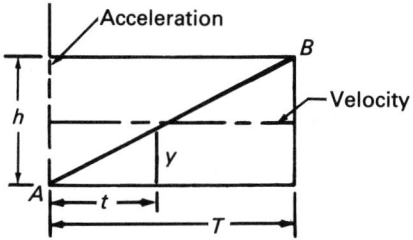

Fig. 8-3. Constant velocity motion displacement diagram.

To calculate the pressure angles for uniform velocity motion, the following formulas are used:

$$\alpha = \arctan\left[\frac{180°h}{\pi\beta R_\alpha}\right]$$

$$\alpha_{MAX} = \arctan\left[\frac{180°h}{\pi\beta R_{MIN}}\right]$$

Here, α is at radius R_α to the pitch curve, α_{MAX} is at radius R_{MIN} to the pitch curve, and $\phi = 0°$. If α_{MAX} is given, then:

$$R_{MIN} = \frac{180°h}{\pi\beta \tan \alpha_{MAX}}$$

PARABOLIC MOTION

A review of the parabolic motion displacement diagram in Figure 8-4 illustrates that when $t=0$, $v=0$. The basic formulas presented here assume that t will range from 0 to $T/2$ in the first half of the curve. When the midpoint is reached, $y=h/2$ and the curve will be inverted for the second half, and $v=0$ when $t=T$ and $y=h$.

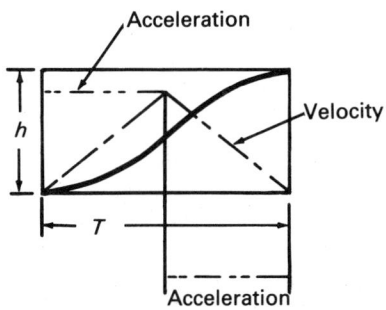

Fig. 8-4. Parabolic motion displacement diagram.

The basic parabolic motion formulas used in cam design are:

$$y = 2h\left(\frac{t}{T^2}\right) \text{ or}$$

$$y = 2h\left(\frac{\phi}{\beta}\right)^2$$

$$v = \frac{4ht}{T^2} \text{ or}$$

$$v = \frac{4h\omega\phi}{\beta^2}$$

$$a = \frac{4h}{T^2} \text{ or}$$

$$a = \frac{4h\omega^2}{\beta^2}$$

Formulas used to find pressure angles are:

$$\alpha = \arctan\left[\frac{720°h\phi}{\pi\beta^2 R_\alpha}\right]$$

When $\phi = \beta/2$ and $R_\alpha = R_{MIN} + (h/2)$, then:

$$\alpha_{MAX} = \arctan\left[\frac{360°h}{\pi\beta R_\alpha}\right]$$

If α_{MAX} is specified, then:

$$R_{MIN} = \left(\frac{360°}{\pi\beta\tan\alpha_{MAX}}\right) - \left(\frac{h}{2}\right)$$

where $\phi = 0°$.

SIMPLE HARMONIC MOTION

Presented in Figure 8-5 is the displacement diagram for a simple harmonic motion cam. The basic assumptions here are that t will be equal to or greater than 0, and T is equal to or greater than t. The primary advantage to this design is the smoothness in velocity and acceleration during the stroke of the follower. The basic formulas used here are:

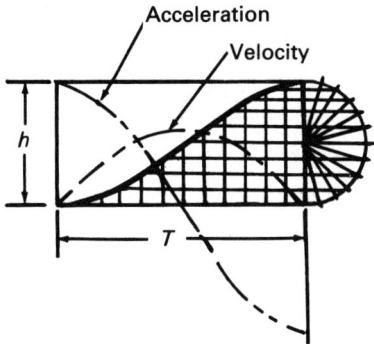

Fig. 8-5. Simple harmonic motion displacement diagram.

$$y = \left[\frac{h}{2}\right]\left[1 - \cos\left(\frac{180°t}{T}\right)\right] \text{ or}$$

$$y = \left[\frac{h}{2}\right]\left[1 - \cos\left(\frac{180°\phi}{\beta}\right)\right]$$

$$v = \left[\frac{h}{2}\right]\left(\frac{\pi}{T}\right) \sin\left(\frac{180°t}{T}\right) \text{ or}$$

$$v = \left[\frac{h}{2}\right]\left(\frac{\pi\omega}{\beta}\right) \sin\left(\frac{180°\phi}{\beta}\right)$$

$$a = \left[\frac{h}{2}\right]\left(\frac{\pi^2}{T^2}\right) \cos\left(\frac{180°t}{T}\right) \text{ or}$$

$$a = \left[\frac{h}{2}\right]\left(\frac{\pi\omega}{\beta}\right)^2 \cos\left(\frac{180°\phi}{\beta}\right)$$

The formulas used to calculate pressure angles for simple harmonic motion are:

$$\alpha = \arctan\left[\left(\frac{90° h}{\beta R_\alpha}\right) \sin\left(\frac{180°\phi}{\beta}\right)\right]$$

at radius R_α to the pitch of the curve at angle ϕ, and:

$$\phi_P = \left(\frac{\beta}{180°}\right)\left[\arccot\left(\frac{\beta}{180°} \tan \alpha_{max}\right)\right]$$

where ϕ_P is the value of ϕ where pressure angle α_{max} is located.

When $\alpha = \alpha_{max}$ and $\phi = \phi_P$, then:

$$R_{\alpha max} = \frac{h[\sin(180°\phi_P/\beta)]^2}{2\cos(180°\phi_P/\beta)}, \text{ and}$$

$$R_{min} = R_{\alpha max} - \frac{h}{2}\left[1 - \cos\left(\frac{180°\phi_P}{\beta}\right)\right]$$

CYCLOIDAL MOTION

A cycloidal motion displacement diagram (Figure 8-6) has no sudden change associated with the acceleration curve, though it is frequently used in the design of cams for high-speed machinery. The reason for this is that it minimizes the levels of noise and vibration during machine operation. The basic formulas used here are:

$$y = h\left[\frac{t}{T} - \left(\frac{1}{2\pi}\right)\sin\left(\frac{360°t}{T}\right)\right] \text{ or}$$

$$y = h\left[\frac{\phi}{\beta} - \left(\frac{1}{2\pi}\right)\sin\left(\frac{360°\phi}{\beta}\right)\right]$$

$$v = \left[\frac{h}{T}\right]\left[1 - \cos\left(\frac{360°t}{T}\right)\right] \text{ or}$$

$$v = \left[h\frac{\omega}{\beta}\right]\left[1 - \cos\left(\frac{360°\phi}{\beta}\right)\right]$$

$$a = \left(\frac{2\pi h}{T^2}\right)\sin\left(\frac{360°t}{T}\right) \text{ or}$$

$$a = \left(\frac{2\pi h\omega^2}{\beta^2}\right)\sin\left(\frac{360°\phi}{\beta}\right)$$

The formulas used for calculating pressure angles for cycloidal motion are:

$$\alpha = \arctan\left\{\left[\frac{180°h}{\pi\beta R_\alpha}\right]\left[1 - \cos\left(\frac{360°\phi}{\beta}\right)\right]\right\}$$

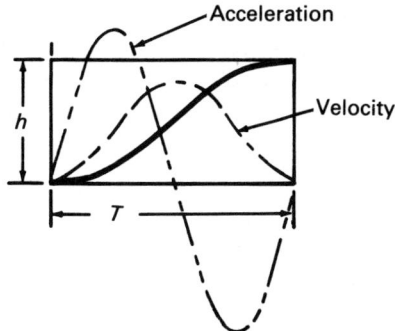

Fig. 8-6. Cycloidal motion displacement diagram.

at a radius of R_α and at an angle of ϕ to the pitch curve, and:

$$\phi_P = \left[\frac{\beta}{180°}\right]\left[\text{arccot}\left(\frac{\beta \tan \alpha_{MAX}}{360°}\right)\right]$$

when the value of ϕ_P is equal to ϕ where the pressure angle α_{MAX} occurs.

$$R_{MAX} = \frac{h\,[(1 - \cos(360°\phi_P/\beta)]^2}{2\pi \sin(360°\phi_P/\beta)}$$

at a position when $\alpha = \alpha_{MAX}$ and $\phi = \phi_P$.

$$R_{MIN} = R_{\alpha_{MAX}} - h\left[\left(\frac{\phi_P}{\beta}\right) - \left(\frac{1}{2\pi}\right)\sin\left(\frac{360°\phi_P}{\beta}\right)\right]$$

Cam Problems

A number of cam design problems are solved by applying the above formulas and their derivatives. The first type of problem deals with synthesizing the displacement diagram itself. This process usually consists of matching both constant velocity and parabolic motion curves to reduce the acceleration to a small constant value.

238 • CAMS AND GEARS

Figure 8-7 is a displacement diagram with three rise distances (i.e., $y_1 = 0.250$ m, $y_2 = 1.250$ m, and $y_3 = 0.500$ m), with the first two rises at angles of over 50°. With this arrangement, it is possible to calculate angles ϕ_1 and ϕ_3 with the following calculations:

$$\frac{CM_1}{\phi_2} = \frac{y_1}{y_2}$$

$$\frac{0.500\ \phi_1}{50°} = \frac{0.250}{1.250}$$

$$\phi_1 = \frac{[(50)(0.250)/1.25]}{0.500}$$

$$= 20°$$

$$\frac{DM_2}{\phi_2} = \frac{y_3}{y_2}$$

$$\frac{0.500\ \phi_3}{50°} = \frac{0.500}{1.250}$$

$$\phi_3 = \frac{[(0.500)(50)/1.250]}{0.500}$$

$$= 40°$$

To calculate follower displacement for the same type of displacement diagram as shown in Figure 8-7, the following formula (derived from the parabolic curve formula) is used:

$$y = \frac{(\phi^2)(2)(2y_1)}{(2\phi_1)^2}$$

An example of this problem would be to calculate follower displacement with a rise angle (ϕ_1) of 20° if $\phi = 10°$ and $h = 0.500$ in.

$$y = \left[\frac{(2)(h)}{(2\phi_1)^2}\right]\phi^2$$

$$= \left[\frac{(2)(0.500)}{[(2)(20)]^2}\right]10^2$$

$$= 0.0625 \text{ in.}$$

It should be noted that to find the follower displacement for the constant velocity portion of the curve, simply divide that seg-

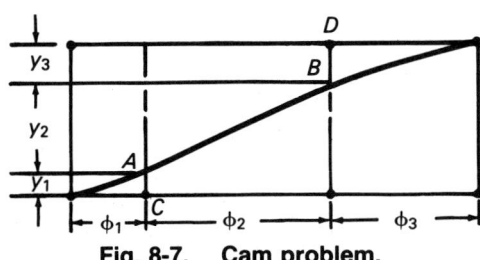

Fig. 8-7. Cam problem.

ment into the appropriate number of uniform divisions (e.g., 1.250 in. into 10 divisions). Furthermore, when there is a cam fall, the value of the displacement is found by subtracting the calculated value from the total rise of the cam $(y_1 + y_2 + y_3)$.

Exercises

Referring to Figure 8-7, solve the following problems:

1. A cam follower is designed to rise 0.750 in. with a constant acceleration, 2.000 in. with constant velocity over a 72° rotation, and then 1.000 in. with constant deceleration. Calculate for ϕ_1.
2. What is ϕ_2 for problem 1?
3. What is the total rise for the cam in problem 1?

For the development of a modified constant velocity cam with parabolic matching (using values given for Figure 8-7), solve the following problems:

4. If rise angle $\phi_1 = 20°$, what will be the follower displacement for the parabolic section with the following values: 0, 5, 10, and 15?
5. When the rise angle $\phi_2 = 50°$, what will be the follower displacement for the constant velocity section of the cam for the following ϕ values: 30, 45, and 60?
6. When the rise angle $\phi_3 = 40°$, what will be the follower displacement for the last parabolic section with the following values: 85, 95, and 105? (*Note*: Assume that the total rise of the cam is 2.000 in.)

Gears

Gears are used in machinery and equipment to transmit rotating or reciprocating motion from one part to another. There is no single method used to classify gears. One method of classification is according to the gears' positioning with their shafts, such as parallel, nonparallel, or intersecting shafts. Another method is according to the design of the gear teeth, such as spur, helical, herringbone, bevel, hypoid, or worm and worm gears.

Standards and formulas have been established for each type of gear-tooth system. Knowing specific dimensional parameters, it is possible to design and specify gearing systems to fit a given need. As might be expected, these specifications are derived by using mathematical formulas, which vary from one type of gear to the next.

Spur Gears

Before we can progress with specific spur gear formulas, it is necessary to understand spur gear nomenclature.

SPUR GEAR NOMENCLATURE

In many instances, gear-tooth parts are quantified by mathematical formulas. Where this exists, a symbol or notation will be given. For a clearer understanding of these terms, refer to Figure 8-8.

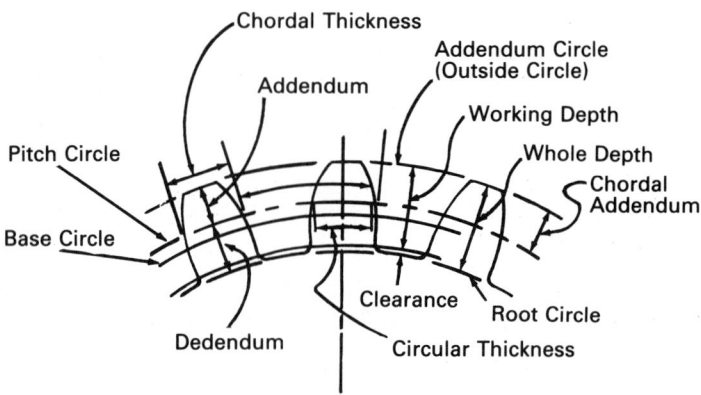

Fig. 8-8.   Spur gear nomenclature.

Addendum (*a*). This is the height of the tooth above the pitch circle, or the radial distance from the pitch circle to the top of the tooth.

Base circle diameter (D_B). The diameter of the circle from which the involute tooth curve is generated.

Center distance (*C*). The distance from the parallel axes of spur gears to parallel helical gears, or the crossed axes of crossed helical gears and worm gears. This distance is also the distance between the centers of the pitch circles.

Circular pitch (*p*). The distance measured along the pitch circle from a given point on one tooth to a corresponding point on an adjacent tooth, or the arc length of the pitch circle between the centers of adjacent teeth.

Circular thickness (*t*). The thickness of the gear tooth as measured along the pitch circle, or the arc length between two sides of a gear tooth on the pitch circle.

Dedendum (*b*). The tooth depth space below the pitch circle, or the radial distance from the pitch circle to the bottom of the tooth space.

Diametral pitch (*P*). The number of teeth on the gear per inch of pitch diameter.

Face width (*F*). That portion of the tooth between the pitch circle and the top of the tooth. The *effective face width* is that portion of the face width that has direct contact with mating teeth.

Gear ratio (m_G). The ratio of the number of teeth in one gear to the number of teeth in a mating gear.

Number of teeth (*N*). The total number of teeth in a gear (e.g., N_G/N_P, where N_G is the number of teeth in the gear and N_P is the number of teeth in the pinion).

Outside diameter (D_O). The diameter of the addendum circle.

Pitch diameter (*D*). The diameter of the pitch circle, where D_G is the pitch diameter of the gear and D_P is the pitch diameter of the pinion.

Pressure angle (ϕ). The angle that determines the direction of the pressure between contacting teeth, or the angle between the tooth profile and a radial line through its pitch point (the pitch point is the point of tangency of two pitch circles).

Root diameter (D_R). The diameter of the root circle tangent to the bottoms of the tooth spaces.

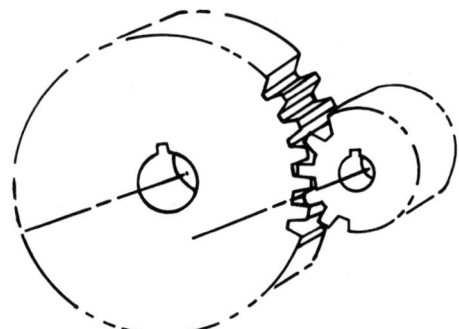

Fig. 8-9. Use of spur gear to transmit motion and power to parallel shaft.

Whole depth of tooth (h_T). The distance of the addendum and dedendum, or the total distance of the depth of the tooth.

Working depth of tooth (h_K). The depth of contact between two gears, or the sum of their addendums.

SPUR GEAR FORMULAS

As shown in Figure 8-9, spur gears transmit motion and power from one shaft to another, parallel shaft. Spur gear design enables the transmission of motion with no slippage or undue pressure. The teeth of mating spur gears must be equal in spacing and width. Thus, the number of teeth (N) in each gear will be proportional to its pitch diameter.

The basic formulas used in spur gear design are presented in Table 8-1.

A common spur gear type is the American standard coarse pitch spur gear; it is illustrated in Figure 8-10. The basic formulas used for this gear are presented in Table 8-2.

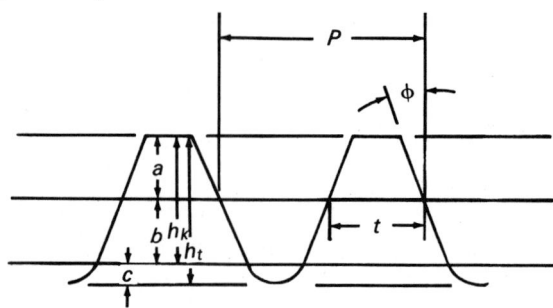

Fig. 8-10. Typical rack of 20 and 25 standard coarse pitch gear.

Table 8-1. General Spur Gear Formulas

Unknown	Formulas
Base circle diameter	$D_B = D \cos \phi$
Circular pitch	$p = \dfrac{\pi D}{N}$
	$p = \dfrac{\pi}{P}$
Clearance preferred	$c = \dfrac{0.250}{P}$
	$c = 0.0796 p$
Shaved or ground	$c = \dfrac{0.350}{P}$
	$c = 0.1114 p$
Center distance	$C = \dfrac{N_P(m_G + 1)}{2P}$
	$C = \dfrac{D_P + D_G}{2}$
	$C = \dfrac{N_G + N_P}{2P}$
	$C = \dfrac{(N_G + N_P)p}{2\pi}$
Dedendum	$b = \dfrac{1.157}{P}$
Diametral pitch	$P = \dfrac{\pi}{p}$
	$P = \dfrac{N}{D}$
	$P = \dfrac{N_P(m_G + 1)}{2C}$
Number of teeth	$N = PD$
	$N = \dfrac{\pi D}{p}$
Outside diameter	$D_o = D + 2a$
Full-depth teeth	$D_o = \dfrac{N + 2}{P}$
	$D_o = \dfrac{(N + 2)p}{\pi}$

(*continued*)

Table 8-1. (continued)

Unknown	Formulas
American standard stub teeth	$D_o = \dfrac{N + 1.6}{P}$
	$D_o = \dfrac{(N + 1.6)p}{\pi}$
Pitch diameter	$D = \dfrac{N}{P}$
	$D = \dfrac{Np}{\pi}$
Root diameter	$D_R = D - 2b$
Whole depth	$h_T = a + b$
Working depth	$h_K = a_G + a_P$

One of the most common forms of spur gears is the ANSI 20° standard gear. The formulas developed for this gear are also applicable for spur gears having 14.5° and 25° pressure angles. Presented in Table 8-3 are the formulas applicable to this gear form. It should be noted that the subnotation N pertains to normal dimensions; hence, P_N is the normal diametral pitch. Also of importance is the fact that P is the transverse diametral pitch and the Greek letter psi (ψ) represents the helix angle.

Table 8-2. American Standard Coarse Pitch Spur Gear Formulas

Unknown	Formulas
Addendum	$a = \dfrac{1.000}{P}$
	$a = 0.3183p$
Circular thickness	$t = \dfrac{1.5708}{P}$
	$t = \dfrac{p}{2}$
Clearance preferred	$c = \dfrac{2.250}{P}$
	$c = 0.0796p$

(continued)

Table 8-2. (continued)

Unknown	Formulas
Shaved or ground	$c = \dfrac{0.350}{P}$ $c = 0.1114p$
Dedendum preferred	$b = \dfrac{1.250}{P}$ $b = 0.3979p$
Shaved or ground	$b = \dfrac{1.350}{P}$ $b = 0.4297p$
Outside diameter	$D_o = \dfrac{N+2}{P}$ $D_o = 0.3183(N+2)p$
Pitch diameter	$D = \dfrac{N}{P}$ $D = 0.3183N_P$
Root diameter preferred	$D_R = \dfrac{N-2.5}{P}$ $D_R = 0.3183(N-2.5)p$
Shaved or ground	$D_R = \dfrac{N-2.7}{P}$ $D_R = 0.3183(N-2.7)p$
Whole depth preferred	$h_t = \dfrac{2.250}{P}$ $h_t = 0.7162p$
Shaved or ground	$h_t = \dfrac{2.350}{P}$ $h_t = 0.7480p$
Working depth	$h_K = \dfrac{2.000}{P}$ $h_k = 0.6366p$

Bevel Gears

Bevel gears differ from spur gear systems in that they are conical in design. The primary function of bevel gears is to connect shafts at intersecting points or axes. Shown in Figure 8-11 is an illustration of a bevel gear.

Table 8-3. Formulas for Spur and Helical Gears for ANSI Standard Gear Formulas

Unknown	Spur	Helical
Addendum	$a = \dfrac{1.000}{P}$	$a = \dfrac{1.000}{P_n}$
Center distance	$C = \dfrac{N+n}{2P}$	$C = \dfrac{N+n}{2P_N \cos \psi}$
Circular pitch	$p = \dfrac{\pi D}{D}$ $p = \dfrac{\pi d}{n}$ $p = \dfrac{\pi}{P}$	$P_N = \dfrac{\pi}{P_n}$
Clearance standard	$c = \dfrac{0.200}{P} + 0.002$	$c = \dfrac{0.002}{P_N} + 0.002$
Shaved or ground	$c = \dfrac{0.350}{P} + 0.002$	$c = \dfrac{0.350}{P_n} + 0.002$
Dedendum	$b = \dfrac{1.200}{P} + 0.002$	$b = \dfrac{1.200}{P_n} + 0.002$
Outside-diameter gear	$D_o = \dfrac{N+2}{P}$	$D_o = \dfrac{1}{P}\left[\left(\dfrac{N}{\cos \psi}\right) + 2\right]$
Pinion	$d_o = \dfrac{n+2}{P}$	$d_o = \dfrac{1}{P_N}\left[\left(\dfrac{n}{\cos \psi}\right) + 2\right]$
Pitch-diameter gear	$D = \dfrac{N}{P}$	$D = \dfrac{N}{P_n \cos \psi}$
Pinion	$d = \dfrac{n}{p}$	$d = \dfrac{n}{P_n \cos \psi}$
Tooth thickness	$t = \dfrac{1.5708}{P}$	$t_N = \dfrac{1.5708}{P_n}$
Whole depth	$h_T = \dfrac{2.200}{P} + 0.002$	$h_T = \dfrac{2.200}{P_N} + 0.002$
Working depth	$h_K = \dfrac{2.000}{P}$	$h_K = \dfrac{2.000}{P}$

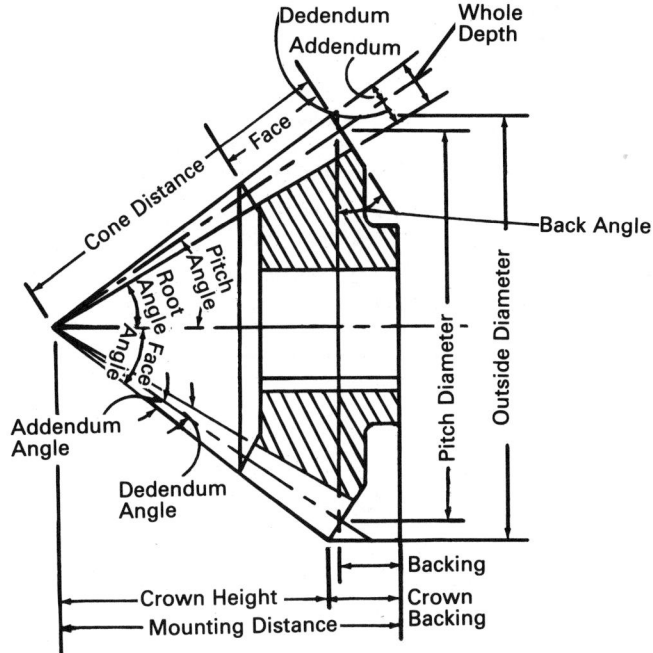

Fig. 8-11. Bevel gear nomenclature.

There are three general types of bevel gear systems. In the first, known as a *straight bevel gear*, the teeth are straight but tapered on the side. Because they are the easiest to design and produce, straight bevel gears are the most commonly used. A second type is the *spiral bevel gear*, which has curved oblique teeth. The primary advantage of this design is that it allows for complete control of tooth contact. The last design is the *hypoid gear*. This gear appears similar to the spiral bevel gear except that the pinion and gear axes are offset. Presented in Table 8-4 is a listing of bevel gear applications.

BEVEL GEAR NOMENCLATURE

Most of the terms used in bevel gear systems have the same meaning as in spur gears. There are, however, some slight differences. These are presented below.

248 • CAMS AND GEARS

Table 8-4. Applications of Bevel Gears

Gear Design	Use
Straight	Peripheral speeds that do not exceed 1000 fpm and where noise and smoothness are unimportant. Small lot gear production.
Spiral	Peripheral speeds that exceed 1000 fpm or rpm. Large reduction ratios.
Hypoid	Where maximum smoothness is desired. High reduction ratios. Nonintersecting shafts.

Addendum angle (α). This is the angle subtended by the addendum and is calculated the same for both gear and pinion.

Backing (Y). The distance measured between the base of the pitch cone and the rear of the hub.

Cone distance (A). The cone distance is the slant height of the pitch cone and is the same for both gear and pinion.

Crown backing (Z). Same as the backing, but used strictly for manufacturing purposes.

Crown height (X). The crown height is measured parallel to the gear axis from the cone apex to the gear crown.

Dedendum angle (δ). The angle subtended by the dedendum. It is the same for both gear and pinion.

Face angle (Γ_O). This is found by measuring the angle formed between the top of the gear teeth and the gear axis. The gear face angle is noted by the symbols Γ_O for the gear and γ_O for the pinion.

Mounting distance (M). A measurement used for assembly purposes.

Pitch diameter (D). This is the diameter of the base of the pitch cone; hence, the relationship of the circular pitch and diametral pitch is the same as for spur gears.

Root angle (Γ_R). The angle formed between the root of the gear teeth and the gear axis.

BEVEL GEAR FORMULAS

Presented in Table 8-5 are the basic formulas used for straight bevel gears. The system of formulas used here are for *Gleason bevel gears*, a universal system for straight bevel gears (see ANSI/AGMA 208.03).

Worm Gears

A gear design used to transmit power between nonintersecting shafts that are at 90° to each other is the worm gear. The *worm* is a threaded screw whose thread is the same shape as a rack tooth. The *worm wheel* is similar in appearance to a spur gear, except its teeth have been shaped and curved to conform to the shape of the worm. Presented in Figure 8-12 is an illustration of a worm and worm gear.

WORM GEAR NOMENCLATURE

Several gear terms have special meaning in worm and worm gear designs.

- *Axial pitch* (P_X). This is the distance from a point on one thread to the corresponding point on the next thread. The axial pitch should be measured parallel to the worm axis. The axial pitch is equal to the circular pitch of the gear.
- *Lead* (l). The lead of a worm gear is the distance that the thread advances in one complete rotation; it is measured axially. Leads are always multiples of the pitch (e.g., in a double-thread worm, the lead will be equal to two times the pitch).
- *Lead angle* (λ or L_A). This is the angle between a tangent to the helix at the pitch diameter and a perpendicular to the worm's axis.

WORM GEAR FORMULAS

Most formulas involving worm gears will employ circular pitch measures rather than the diametral pitch, because the circular pitch is a measure that must be known for machining the gear. Table 8-6 presents formulas used for American standard fine-pitch worm

Table 8-5. Straight Bevel Gear Formulas

Unknown	Formulas
Gear	
Addendum	$a_G = \dfrac{0.540}{P} + \dfrac{0.460}{P(N/n)^2}$
Chordal addendum	$a_{CG} = a_G + \dfrac{T^2 \cos \Gamma}{4D}$
Chordal thickness	$T_C = T - \dfrac{T^3}{6D^2} - \dfrac{B}{2}$
Circular pitch	$p = \dfrac{\pi}{P}$
Circular thickness	$T = \dfrac{p}{2} - (a_P - a_G) \tan \phi - \dfrac{K}{P}$
Clearance	$c = h_T - h_K$
Cone distance	$A_o = \dfrac{D}{2 \sin \Gamma}$
Dedendum	$b_G = \dfrac{2.188}{P} - a_G$
Dedendum angle	$\delta_G = \arctan \dfrac{b_G}{A_O}$
Outside diameter	$D_O = D + 2a_G \cos \Gamma$
Pitch angle	$\Gamma = 90° - \gamma$
Pitch apex to crown	$x_O = \dfrac{d}{2} - a_G \sin \Gamma$
Root angle	$\Gamma_R = \Gamma - \delta_G$
Tooth angle	$\theta = \dfrac{3438}{A_O} \left(\dfrac{T}{2} + b_G \tan \phi \right)$ minutes
Whole depth	$h_T = \dfrac{2.188}{P} + 0.002$
Working depth	$h_K = \dfrac{2.000}{P}$

(continued)

Table 8-5. (*continued*)

Unknown	Formulas
Pinion	
Addendum	$a_P = h_K - a_G$
Chordal addendum	$a_{CP} = a_P + \dfrac{t^2 \cos \gamma}{4d}$
Chordal thickness	$t_C = t - \dfrac{t^3}{6d^2} - \dfrac{B}{2}$
Circular pitch	same as gear
Circular thickness	$t = p - T$
Clearance	same as gear
Cone distance	same as gear
Dedendum	$b_P = \dfrac{2.188}{P} - a_P$
Dedendum angle	$\delta_P = \arctan \dfrac{b_P}{A_O}$
Outside diameter	$d_O = d + 2a_P \cos \gamma$
Pitch angle	$\gamma = \arctan \dfrac{n}{N}$
Pitch apex to crown	$x_O = \dfrac{D}{2} - a_P \sin \gamma$
Root angle	$\gamma_R = \gamma - \delta_P$
Tooth angle	$\dfrac{3438}{A_O} \left(\dfrac{T}{2} + b_P \tan \phi \right)$ minutes
Whole depth	same as gear
Working depth	same as gear

252 • CAMS AND GEARS

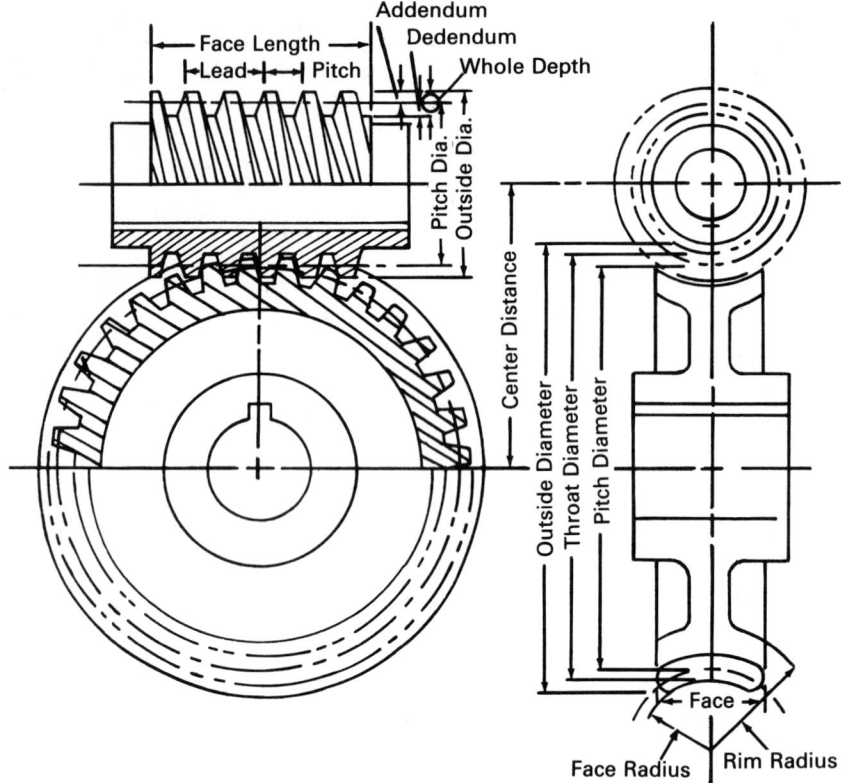

Fig. 8-12. Worm and worm gear nomenclature.

gears. In these formulas, n = number of threads in the worm, N = number of teeth in the worm gear (and is the same as nm_G), and m_G = ratio of the gear as a proportion of $N:n$.

Industrial worm gearing includes heavier gear designs that are primarily used to transmit power efficiently and at a slower velocity, with the ability to handle extremely heavy loads or high resisting forces. Gear designs for these systems have relatively coarse pitch. As a result, the formulas used in this type of gearing will be somewhat different from those used in other worm gear systems. Table 8-7 presents formulas used in industrial worm gearing.

Table 8-6. Fine-Pitch Worm Gear Formulas

Unknown	Formulas
Worm	
Lead	$l = nP_x$
Pitch diameter	$d = \dfrac{1}{\pi \tan \lambda}$
Outside diameter	$d_O = d + 2a$
Gear	$D = \dfrac{NP}{\pi}$
Pitch diameter	$D = \dfrac{NP_x}{\pi}$
Outside diameter	$D_O = 2C - d + 2a$
Face width	$F_{G_{MIN}} = 1.125\sqrt{(d_O + 2c)^2 - (d_O - 4a)^2}$
Worm and gear	
Addendum	$a = 0.3182 P_N$
Center distance	$C = 0.5(d + D)$
Clearance	$c = h_T - h_K$
Tooth thickness	$t_N = 0.5 P_N$
Whole depth	$h_T = 0.7003 P_N + 0.002$
Working depth	$h_K = 0.6366 P_N$

Helical Gears

Helical gears are used for two major functions: driving parallel shafts and driving skew shafts. Shown in Figure 8-13 is an illustration of common helical gear systems. Presented in Table 8-8 are the basic formulas used with helical gearing.

Gear Problems

Most gear problems require that one determine design specifications with certain givens. For example, given an ANSI 20° spur gear with

Table 8-7. Industrial Worm Gearing Formulas

Unknown	Formula
Addendum	
Single and double thread	$a = 0.318P$
Triple and multiple threads	$a = 0.286P$
Center distance	$C = \dfrac{(D+d)}{2}$
	$C = \dfrac{P}{6.2832}\left(\dfrac{t}{\tan L_A} + T\right)$
Clearance	$c = 0.2m \cos L_A$
Face width, worm gear	
Single and double threads	$F = 2.38P + 0.25$
Triple and quadruple threads	$F = 2.15P + 0.2$
Worm threads are part of shaft	$F = \dfrac{(C^{0.875})}{3}$
Lead of worm thread	$L = tP$
	$L = \pi d \tan L_A$
	$L = \dfrac{\pi D}{R}$
Lead angle of worm	$\tan L_A = \dfrac{L}{\pi d}$
Outside diameter of worm	$d_O = d + 2a$
Pitch of worm and worm gear	$P = \dfrac{L}{t}$
	$P = \dfrac{(2C-d)(\pi)}{T}$
Pitch of worm, normal	$P_N = P \cos L_A$
Pitch diameter	
Worm	$d = 2C - D$
	$d = d_O - 2a$
	$d = L \cot \dfrac{L_A}{\pi}$

(*continued*)

Table 8-7. *(continued)*

Unknown	Formula
Worm gear	$D = 2C - d$
	$D = \dfrac{TP}{\pi}$
Ratio	$R = T : t$
Rubbing speed (fpm)	$V = 0.262n \sqrt{d^2 + (D/R)^2}$
	$V = 0.262dn \sec L_A$
Throat diameter of worm gear	$D_1 = D + 2A$
Throat radius of worm gear	$U = \dfrac{d_O}{2} - 2a$
Whole depth	
Single and double threads	$W = 0.686P$
Triple and quadruple threads	$W = 0.623P$
Worm gear length	$G = P(4.5 + 0.02T)$
	$G = \sqrt{D_O^2 - D^2}$

Note: In these formulas, m is a module equal to $0.3183 \times$ axial pitch, and n is the rpm of the worm.

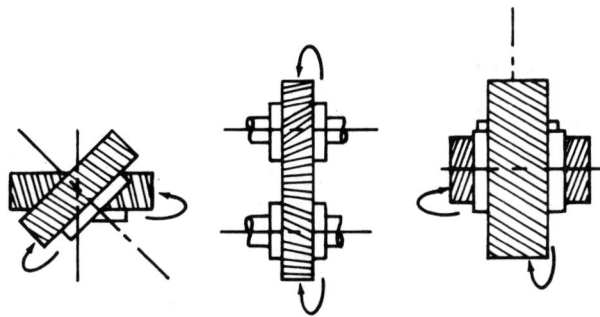

Fig. 8-13. Three common arrangements for helical gear systems.

Table 8-8. Helical Gear Formulas

Unknown	Formula
Addendum	$S = \dfrac{1}{P_N}$
Center distance	$C = \dfrac{D_A + D_B}{2}$
Lead of tooth helix	$L = \pi D \cot\alpha$
Outside diameter	$O = D + 2S$
Pitch diameter	$D = \dfrac{N}{P_N \cos\alpha}$
Whole depth	$W = \dfrac{2.157}{P_N}$
Tooth thickness (at pitch line)	$T_N = \dfrac{1.571}{P_N}$

Note: To check calculations for pitch diameter and center distance, the following formula is used:

$$N_B + (N_A \tan\alpha_A) = 2CP_N \sin\alpha_A$$

48 teeth and a diametral pitch of 5 in., the following information must be calculated to machine this gear: whole depth, working depth, clearance, outside diameter, pitch diameter, root diameter, and tooth thickness. To accomplish this, the following calculations are made:

$$h_T = \frac{2.200}{P} + 0.002$$
$$= \frac{2.200}{5} + 0.002$$
$$= 0.442 \text{ in.}$$

$$h_K = \frac{2.000}{P}$$
$$= \frac{2.000}{5}$$
$$= 0.400 \text{ in.}$$

$$c = \frac{0.200}{P} + 0.002$$
$$= \frac{0.200}{5} + 0.002$$
$$= 0.042 \text{ in.}$$

$$D_o = \frac{N+2}{P}$$
$$= (48+2)5$$
$$= 10 \text{ in.}$$

$$D = \frac{N}{P}$$
$$= \frac{48}{5}$$
$$= 9.6 \text{ in.}$$

Note: To use the following formula, the dedendum, b, must be determined. Hence the following procedures are used:

$$b = \frac{1.200}{P} + 0.002$$
$$= \frac{1.200}{5} + 0.002$$
$$= 0.242 \text{ in.}$$

Thus:

$$D_R = D - 2b$$
$$D_R = 9.6 - 2(0.242)$$
$$= 9.176 \text{ in.}$$

$$t = \frac{1.5708}{P}$$
$$= \frac{1.5708}{5}$$
$$= 0.31416 \text{ in.}$$

As a result of these calculations, it will now be possible to machine this spur gear. It should be noted that when a specific formula is not available for a given gear form (i.e., root diameter formula for an ANSI 20° spur gear), then the general formula might be used. This practice, however, should be checked against company policies and design procedures.

Exercises

7. If a gear has 48 teeth and its pinion has 12, what will be the pitch diameter of that gear if the pinion's pitch diameter is 4 in.?

8. What will be the diametral pitch of a coarse-pitch spur gear with 54 teeth and a pitch diameter of 38 in.?

9. Compute the following specifications for an ANSI 25° spur gear that has 28 teeth with an outside diameter of 45 cm:

   diametral pitch   tooth thickness
   addendum   whole depth
   dedendum   working depth
   circular pitch   clearance standard

10. Compute the following specifications for a straight bevel gear that has a pitch diameter of 7 in., 126 teeth with a mating pinion of 63 teeth, a pitch angle of 16°, and diametral pitch of 18 in.:

    addendum   whole depth
    dedendum   working depth
    circular pitch
    outside diameter

11. Compute the following specifications for the mating pinion of the gear in problem 10. It has a pitch diameter of 4.000 in. and a pitch angle of 16°:

    addendum   pitch angle
    dedendum   circular thickness
    clearance   whole depth

12. What is the outside diameter of a fine-pitch worm gear that is mated to a worm with a pitch diameter of 56 in., where their normal diametral pitch is 18 in., center distance is 12 in., and pitch diameter is 34 in.?

13. Find the whole depth and working depth for the worm and gear in problem 12.
14. Calculate the lead angle for an industrial worm gear that has a lead of 2.4 m and a pitch diameter of 0.184 m.
15. If the normal diametral pitch for a helical gear is 14.250 in., calculate its addendum, tooth thickness, and whole depth.

CHAPTER 9

Machine Elements

- Screws, Rivets, and Splines
- Springs and Bearings

Within the context of this book, machine elements pertain to those parts, devices, units, or mechanisms that are used for the operation of mechanical equipment, tools, or machines. As might be expected, a number of different elements are found in industrial equipment and machinery: screws, rivets, splines, couplings, clutches, bearings, brakes, chain and belt drives, springs, and lines. Presented in this chapter, however, will be those basic machine elements, and their formulas, that are most frequently found in mechanical equipment.

Screws, Rivets, and Splines

Almost all screws, rivets, and splines used as machine elements are selected from standardized sizes and materials. Only in a customized situation might it be necessary to machine these elements. Most formulas used for screws, rivets, and splines pertain to the selection of standardized products on the market.

Screws

Screw fasteners make use of a threaded shaft to secure one part to another. These fasteners come in various forms such as screws, bolts, and nuts. Figure 9-1 illustrates the common screw-thread nomenclature used in the profession.

The American standard thread is used for machine screws and is proportionally manufactured by the formula $p = 1/n$ and $f = p/8$, where p is the pitch, n is the number of threads per inch, and f is the flat. The dimensions for these screws will vary according to the thread *series* used: coarse thread, fine thread, special pitch, and 8-, 12-, and 16-pitch series. For example, a ¼-in. coarse-thread series has 20 threads per inch, while a fine-thread series has 28 threads per inch. The pitch for each thread will be:

$$p = \frac{1}{n} = \frac{1}{20} = 0.05 \text{ in.}$$

$$p = \frac{1}{n} = \frac{1}{28} = 0.0357 \text{ in.}$$

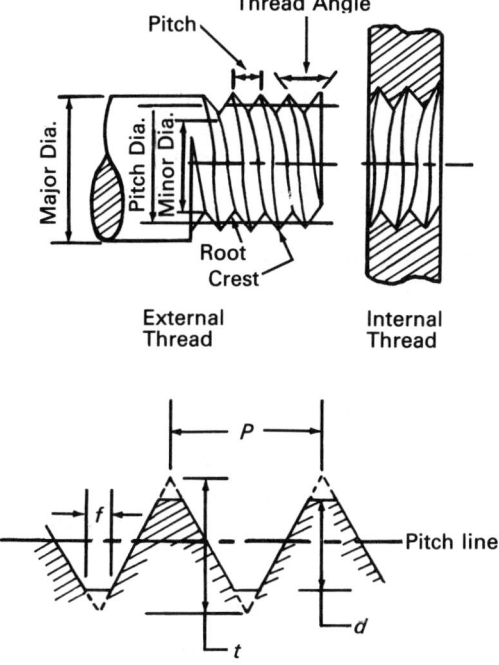

Fig. 9-1. Screw-thread nomenclature.

Table 9-1. Formulas for Foreign Standard Screw Threads

Thread Form	Unknown	Formulas
Whitworth standard thread	Thread depth	$d = 0.6403p$
	Radius	$r = 0.1373p$
French metric screw thread	Pitch	p = mm
	Thread depth	$d = 0.6495p$
	Flat	$f = \dfrac{p}{8}$
International metric standard screw thread	Pitch	p = mm
	Thread depth (actual)	$d = 0.7036p$
	Thread depth (theoretical)	$t = 0.866p$

Each year, there has been increasing usage of foreign standard screw threads, especially for companies competitive in the international market. This is also the case for companies that are converting over to the international metric standard screw thread. Table 9-1 presents formulas used for foreign standard thread forms (see Figure 9-2).

Perhaps one of the most important factors to be determined for screw-thread fasteners pertains to their performance under working loads and stresses. As a result, several formulas have been derived to calculate the performance of bolts and screw threads.

WORKING STRENGTH OF BOLTS

There are two basic formulas used to calculate the working strength of bolts. The first formulas should be used for bolts employed for *packed joints* or where the gasket's elasticity exceeds that of the studs or bolts:

$$W = S_T(0.55d^2 - 0.25d)$$

Here, W is the working strength of the bolt, S_T is the allowable working stress in tension, and d is the nominal outside diameter of the bolt.

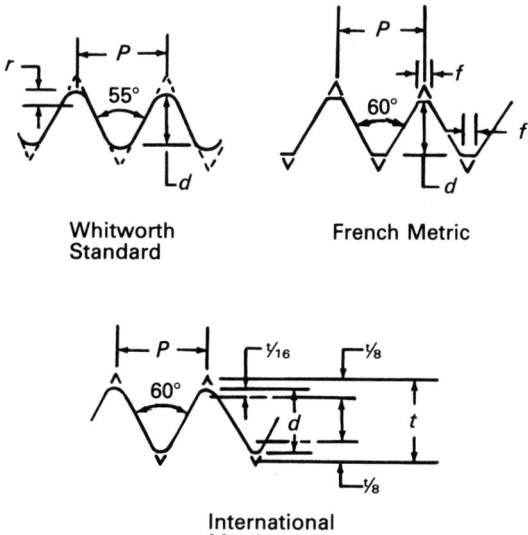

Fig. 9-2. Foreign standard screw threads.

The second formula is considered to be a more convenient form for most applications where the bolt is not in a packed joint:

$$W = S_T(A - 0.25d)$$

where A is the area of the bolt at the root of the thread.

For example, if a 0.75-in. bolt with an allowable working stress of 8500 psi is screwed tightly into a packed joint, its working strength would be calculated as:

$$\begin{aligned} W &= S_T(0.55d^2 - 0.25d) \\ &= 8500[(0.55)(0.75)^2 - (0.25)(0.75)] \\ &= (8500)(0.121875) \\ &= 1035.94 \text{ lb (approx.)} \end{aligned}$$

STRESS AND LENGTH OF ENGAGEMENT

An important factor when specifying screw threads is the amount of load or stress that it can withstand before its threads are stripped. There are several basic formulas used here. The first is for mating internal and external threads that have the same tensile strength.

To prevent thread stripping of the external thread, the length of engagement must not be less than the value of the following formula:

$$L_E = \frac{2A_T}{\pi K_N \max[0.57735n(E_S\min - K_N\max/2)]}$$

Here, L_E = length of engagement, A_T = tensile stress area of the screw thread, n = number of threads per inch, $K_N\max$ = internal thread's minor diameter, $E_S\min$ = external thread's pitch diameter. When the tensile stress area of the screw thread is not given, it may be calculated by using the one of two formulas. The first is for screw threads made out of steels with an ultimate tensile strength of up to 100,000 psi, and the second is for steels that exceed 100,000 psi in tensile strength. The respective formulas are (D is the major diameter of the threads):

$$A_T = 0.7854\left(D - \frac{0.9743}{n^2}\right)$$

$$A_T = \pi\left(\frac{E_S\min}{2} - \frac{0.16238}{n}\right)^2$$

In some situations, the internal thread will be made of material that has a lower strength than the external thread. To find out whether or not this condition exists, a J factor must be computed. If the J factor is equal to or less than 1, then the L_E will be adequate to prevent stripping. Thus:

$$J = \frac{(A_S)(\text{tensile strength of external thread})}{(A_N)(\text{tensile strength of internal thread})}$$

Within this formula, A_S is the shear area of the external thread and A_N is the shear area of the internal thread. If the J factor is calculated as 1 or greater, then the required length of engagement L_Q is determined by:

$$L_Q = JL_E$$

THREAD BREAKS

The amount of direct tensile load (P) that can break the threaded section of a screw or bolt is found by using the following formula:

$$P = SA_T$$

where S is the ultimate tensile strength of the screw or bolt material (psi) and A_T is the tensile stress area.

Exercises

1. Calculate the pitch and flat measures for an American standard thread where the number of teeth per inch are 8, 14, 26, and 38.

2. The pitches for three international metric standard screw threads are 1.2500, 0.0458, and 0.2445 mm, respectively. Calculate the actual and theoretical thread depth for each.

3. What is the working strength of 1- and 1.5-in. bolts that are screwed tightly into a packed joint if both their allowable working stresses equal 12,500 psi?

4. If the area of a bolt at its root is 0.040 in.^{2} and it has an outside diameter of 9/64 in., what will its working strength be with an allowable working stress of 5400 psi?

5. What load is required to break the threaded portion of a screw that has an ultimate tensile strength of 7200 psi and a stress area of 0.1223 in.^{2}?

Rivets

A rivet is a single piece fastener that can be described as a screw with no threads. All rivets consist of a head and body which are used to fasten *permanently* two or more pieces of sheet material. Figure 9-3 illustrates a common rivet and rivet-joint terminology.

The effectiveness and efficiency of a riveted joint is calculated as a ratio of the strength of the joint to that of the thinnest sheet material being joined. Thus, the strength of the rivet joint will be determined by where failure can be expected in bearing, shearing, and/or tension. Formulas used to calculate the strength of riveted joints are presented in Table 9-2.

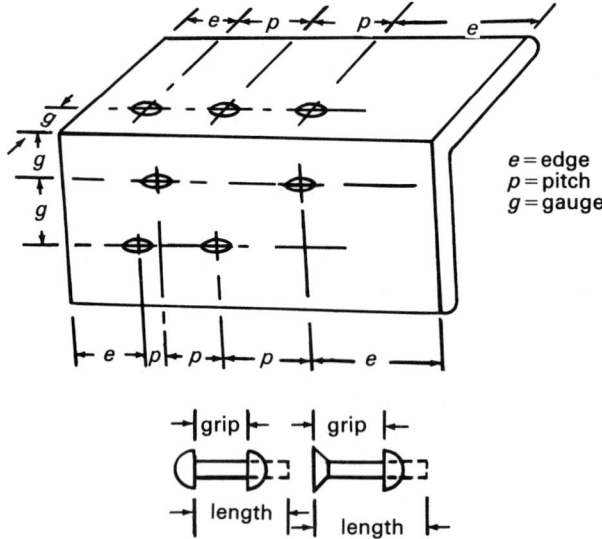

Fig. 9-3. Rivet and rivet-joint nomenclature.

Exercises

6. What is the shear strength of a rivet joint that has a cross-sectional area of 4.6 in.^{2}, six shear planes, and a specified rivet shear strength of 2400 psi?

7. What thickness plate must be used to produce a tensile strength rivet joint of 100,000 psi if the rivet holes are 0.500 in. in diameter and are spaced every 2.750 in., with a plate ultimate shear strength of 60,000 psi?

Splines

Splines are used on shafts to mate with corresponding hubs. These shafts have a series of parallel keys that are an integral part of the shaft. The keys, in turn, are mated with corresponding grooves that are cut into a hub, unit, or fitting. One of the most common forms of splines used as machine elements is the *involute spline*, which has multiple keys similar in design to involute gears (see Figure 9-4).

Table 9-2. Strength of Rivet Joint Formulas

Calculated Strength	Formulas	Factors
Shear strength	$Ps = (Ss)AN$	Ss = shear strength of sheet material A = cross-sectional area of rivet N = number of shear planes
Tensile strength	$Pu = (Su)(p - d)t$	Su = ultimate shear strength of the sheet material p = pitch (spacing) of rivets d = diameter of rivet hole t = thickness of plate
Bearing strength	$Pb = (Sb)(Ac)$	Sb = ultimate bearing strength of the sheet material Ac = projected bearing area of rivet

Involute splines are available in three standard pressure angles: 30°, 37.5°, and 45°. In addition, there are two classes of spline fits. The first is the *side fit*, where the mating sections are in contact only on the sides of the teeth. The second is the *major diameter fit*, where the mating parts are in contact along the major diameter. In addition, involute splines will have either flat or fillet roots.

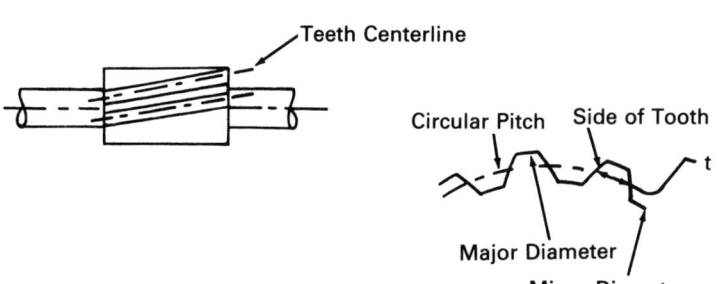

Fig. 9-4. Involute spline.

There are a number of dimensional terms used in specifying involute splines. Most terms are found by using the diametral pitch (P), which is the number of spline teeth per inch of pitch diameter. Their definitions and formulas follow:

Stub pitch (P_S). This is a value used to note the radial distance from the pitch circle to the major circle of the male (external) and female (internal) spline. The formula used for its calculation is:

$$P_S = 2P$$

Pitch diameter (D). The pitch diameter is the diameter dimension of the pitch circle and is found by using the following formula:

$$D = \frac{N}{P}$$

Base diameter (D_B). This is the diameter of the base circle and uses the formula:

$$D_B = D \cos \phi_D$$

where ϕ = pressure angle.

Circular pitch (p). The distance between two corresponding points on adjacent teeth. To find the circular pitch, the formula used is:

$$p = \frac{\pi}{P}$$

Minimum effective space width (s_v). This is basically the circular tooth width and is used to determine optimal fit between mating splines. There are three different formulas used here. For all 30° pressure angle splines, use the formula:

$$s_v = \frac{\pi}{2P}$$

When a 37.5° pressure angle is employed, the following formula is used:

$$s_v = \frac{0.5\pi + 0.1}{P}$$

Finally, for all 45° pressure angle splines, use the following formula:

$$s_v = \frac{0.5\pi + 0.2}{P}$$

Internal major diameter (D_{RI}). This is the diameter of the major circle for the internal spline. The specific formula used here will depend upon the characteristic of the spline. For 30° pressure angle splines with a flat root side fit and a 2.5/5 to 32/64 pitch, the following formula is used:

$$D_{RI} = \frac{N + 1.35}{P}$$

For 30° pressure angle splines with flat root major diameter fits and 3/6 to 16/32 pitch, use the formula:

$$D_{RI} = \frac{N + 1}{P}$$

For 30° pressure angle splines with fillet root side fits and pitches ranging between 2.5/5 and 48/96, the following formula is employed:

$$D_{RI} = \frac{N + 1.8}{P}$$

For 37.5° pressure angle splines, use the following formula:

$$D_{RI} = \frac{N + 1.6}{P}$$

45° pressure angle splines require use of the following formula:

$$D_{RI} = \frac{N + 1.4}{P}$$

(*Note*: Pitch is the ratio P/P_s.)

External major diameter (D_0). The diameter of the major circle for the external spline. Unlike for the internal major diameter, only one formula is used here:

$$D_0 = \frac{N + 1}{P}$$

Internal minor diameter (D_I). This is the diameter of the minor circle for the internal spline. Again, the formula used will depend on the type of spline specified. For all 30° pressure angle splines use:

$$D_I = \frac{N - 1}{P}$$

For all 37.5° pressure angle splines, the following formula is used:

$$D_I = \frac{N - 0.8}{P}$$

All 45° pressure angle splines require:

$$D_I = \frac{N - 0.6}{P}$$

External minor diameter (D_{RE}). This is the diameter of the minor circle for the external spline. The formulas used here will vary according to spline specifications. For all 30° pressure angle splines with either flat root side fit or flat root major diameter fit, use the formula:

$$D_{RE} = \frac{N - 1.35}{P}$$

For all 30° pressure angle splines with fillet root side fits and pitches 2.5/5 to 12/24, use:

$$D_{RE} = \frac{N - 1.8}{P}$$

If these splines have pitches between 16/32 and finer, use:

$$D_{RE} = \frac{N - 2}{P}$$

For all 37.5° pressure angle splines, use:

$$D_{RE} = \frac{N - 1.3}{P}$$

For all 45° pressure angle splines with pitches of 10/20 or finer, use:

$$D_{RE} = \frac{N - 1}{P}$$

Internal form diameter (D_{FI}). The diameter of the form circle for the internal spline. The formula used here for all splines, except a 30° pressure angle spline with a flat root major diameter fit, is:

$$D_{FI} = \frac{N + 1}{P} + c_F$$

The formula for the exception is:

$$D_{FI} = \frac{N+0.8}{P} - 0.004 + 2c_F$$

External form diameter (D_{FE}). This is the diameter of the form circle for the external spline. The formula used for all 30° pressure angle splines is:

$$D_{FE} = \frac{N-1}{P} - 2c_F$$

For all 37.5° pressure angle splines, use:

$$D_{FE} = \frac{N-0.8}{P} - 2c_F$$

For all 45° pressure angle splines, use:

$$D_{FE} = \frac{N-0.6}{P} - 2c_F$$

Form clearance (c_F). This is the radial depth of the involute's profile that extends beyond the engagement depth with the mating part and allows for looseness between the splines. The formula used here is:

$$c_F = 0.001D$$

with an acceptable range of 0.002 to 0.010.

The use of involute spline formulas is similar in concept to that of gear formulas. For example, if a 37.5° pressure angle involute spline has a pitch of 24/48 with six teeth, and we need to find its circular pitch, stub pitch, pitch diameter, and minimum effective space width, the following procedures would be used:

$$\begin{aligned} p &= \frac{\pi}{P} \\ &= \frac{3.14159}{24} \\ &= 0.13090 \end{aligned}$$

If pitch $= P/P_S$, then $P_S = 48$.

$$D = \frac{N}{P}$$
$$= \frac{6}{24}$$
$$= 0.25$$
$$s_v = \frac{(0.5\pi) + 0.1}{P}$$
$$= \frac{(0.5)(3.14159) + 0.1}{24}$$
$$= 0.0696$$

Exercises

8. Compute the pitch diameter and form clearance for a 48/96 pitch involute spline with six teeth.

9. What is the base diameter of a 45° pressure angle 48/96 pitch involute spline with six teeth?

10. Given a 12/24 pitch spline, find its diametral pitch, stub pitch, circular pitch, and minimum effective space width for pressure angles of 30°, 37.5° and 45°.

Springs and Bearings

Two machine elements that have a significant function in the operation of industrial equipment and devices are springs and bearings. Spring calculations are often used to help identify spring material, allowable spring stresses, and spring characteristics. Bearings, on the other hand, are used to reduce friction and wear and play an important role in maintaining and extending the life of equipment and machines.

Springs

When dealing with springs, design and specification formulas assume that the springs will not be stressed beyond their plastic limits.

Otherwise, they would permanently deform and be useless. Generally, most springs can be categorized into two major groups: flat and coiled.

FLAT SPRINGS

In order to use the formulas designed for flat springs, it is first necessary to become familiar with the notations used. They are presented as follows:

P = safe load
E = modulus of elasticity
f = the amount of deflection under a load P
L = length of the spring
V = volume of spring
S_s = safe stress due to bending
S_v = safe shearing stress
U = resilience.

The basic relationship of work performed by a deflecting spring (from 0 to f) can be expressed in the following relationship:

$$U = \frac{Pf}{2}$$

This formula is applicable only when the amount of deflection is proportional to the applied load. Other formulas for particular spring designs are presented as follows:

Rectangular plate springs. Shown in Figure 9-5 is a rectangular plate spring with appropriate notations. The formulas used here are:

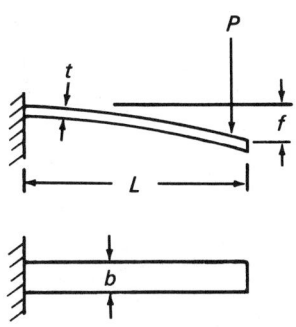

Fig. 9-5. Rectangular plate spring.

MACHINE ELEMENTS

$$P = \frac{bt^2 S_s}{6L}$$

$$U = \frac{Pf}{2} \text{ or}$$

$$U = \frac{S_s^2 V}{18E}$$

$$f = \frac{PL^3}{3EI} \text{ or}$$

$$f = \frac{4PL^3}{bt^3 E} \text{ or}$$

$$f = \frac{2L^2 S_s}{tE}$$

Where:

$$I = \frac{bt^3}{12}$$

$$t = \frac{2S_s L^2}{3Ef} \text{ or}$$

$$t = \sqrt[3]{4PL^3/Ebf}$$

If this spring has a tapered end, then:

$$U = \frac{Pf}{2} \text{ or}$$

$$U = \frac{S_s^2 V}{9E}$$

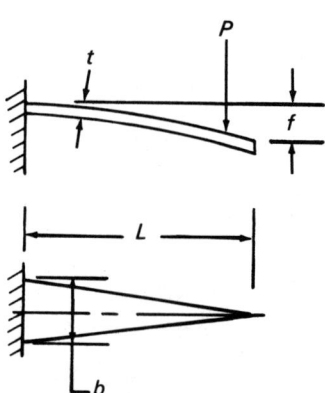

Fig. 9-6. Triangular plate spring.

Triangular plate spring. This spring is illustrated in Figure 9-6, with its formulas as follows:

$$P = \frac{bt^2}{6L}$$

$$U = \frac{Pf}{2} \text{ or}$$

$$U = \frac{S_s^2 V}{6E}$$

$$f = \frac{PL^3}{2EI} \text{ or}$$

$$f = \frac{6PL^3}{bt^3 E} \text{ or}$$

$$f = \frac{L^2 S_s}{tE}$$

Where:

$$I = \frac{bt^3}{12}$$

$$t = \frac{S_s L^2}{Ef} \text{ or}$$

$$t = \sqrt[3]{6PL^3/Ebf}$$

Semielliptic springs. These springs are often used for trucks, locomotives, and other vehicles (Figure 9-7). In this situation:

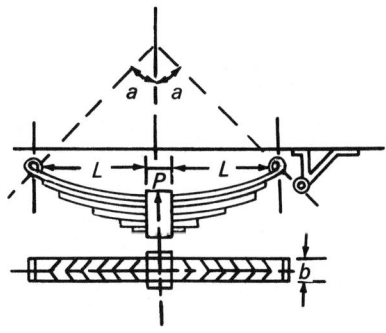

Fig. 9-7. Semielliptic spring.

276 • MACHINE ELEMENTS

$$\text{bearing force} = 2P \text{ or}$$
$$= \left[\frac{2nbt^2}{6}\right]\left[\frac{S_s}{(L+p \tan a)}\right]$$
$$\text{deflection } f = \frac{L^2 S_s}{tE} \text{ or}$$
$$= \frac{(6L^2/nbt^3)P(L+p \tan a)}{E}$$

And:

$$\text{bending moment } M = P(L + p \tan a)$$

Other formulas for flat springs are presented in Table 9-3.

COILED SPRINGS

There are generally two types of coiled springs: compression and extension (Figure 9-8). Because there is a wide variety of spring designs on the market, only the more commonly used will be covered here. Critical to the successful use of springs is their ability to withstand working stresses. Safe working stress will be based on factors such as the type and size of spring, spring material, stress range, type of load and service, and spring design.

The specific formula used for compression springs will depend upon the type being designed. These are generally based upon how their ends appear. In all, there are four: open or plain ends not ground, open or plain ends ground, squared or closed ends not ground, and closed ends ground. In each case, the spring may have either a left- or right-hand helix. These are shown in Figure 9-9.

(a) Compression Spring

(b) Extension Spring

Fig. 9-8. Two types of coiled springs.

Table 9-3. Flat Spring Formulas

Design	Unknown	Formula
	Deflection	$f = \dfrac{PL^3}{4Ebt^3}$ $f = \dfrac{S_s L^2}{6Et}$
	Load	$P = \dfrac{2S_s bt^2}{3L}$ $P = \dfrac{4Ebt^3 f}{L^3}$
	Stress	$S_s = \dfrac{3PL}{2bt^2}$ $S_s = \dfrac{6Etf}{L^2}$
	Spring thickness	$t = \dfrac{S_s L^2}{6Ef}$ $t = \sqrt[3]{PL^3/4Ebf}$
	Deflection	$f = \dfrac{5.22 PL^3}{Ebt^3}$ $f = \dfrac{0.87 S_s L^2}{Et}$
	Load	$P = \dfrac{S_s bt^2}{6L}$ $P = \dfrac{Ebt^3 f}{5.22 L^3}$
	Stress	$S_s = \dfrac{6PL}{bt^2}$ $S_s = \dfrac{Etf}{0.87 L^2}$
	Spring thickness	$t = \dfrac{0.87 S_s L^2}{Ef}$ $t = \sqrt[3]{5.22 PL^3/Ebf}$

Right Helix
Open Ends Not
Ground

Right Helix
Closed Ends
Not Ground

Left Helix
Closed Ends Ground

Left Helix
Open Ends Ground

Fig. 9-9. Types of helical compression springs.

Before the formulas for compression and extension springs can be given, it is necessary to list their notations. These are:

d = wire diameter or side of square
D = mean coil diameter $(OD - d)$
f = deflection for N coils
h_s = solid height
L = free length for unloaded spring
G = modulus of elasticity in torsion
T_I = initial torsion
N = number of active coils
P = load
S = stress
C = total coils

The formulas used for compression and extension springs are given in Tables 9-4 and 9-5.

Exercises

11. Given four plate springs, calculate each one's resilience if, when subjected to a safe load of 225 lb, they will have deflections of 1.25, 2.225, 3.0, and 5.75 in., respectively.

12. For a semielliptic spring, similar to the one illustrated in Figure 9-7, calculate the bearing force and bending moment if the applied safe load is 1250 lb. With the spring having a length of 36 in., angle a will be 42° and deflection p will be 8 in. when the loan is applied.

Table 9-4. Formulas for Compression Spring Design

	Type of Spring End			
Unknown	Open or Plain Not Ground	Open or Plain Ground	Closed or Squared Not Ground	Closed Ground
Pitch (p)	$\dfrac{(L-d)}{N}$	$\dfrac{L}{C}$	$\dfrac{L-3d}{N}$	$\dfrac{L-2d}{N}$
Solid height (h_s)	$(C+1)d$	$C \times d$	$(C+1)d$	$C \times d$
Number of active coils (N)	C or $\dfrac{L-d}{p}$	$C-1$ or $\dfrac{L}{p}-1$	$C-2$ or $\dfrac{L-3d}{p}$	$C-2$ or $\dfrac{L-2d}{p}$
Total coils (C)	$\dfrac{L-d}{p}$	$\dfrac{L}{p}$	$\dfrac{(L-3d)}{p}+2$	$\dfrac{(L-2d)}{p}+2$
Free length (L)	$pC+d$	pC	$pN+3d$	$pN+2d$

13. Calculate the pitch for four springs, each of which has a free length of 1.185 in., 22 coils, and a diameter of 0.0156 in. Assume that each coil has a different type of end.

14. What will be the free lengths of an open and a squared end spring if they are not ground and they each have a pitch of

Table 9-5. Design Formulas for Compression and Extension Springs

	Type of Spring Material	
Unknown	Round Wire	Square Wire
Load (P)	$\dfrac{0.393Sd^3}{D}$ or $\dfrac{Gd^4f}{\pi ND^3}$	$\dfrac{0.416Sd^3}{D}$ or $\dfrac{Gd^4f}{5.58ND^3}$
Stress, torsional (S)	$\dfrac{Gdf}{\pi ND^2}$ or $\dfrac{PD}{0393d^3}$	$\dfrac{Gdf}{2.32ND^2}$ or $0.416d^3$
Deflection (f)	$\dfrac{8PND^3}{Gd^4}$ or $\dfrac{\pi SND^2}{Gd}$	$\dfrac{5.58PND^3}{Gd^4}$ or $\dfrac{2.32SND^2}{Gd}$
Number of active coils (C)	$\dfrac{Gd^4f}{8PD^3}$ or $\dfrac{Gdf}{\pi SD^2}$	$\dfrac{Gd^4f}{5.58PD^3}$ or $\dfrac{Gdf}{2.32SD^2}$
Diameter of wire (d)	$\dfrac{\pi SND^2}{Gf}$ or $\sqrt[3]{2.55PD/S}$	$\dfrac{2.32SND^2}{Gf}$ or $\sqrt[3]{PD/0.416S}$
Stress resulting from initional tension (S_{IT})	$\dfrac{ST_I}{P}$	$\dfrac{ST_I}{P}$

0.150 in., 72 active and 84 total coils, and a diameter of 0.0125 in?

15. Find the torsional stress in a 0.125-in.-square extension spring with an outside diameter of 0.750 in. when a load of 54 lb is applied.

Bearings

As is the case with other machine elements, bearings come in various sizes and designs. Within the industry, bearings are categorized as being either plain or ball and roller. The subcategories for each are based primarily on function.

PLAIN BEARINGS

There are three general types of plain bearings. The first, *journal bearings*, are cylindrical in shape and are used for carrying a rotating shaft or journal. The second are *thrust bearings*, and these function to prevent rotating shafts from moving lengthwise. Last, *guide bearings* are used to guide machine elements in a lengthwise motion. Shown in Figure 9-10 are examples of each type.

Most calculations made for bearings pertain to lubrication analysis. Many values can be found in standardized tables, though there are times when specific calculations must be made. The notations used in these formulas are somewhat different from those found in other machine elements and must therefore be clearly recognized:

c_D = diametrical clearance
C_N = bearing capacity number
d = diameter of the journal
K = constants
L = bearing length
m = clearance modulus
N = rpm
p_B = unit load (psi)
P_F = friction horsepower
P' = bearing pressure parameter
p_S = oil supply pressure
q = flow factor
Q_R = total flow required (gpm)

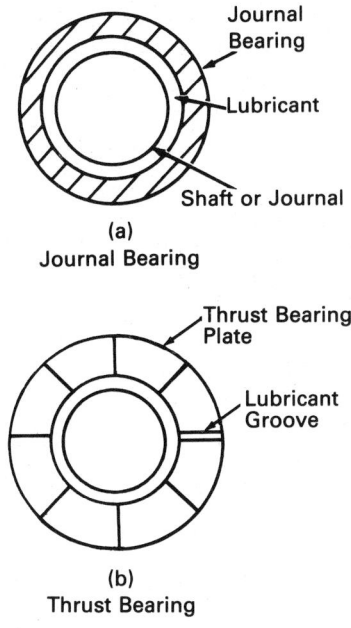

Fig. 9-10. Types of plain bearings.

- r = journal radius
- Δt_A = temperature rise of bearing oil
- t_B = bearing operating temperature
- t_{IN} = t = inlet temperature of oil
- T' = torque parameter
- W = load
- X = factor
- Z = viscosity

The formulas commonly used for lubricant analysis in journal bearing systems are presented here:

Bearing pressure. Though the bearing pressure is often obtained from standardized tables (Table 9-6), calculations are often made as a safeguard against errors. The unit load formula is:

$$p_B = \frac{W}{KLd}$$

Here, K will have a value of 1 for a single oil hole and 2 for a central groove configuration.

Clearance modulus. This modulus is calculated by using the following formula:

$$m = \frac{c_D}{d}$$

Bearing pressure parameter. The bearing pressure parameter is used to calculate other factors (e.g., eccentricity ratio) and uses the formula:

$$P' = \frac{6.9(1000\ m)^2 p_B}{ZN}$$

Friction torque. This is calculated by:

$$T = \frac{T' r^2 ZN}{6900(1000m)}$$

Friction horsepower. The basic formula used to find the friction horsepower for a journal bearing is:

$$P_F = \frac{KTNL}{63,000}$$

Table 9-6. Allowable Bearing Pressures

Type of Bearing/Service	Pressure (psi)
Electric motor and generator bearings	100–200
Heavy-line shafting	100–150
Light-line shafting	15–35
Locomotive axles	300–350
Diesel engine, main	800–1500
Diesel engine, rod	1000–2000
Automotive, main bearings	500–700
Automotive, rod bearings	1500–2500

Table 9-7. X Factor for Mineral Oils

X Factor	Temperature
12.9	100
12.4	150
12.1	200
11.8	250
11.5	300
11.1	350

X factor. This factor is usually obtained by use of a table (Table 9-7).

Total flow of lubricant required. The formula used here is:

$$Q_R = \frac{X(P_F)}{\Delta t_A}$$

Assumed operating temperature. There is always an assumed temperature rise of oil as it is used in a bearing system. The formula used in journal bearings is:

$$t_B = t_{IN} + \Delta t_A$$

Bearing capacity number. The bearing capacity number is a value that must be obtained to determine the flow factor. It is found by:

$$C_N = \frac{(L/d)^2}{60P'}$$

In addition to these formulas, five others are used for journal-bearing lubricant analysis. These are presented in Table 9-8.

THRUST BEARINGS

Presented in Table 9-9 is a listing of commonly used thrust bearing design formulas. These formulas use a somewhat different set of notations and symbols. Those of importance are listed below (for clarification refer to Figure 9-11):

- a = width of pad (radial)
- b = circumference length of pad at pitch line
- B = pitch circle circumference

Table 9-8. Lubricant Analysis Formulas for Journal Bearings

Unknown	Formula
Actual hydrodynamic lubricant flow	$Q_1 = \dfrac{NLc_Dqd}{194}$
Actual pressure flow of lubricant*	$Q^2 = \dfrac{Kp_s c_D^3 d(1 + 1.5\epsilon^2)}{ZL}$
Actual total lubricant flow	$Q = Q_1 + Q_2$
Actual bearing temperature	$t = \dfrac{XP_F}{Q}$
Minimum lubricant film thickness**	$h_O = \dfrac{c_D(1-\epsilon)}{2}$

*K will be equal to 164,000 for a single oil hole, and 235,000 for a central groove.
**ϵ = eccentricity ratio (eccentricity : radial clearance).

c = lubricant specific heat
D = diameter (D_1 = inside diameter and D_2 = outside diameter)
h = film thickness
i = number of pads
K_G = that fraction of the circumference that is occupied by the pad (typically 0.8)
L = length of the chamfer

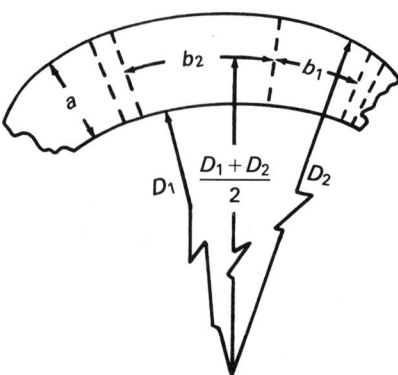

Fig. 9-11. Basic elements of thrust bearings.

Table 9-9. Thrust Bearing Design Formulas

Unknown	Formula
Outside diameter (*Note:* Inside diameter will be determined by shaft size and clearance.)	$D_2 = \left(\dfrac{4W}{\pi K_G P + D_1^2}\right)^{1/2}$
Radial pad width	$a = 0.5(D_2 - D_1)$
Pitch line circumference	$B = \pi(D_2 - a)$
Number of pads	$i = \dfrac{B}{a+s}$
Length of pad	$b = \dfrac{B - si}{i}$
Actual unit load	$p = \dfrac{W}{iab}$
Pitch line velocity	$U = \dfrac{BN}{12}$
Friction power loss	$P_F = iabM$
Oil flow required	$Q = \dfrac{42.4 P_F}{c \Delta t}$
Oil flow (*Note:* Z_2 is at outlet temperature, and value h at 0.002 in.)	$Q_F = \dfrac{150{,}000\, iVh^3 p_s}{Z_2}$
Required flow per chamfer	$Q_c = \dfrac{Q}{i}$
Uncorrected required flow per chamfer	$Q_c^0 = \dfrac{Q_c}{\xi}$
Depth of chamfer	$g = \sqrt[4]{(Q_c^0 L Z_2)/47{,}400 p_s}$

M = horsepower per area
N = rpm
p = bearing unit load
p_s = oil supply pressure
p_F = friction horsepower
Q = total flow
Q_c = flow required per chamfer
Q_c^0 = uncorrected required flow per chamfer

Q_F = film flow
s = width of oil groove
Δt = temperature increase
U = velocity
V = effective ratio of pad's width to its length
W = applied load
Z = viscosity
ξ = kinetic energy (correction factor)

BALL AND ROLLER BEARINGS

Most formulas used in ball and roller bearing work address problems associated with working life and maximum work loads. Ball bearing rating life is established by the Anti-Friction Bearing Manufacturers Association and is noted as the Rating Life L10 of ball bearings (the magnitude is expressed in terms of millions of revolutions).

These calculations are made by using a formula that establishes a ratio between the basic load ratings (C) and equivalent loads (P), usually expressed in terms of newtons or pounds. The notations used in these formulas are:

f_c = a standardized factor determined by bearing design and geometry
i = number of rows of balls within the bearing
α = nominal contact angle
L = effective length
Z = number of balls per row in the bearing
D = diameter of the ball
F_R = applied radial load
F_A = applied axial load
X = radial load factor
Y = axial load factor

The particular design of the bearing will influence the nature of this formula. A listing of common ball and roller bearing rating life formulas are given as follows:

Radial and angular contact ball bearings.

$$L_{10} = \left(\frac{C}{P}\right)^3$$

When balls are no larger than 25.4 mm or 1 in. diameter, then:
$$C = f_c(i \cos \alpha)^{0.7} Z^{2/3} D^{1.8}$$
If the balls exceed 25.4 mm or 1 in. diameter, then:
$$C \text{ (metric)} = 3.647 f_c(i \cos \alpha)^{0.7} Z^{2/3} D^{1.4}$$
and:
$$C \text{ (inch)} = f_c(i \cos \alpha)^{0.7} Z^{2/3} D^{1.4}$$
The equivalent radial load will be:
$$P = XF_R + YF_A$$

Thrust ball bearings.
$$L_{10} = \left(\frac{C_A}{P_A}\right)^3$$

When balls are no larger than 25.4 mm or 1 in. in diameter, and $\alpha = 90°$, then
$$C_A = f_c Z^{2/3} D^{1.8}$$
If α does not equal 90°, then:
$$C_A = f_c(\cos \alpha)^{0.7} Z^{2/3} D^{1.8} \tan \alpha$$
When balls are larger than 25.4 mm or 1 in. in diameter, and $\alpha = 90°$, then:
$$C_A \text{(metric)} = 3.647 f_c Z^{2/3} D^{1.4}$$
and:
$$C_A \text{ (inch)} = f_c Z^{2/3} D^{1.4}$$
If α does not equal 90°, then:
$$C_A \text{ (metric)} = 3.647 f_c(\cos \alpha)^{0.7} Z^{2/3} D^{1.4} \tan \alpha \text{ and}$$

$$C_A \text{ (inch)} = f_c(\cos \alpha)^{0.7} Z^{2/3} D^{1.4} \tan \alpha$$
The equivalent thrust load will be:
$$P_A = XF_R + YF_A$$

288 • MACHINE ELEMENTS

Radial roller bearings.

$$L_{10} = \left(\frac{C}{P}\right)^{10/3}$$

For all radial roller bearings:

$$C = f_c(iL\cos\alpha)^{7/9} Z^{3/4} D^{29/27}$$

The equivalent radial load will be:

$$P = XF_R + YF_A$$

Thrust roller bearings.

$$L_{10} = \left(\frac{C}{P}\right)^{10/3}$$

When α is equal to 90°, then:

$$C_A = f_c L^{7/9} Z^{3/4} D^{29/27} \tan\alpha$$

If α is not equal to 90°, then:

$$C_A = f_c (L\cos\alpha)^{7/9} Z^{3/4} D^{29/27} \tan\alpha$$

The size of the equivalent thrust load will be:

$$P_A = XF_R + YF_A$$

Exercises

16. Calculate the clearance modulus for a journal bearing that has diametral clearance of 0.003 in. and journal diameter of 2.5 in.

17. If the unit load is 1372 psi, viscosity 8.00 centipoise, and the bearing is revolving at a rate of 4800 rpm., what will be the bearing pressure parameter for the journal bearing in problem 16?

18. Given a thrust bearing with an internal diameter of 1.000 in. and an external diameter of 3.000 in., calculate its pitch-line circumference

Springs and Bearings • 289

19. If the thrust bearing in problem 18 is rotating at a rate of 9500 rpm, what will be the pitch-line velocity?

20. Find the rating life for a radial contact ball bearing that has a basic load rating of 750 lb and an equivalent radial load of 718 lb. Express this in terms of the number of revolutions that one can expect from the ball bearing.

CHAPTER 10

Machining Operations

- **Cutting Speeds and Feeds**
- **Basic Machining Operations**
- **Power and Force Calculations in Machining**

Machining operations are central to many industrial settings. These include the use of traditional machine shop equipment plus many production units. It is the purpose of this chapter to present formulas used to determine material feeds and speeds during machining operations. The machining operations covered here will include only those commonly used in the work setting. For specialized machining operations, the reader is encouraged to review the operating manuals for each piece of equipment.

Cutting Speeds and Feeds

The primary function of a machining operation is the removal of material by use of a cutting tool. The exact speed and rate of feed will depend on a number of factors, such as type of operation being performed, work material, cutting tool material, depth of cut, and lubrication. Generally, it has been found that the harder the material, the more difficult it is to cut. Material hardness, then, is a property that must be considered during machining.

An important element of machining operations is the cutting tool, designed for the actual removal of material. Examples are high-speed steels, coated carbides, titanium carbides, cemented carbides, ceramic materials, cast nonferrous alloys, diamonds, cubic boron nitride (CBN), and carbon tool steels.

Speed Tables

There is a wide range of tables available that recommend cutting speeds for different materials. These speeds, however, are based upon *average* or typical conditions and should be considered as a reference point only. As might be expected, it is difficult, if not impossible, to specify cutting and feed speeds that will fit every situation. Too slow a cutting speed will result in an increase in part cost and a low rate of production. Too fast a cutting speed will cause excessive cutting tool wear and increased servicing and maintenance costs.

An example of a partial cutting-speed table is shown in Table 10-1. Here, several steels are listed, along with their Brinell hardness number (BHN), condition (hot rolled, HR; cold rolled, CR; annealed, A; normalized, N; quenched, Q; and tempered, T), and type of cutting tool material (high-speed steel, or HSS, and carbide).

As can be surmised from the review of Table 10-1, the recommended cutting speed will become slower as the material hardness increases. Also, carbide cutting tools can withstand greater cutting speeds than high-speed steels. An example of how this table would be used is to determine the cutting speed that should be selected for turning a 3.500-in.-diameter stock made from AISI 4150 alloy steel that has been quenched and tempered and has a Brinell hardness number of 395. According to the table, a cutting speed of 40 feet per minute (fpm) should be used for high-speed steel cutting tools and 165 fpm for carbide cutting tools.

Cutting-Speed Formulas

The majority of machining operations used in industry employ a rotating spindle that rotates the cutting tool or the workpiece. Here, the cutting speed of feet or meters per minute (fpm or m/min) must be converted to the speed or revolution rate of the spindle (rpm). The two basic formulas used for this conversion are:

$$N = \frac{12V}{\pi D} \text{ (inch units)}$$

Table 10-1. Recommended Cutting Speeds for Turning Plain Carbon and Alloy Steels

AISI and SAE Steels	Hardness BHN	Material Condition	Cutting Speed (fpm)	
			HSS	Carbide
Free-machining plain carbon-steels (resulphurized), 1212, 1213, 1215	100–150 150–200	HR, A CD	150 160	600 625
1108, 1109, 1115, 1117, 1118, 1120, 1126, 1211	110–150 150–200	HR, A CD	130 120	500 525
1132, 1137, 1139, 1140, 1144, 1146, 1151	175–225 275–325 325–375 375–425	HR, A, N, CD Q and T Q and T Q and T	120 75 50 40	400 300 225 200
Free-machining plain carbon-steels (leaded), 11L17, 11L18, 12L13, 12L14	100–150 150–200 200–250	HR, A, N, CD HR, A, N, CD N, CD	140 145 110	550 560 400
Free-machining alloy steels (resulphurized), 4140, 4150	175–200 200–250 250–300 300–375 375–425	HR, A, N, CD HR, N, CD Q and T Q and T Q and T	110 90 65 50 40	400 350 300 220 165
Free-machining alloy steels (leaded), 41L30, 41L40, 41L47, 41L50, 43L47, 51L32, 52L100, 86L20 86L40	150–200 200–250 250–300 300–375 375–425	HR, A, N, CD HR, N, CD Q and T Q and T Q and T	120 100 75 55 50	430 380 275 220 200
Plain carbon-steels, 1006, 1008, 1009, 1010, 1012, 1015 through 1026, 1513, 1514	100–125 125–175 175–225 225–275	HR, A, N, CD HR, A, N, CD HR, N, CD CD	120 110 90 70	450 400 350 300

(continued)

Table 10-1. (continued)

AISI and SAE Steels	Hardness BHN	Material Condition	Cutting Speed (fpm)	
			HSS	Carbide
1027, 1030, 1033	125–175	HR, A, N, CD	100	375
1035, through	175–225	HR, A, N. CD	85	325
1043, 1045, 1046,	225–275	N, CD, Q and T	70	225
1048, 1049, 1050,	275–325	Q and T	60	200
1052, 1524, 1526,	325–375	Q and T	40	160
1527, 1541	375–425	Q and T	30	140
1055, 1060, 1064,	125–175	HR, A, N, CD	100	370
1065, 1070, 1074,	175–225	HR, A, N, CD	80	320
1708, 1080, 1084,	225–275	N, CD, Q and T	65	220
1086, 1090, 1095	275–325	Q and T	50	180
1548, 1551, 1552,	325–375	Q and T	35	150
1561, 1566	375–425	Q and T	30	130
Alloy steels, 4012,	125–175	HR, A, N, CD	100	400
4023, 4024, 4028,	175–225	HR, N, CD	90	350
4118, 4320, 4419,	225–275	CD, N, Q and T	70	300
4422, 4427, 4615,	275–325		60	250
4620, 4621, 4626,	325–375	Q and T	50	200
4718, 4720, 4815,	375–425	Q and T	35	175
4817, 4820, 5015,		Q and T		
5117, 5120, 6118,				
8115, 8615, 8617,				
8620, 8622, 8625,				
8627, 8720, 8822,				
94B17				
Alloy steels E51100,	175–225	HR, A, CD	70	310
E53100	225–275	N, CD, Q and T	65	260
	275–325		50	220
	325–375	N, Q and T	30	180
	375–425	N, Q and T Q and T	20	140

(continued)

Table 10-1. (*continued*)

AISI and SAE Steels	Hardness BHN	Material Condition	Cutting Speed (fpm)	
			HSS	Carbide
Alloy steels 1330, 1335, 1340, 1345, 4032, 4037, 4042, 4047, 4130, 4135, 4137, 4140, 4142, 4145, 4147, 4150, 4161, 4337, 4340, 50B44, 50B46, 50B50, 50B60, 5130, 5132, 5140, 5145, 5147, 5150, 5160, 51B60, 6150, 81B45, 8630, 8635, 8637, 8640, 8642, 8645, 8650, 8655, 8660, 8740, 9254, 9255, 9260, 9262, 94B30,	175–225 225–275 275–325 325–375 375–425	HR, A, N, CD N, CD, Q and T N, Q and T N, Q and T Q and T	85 70 60 40 30	325 275 230 200 150
Steels Not AISI				
Ultra-high-strength steels AMS 6421, AMS 6422, AMS 6424, AMS 6427, AMS 6428, AMS 6430, AMS 6432, AMS 6433, AMS 6434, AMS 6436, AMS 6442, 300M, D6ac	220–300 300–350 350–400 Rockwell C scale Hardness No.: 43–48 48–52	A N N Q and T Q and T	65 50 35 25 10	270 200 150 120 80
Maraging steels, 18% Ni Grades 200, 250, 300, 350	250–325 Rockwell C scale Hardness No.: 50–52	A Maraged	60 10	300 80
Nitriding steels, Nitrex 1, Nitralloys 125, 135, 135 Mod., 225, 230, N, EZ	200–250 300–350	A N, Q and T	70 30	300 225

$$N = \frac{1000V}{\pi D} \text{ (metric units)}$$

where N is the speed of the spindle in rpm, V is the cutting speed in fpm or m/min, and D is the diameter of the rotating workpiece or cutter.

To calculate the cutting speed to be used, the following formula is used:

$$V = V_O F_F F_D$$

where V is the cutting speed to be determined in fpm or m/min, V_O is the cutting speed obtained from cutting-speed tables, F_f is the feed factor, and F_D is the depth-of-cut factor. Both the feed and depth-of-cut factors are presented in Table 10-2.

For an example of how this formula would be used, let's find the cutting and spindle speeds when using a carbide cutting tool for a 1.250-in. (31.75-mm) turning of 215 BHN AISI 1118 steel using a depth of cut of 0.100 in. (2.54 mm) and a feed rate of 0.015 in. (0.38 mm/revolution). The procedures used here are:

$$\begin{aligned} V &= V_O F_F F_D \\ &= (500)(0.91)(1.03) \\ &= 456 \text{ fpm} \end{aligned}$$

$$\begin{aligned} N &= \frac{12V}{\pi D} \\ &= \frac{(12)(456)}{(3.14159)(1.250)} \\ &= 1393 \text{ rpm} \end{aligned}$$

To solve this problem in metric units, we apply the same principle (there is a slight but insignificant difference in answers due to rounding off during conversions). To convert V_O into metric units:

$$V_O = (500)(0.3048) = 152.4 \text{ m/min}$$

Then:

$$\begin{aligned} V &= V_O F_F F_D \\ &= (152.4)(.091)(1.03) \\ &= 143 \text{ m/min} \end{aligned}$$

$$\begin{aligned} N &= \frac{1000V}{\pi D} \\ &= \frac{(1000)(143)}{(3.14159)(31.75)} \\ &= 1434 \text{ rpm} \end{aligned}$$

MACHINING OPERATIONS

Table 10-2. Feed and Depth-of-Cut Factors

Feed (in./rev.)	(F_F) Feed Factor	Depth of Cut (in.)	(F_D) Depth-of-Cut Factors
0.002	1.50	0.005	1.50
0.003	1.50	0.010	1.42
0.004	1.50	0.016	1.33
0.005	1.44	0.031	1.21
0.006	1.34	0.047	1.15
0.007	1.25	0.062	1.10
0.008	1.18	0.078	1.07
0.009	1.12	0.094	1.04
0.010	1.08	0.100	1.03
0.011	1.04	0.125	1.00
0.012	1.00	0.150	0.97
0.013	0.97	0.188	0.94
0.014	0.94	0.200	0.93
0.015	0.91	0.250	0.91
0.016	0.88	0.312	0.88
0.018	0.84	0.375	0.86
0.020	0.80	0.438	0.84
0.022	0.77	0.500	0.82
0.025	0.73	0.625	0.80
0.028	0.70	0.688	0.78
0.030	0.68	0.750	0.77
0.032	0.66	0.812	0.76
0.035	0.64	0.938	0.75
0.040	0.60	1.000	0.74
0.045	0.57	1.125	0.73
0.050	0.55	1.250	0.72
0.060	0.50	1.375	0.71

Frequently, the spindle speed and workpiece diameter are given parameters in a machining operation. What must be determined is the cutting speed to be used. To find this, the following formulas are used:

$$V = \frac{\pi DN}{12} \text{ (inch units)}$$

$$V = \frac{\pi DN}{1000} \text{ (metric units)}$$

To find the cutting speed in fpm for a 5/8-in. drill rotating at a rate of 750 rpm, the following procedure is used:

$$V = \frac{\pi DN}{12}$$
$$= \frac{(3.14159)(0.625)(750)}{12}$$
$$= 1473 \text{ fpm}$$

Exercises

1. Find the cutting speeds when high-speed steel is used to cut the following metals: 100 BHN AISI 1212; HRC 50; Maraged steel 18% Ni grade 300; 325 BHN AISI 4140; and 200 BHN Nitralloy 225, annealed.

2. Determine the spindle speeds for all the metals in problem 1. Assume that their diameters are 0.250 in. What would their spindle speeds be if their diameters were 5.54 mm?

3. Calculate the cutting speeds for HSS and carbide cutting tools for 375 BHN AISI 1132 plain carbon-steel with a diameter of 1.667 in. What is the spindle speed?

4. What would be the cutting speed for the metal in problem 3 if the stock's diameter were 28.50 mm, feed factor were 1.50, and the depth of cut were 1 mm? (*Hint:* Use conversion factor of 1 mm = 0.03937 in.)

5. Find the cutting speed required for a 19.05-mm drill rotating at 375 rpm.

6. Calculate the cutting speed for problem 5 in m/min using the conversion factor of 1 mm = 0.03937 in.

Basic Machining Operations

The majority of machining operations performed in small shop and production settings are based primarily on basic machining techniques. These operations are turning, milling, shaping and planing, drilling, grinding, and broaching. The formulas used here serve as the foundation for other calculations made in more sophisticated production processing. Presented in this section will be a brief description of these operations and related formulas.

Descriptions of Machining Operations

The selection of specific machining operations is critical for efficient and economical production. To fully appreciate this point, and the use of related formulas, it is necessary to understand the differences and similarities among basic machining operations.

TURNING

The oldest and most common form of turning machine is the lathe, which rotates the part as a cutter is held against it. There are a number of different types of lathes used in industry, each designed for a specific purpose. Examples of these include: speed lathe, engine lathe, bench lathe, toolroom lathe, special-purpose lathe, turret lathe, automatic lathe, vertical lathe, and automatic screw machine.

Operations performed on the lathe are varied (e.g., turning, boring, facing, threading, and tapering). In most cases a single-point cutter is used, though other cutting tools are also used for operations such as drilling and reaming. Figure 10-1 is an illustration of a simple turning operation with a single-point tool.

MILLING

A milling machine works in an opposite way from a lathe in that the cutting tool revolves as the workpiece is fed against it. The workpiece is held on a table that can be moved longitudinally, crosswise, or vertically (some tables also have a rotational movement). The more common types of milling machines are: column and knee, planer

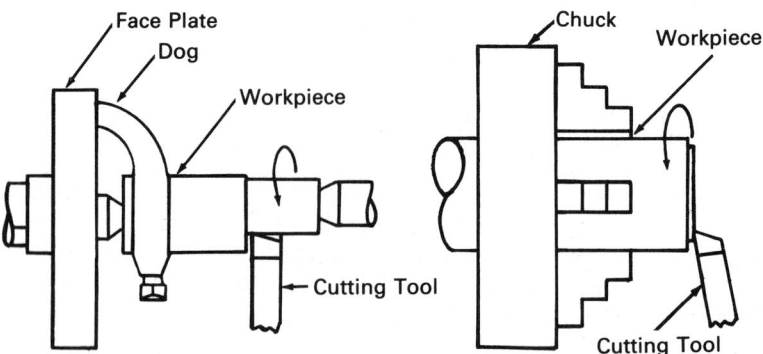

Fig. 10-1. Two typical lathe operations.

milling machine, fixed-bed, and machining centers. There are also special types, such as the rotary table, planetary, profiling, duplicating, and pantograph milling machines.

A great number of operations can be performed on milling machines. These include drilling, boring, angle cutting, side cutting, gang milling, facing, T-slot cutting, gear cutting, and metal sawing. Shown in Figure 10-2 are illustrations of two basic types of milling operations.

SHAPING AND PLANING

Both shapers and planers employ reciprocating motion for tool cutting. Thus, the cutting made is considered to be a straight-line cut. Shapers use a reciprocating tool to cut a stationary workpiece, while planers have stationary cutting tools and reciprocating tables. Shapers are classified according to their design: horizontal, vertical, and

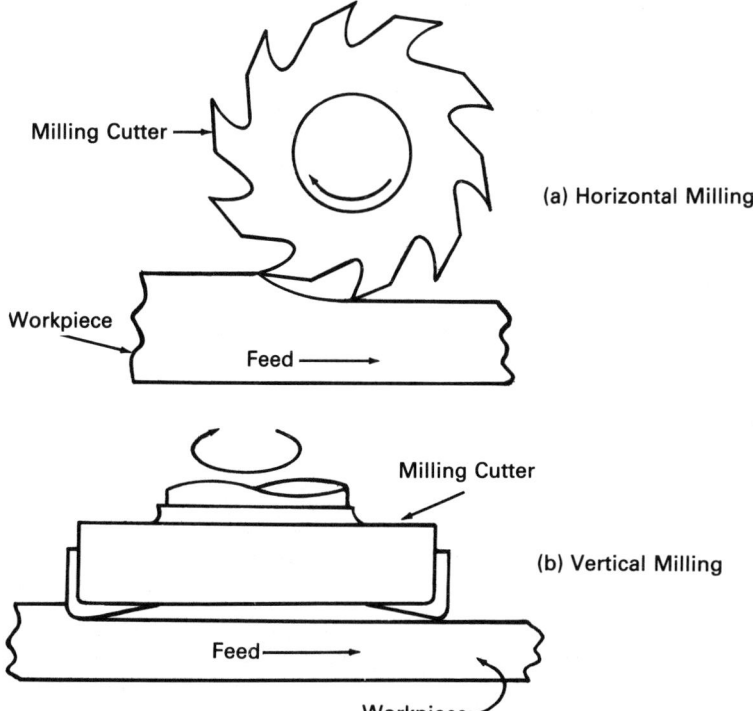

Fig. 10-2. Typical milling operations.

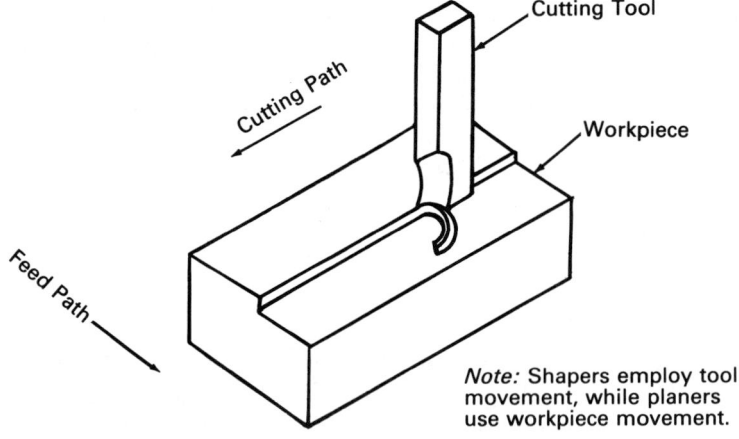

Fig. 10-3. Shaping and planing.

special purpose (e.g., gear cutting). Planers, on the other hand, are grouped according to construction: double housing, open side, pit type, and edge or plate.

In addition to making straight plane cuts, shapers and planers can also make specialty cuts by use of special tooling, attachments, and clamping devices. These include the machining of keyways, spiral grooves, gear racks, and slots. Figure 10-3 is an illustration of a basic shaper operation.

DRILLING

Drilling is one of the simplest operations performed in the production and toolroom setting. It is a process whereby holes are machined by driving a rotating drill into the workpiece. Examples of different types of drilling machines are portable drills, as well as sensitive, upright, radial, turret, multispindle, automatic-production, and deep-hole drilling machines.

In addition to drilling, other operations such as reaming, tapping, boring, hole sawing, and sanding can be performed on drilling machines. Figure 10-4 is a simple illustration of drilling.

GRINDING

The term *grind* implies the use of an abrading force to remove material. Grinding machines are primarily used to finish parts with

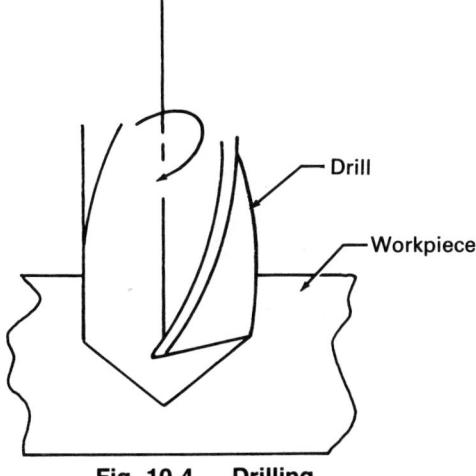

Fig. 10-4. Drilling.

either cylindrical, flat, or internal surfaces. The basic classifications of grinding machines are: cylindrical, internal, surface, universal, tool, and special (e.g., cutting-off, portable, flexible-shaft, abrasive, and surface-preparing) grinders.

In addition to standard grinding of surfaces, other special finishing operations can also be executed. Three of the more common operations are wire brushing, polishing, and buffing. Shown in Figure 10-5 is a simple grinding operation.

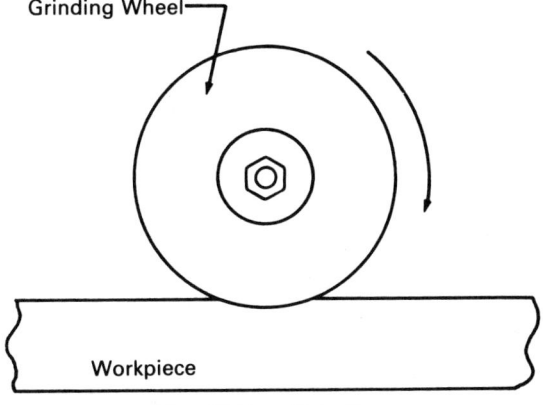

Fig. 10-5. Grinding.

302 • MACHINING OPERATIONS

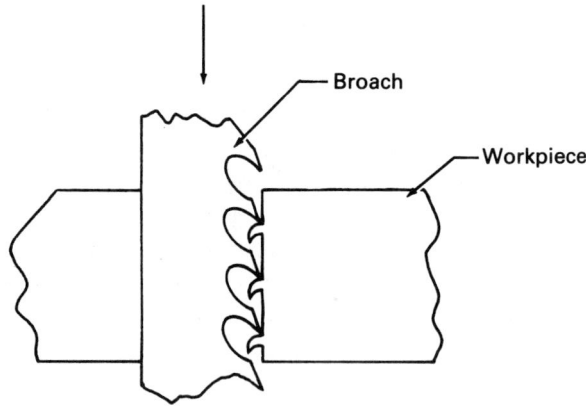

Fig. 10-6. Broaching.

BROACHING

Broaching is a somewhat specialized operation that is used to remove metal by moving a long cutting tool over the workpiece. Unlike many other operations, the part is finished after the completion of one stroke of the broaching machine. In the majority of operations, the broach is moved across the work.

There are four basic types of broaching machines: pull, push, surface, and continuous. Figure 10-6 is an illustration of the basic concept of broaching.

Machining Formulas

Of the numerous formulas used for shop and production work, the first discussed will be those used for turning, milling, drilling, and broaching operations. These are presented in Table 10-3. The symbols used in these formulas are presented below:

D_T = diameter of turned workpiece
D_M = diameter of milling cutter
D_D = diameter of drill
d = depth of cut
d_T = total broaching depth of stroke
E = efficiency of spindle drive
f_M = feed rate per minute
f_R = feed per revolution

f_T = feed per tooth
HP_M = motor horsepower
HP_S = horsepower at spindle
L = length of cut
n = number of teeth in cutter
P = unit power (in.^{3}/min)
Q = rate of material removal
T_S = torque at spindle
t = cutting time
V_C = cutting speed
w = width of cut

As an example of how these formulas are used, let's determine the cutting speed, feed rate, cutting time, rate of metal removal, and motor horsepower required for turning 0.040 in. off a 5.250-in. section of a 1-in. rod, with a spindle speed of 1800 rpm, at a feed rate of 0.020 in./rev (assume that $P = 1.1$ and the efficiency is 0.90). To find these answers, the following calculations are made:

$$V_C = 0.262 D_T (\text{rpm})$$
$$= (0.262)(1)(1800)$$
$$= 471.6 \text{ fpm}$$

$$f_M = f_R(\text{rpm})$$
$$= (0.020)(1800)$$
$$= 36 \text{ in./min}$$

$$t = \frac{L}{f_M}$$
$$= \frac{5.250}{36}$$
$$= 0.1458 \text{ min}$$
$$= 8.75 s$$

$$Q = 12 d f_R V_C$$
$$= (12)(0.040)(0.020)(471.6)$$
$$= 4.53 \text{ in.}^3/\text{min}$$

$$HP_M = \frac{QP}{E}$$
$$= (4.53)(1.1)/0.90$$
$$= 5.54$$

Table 10-3. Basic Formulas for Turning, Milling, Drilling, and Broaching

Unknown	Formulas (inch units)	Operation
Cutting speed	$V_C = 0.262 D_T (\text{rpm})$	Turning
	$V_C = 0.262 D_M (\text{rpm})$	Milling
	$V_C = 0.262 D_D (\text{rpm})$	Drilling
	$V_C = \text{fpm}$	Broaching
RPM	$\text{rpm} = \dfrac{3.82 V_C}{D_T}$	Turning
	$\text{rpm} = \dfrac{3.82 V_C}{D_M}$	Milling
	$\text{rpm} = \dfrac{3.82 V_C}{D_D}$	Drilling
Feed rate	$f_M = f_R (\text{rpm})$	Turning and drilling
	$f_M = f_T n (\text{rpm})$	Milling
Feed per tooth	$f_T = \dfrac{f_M}{n}(\text{rpm})$	Milling
	$f_T = \text{inches}$	Broaching
Cutting time	$t = \dfrac{L}{f_M}$	Turning, milling, and drilling
	$t = \dfrac{L}{12 V_C}$	Broaching
Rate of metal removal	$Q = 12 d f_R V_C$	Turning
	$Q = w d f_M$	Milling
	$Q = \dfrac{\pi D^2 d f_M}{4}$	Drilling
	$Q = 12 w d V_C$	Broaching
Motor horsepower required	$HP_M = \dfrac{QP}{E}$	Turning, milling, drilling, and broaching
Horsepower required at spindle	$HP_S = QP$	Turning, milling, and drilling
Torque at spindle	$T_S = \dfrac{63{,}030\, HP_S}{\text{rpm}}$	Turning, milling, and drilling

Table 10-4. Approximate Average Unit Power Requirements

| | | Unit Power* | | |
| | | Turning | Drilling | Milling |
Material	Hardness (BHN)	HSS and Carbide	HSS	HSS and Carbide
Wrought and cast steels	85–200	1.1	1.0	1.1
Cast irons	110–190	0.7	1.0	0.6
	190–320	1.4	1.6	1.1
Stainless steels	135–275	1.3	1.1	1.4
Precipitation hardening stainless steels	150–450	1.4	1.2	1.5
Titanium	250–375	1.2	1.1	1.1
Nickel and cobalt base high-temp. alloys	200–360	2.5	2.0	2.0
Nickel alloys	80–360	2.0	1.8	1.9
Aluminum alloys	30–150	0.25	0.16	0.32
Magnesium alloys	40–90	0.16	0.16	0.16

*For sharp tools only, dull tooling will increase unit power.

To find the unit power (P) for turning, milling, and drilling, see Table 10-4. The unit power used for broaching operations will be based on the chip load measured in amount of metal removed per tooth (see Figure 10-7).

Another commonly used machining operation is known as abrasive and grinding machining. As the name implies, this involves the use of abrasive wheels, disks, or materials for the removal of material. The symbols used in these formulas are referenced to Figure 10-8 and given as follows:

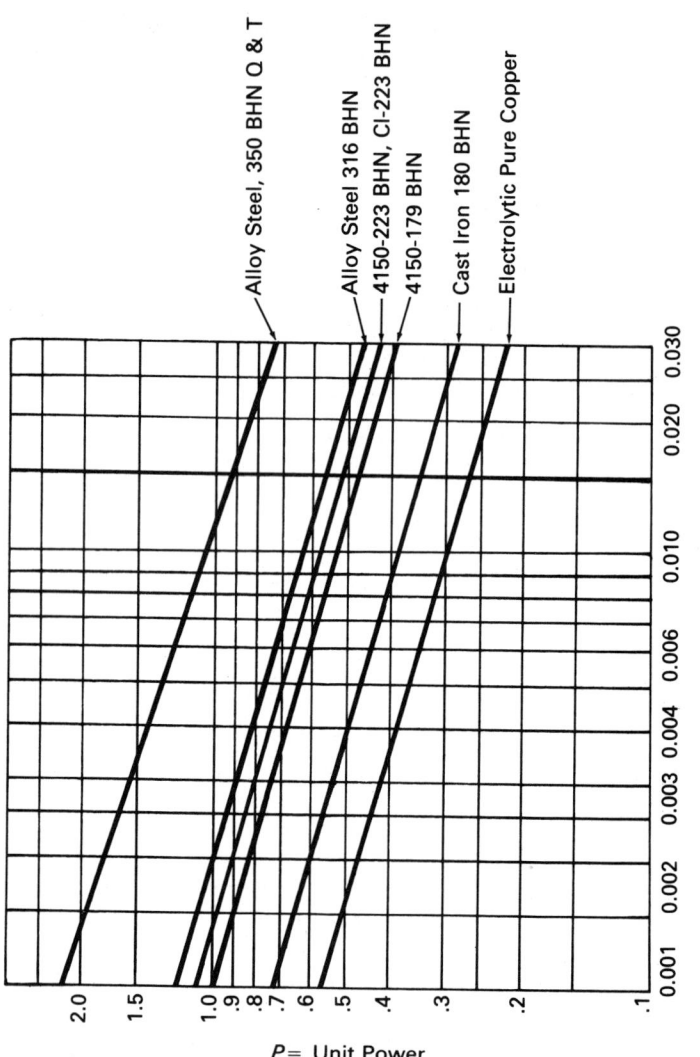

Fig. 10-7. Unit power for broaching.

a = depth of grind per pass
a_T = total depth of grind
b = width of cut (plunge grinding)
b_C = cross-feed distance per pass
b_S = beginning width of the grinding wheel
b_W = workpiece width
d_O = outside diameter of grinding wheel
d_I = inside diameter of grinding wheel
d_R = regulating wheel diameter (centerless grinding)
d_{S_1} = initial grinding diameter
d_{S_2} = final grinding diameter
d_T = mean diameter of workpiece path on a rotary table (vertical spindle surface grinder)
d_{W_1} = original work diameter
d_{W_2} = final work diameter
E = grinding wheel drive efficiency
f_T = traverse feed rate of table, or through-feed of work during centerless grinding
f_P = plunge infeed rate
G = grinding ratio
L = ground workpiece length
hp = horsepower required at motor
n_R = regulating wheel rpm
n_S = grinding wheel rpm
n_T = rotary table rpm
n_W = work rpm
P = unit horsepower (hp/in.^{3}/min)
s = feed per revolution of work
t = axial wear from face of grinding wheel
v_S = peripheral speed of grinding wheel
v_W = peripheral speed of workpiece
w_M = maximum contact width of grinding wheel on work
Z = metal removal rate
α = regulating wheel inclination angle

The formulas most often used for grinding operations are listed in Table 10-5. Note that the four major types of grinding operations listed are surface grinding, center-type cylindrical grinding, centerless grinding, and internal grinding. Furthermore, surface grinding is subdivided into two additional categories of horizontal spindle

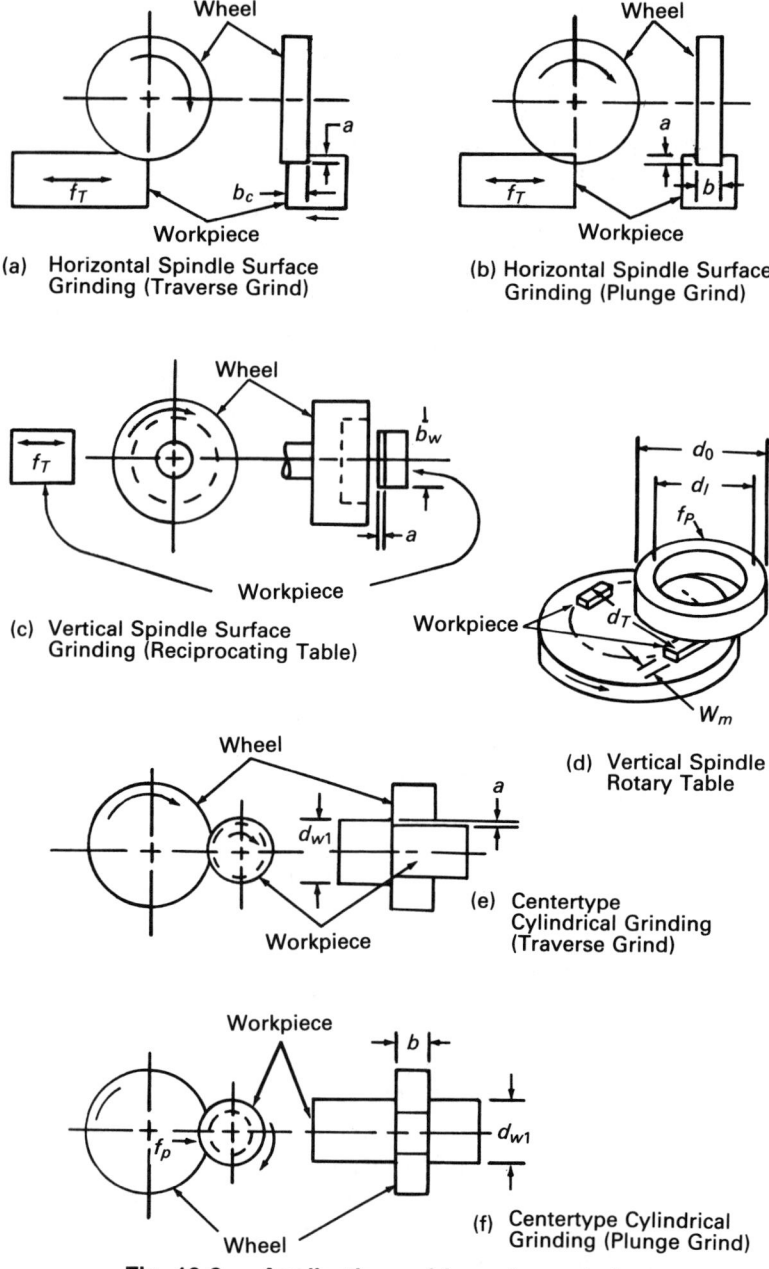

Fig. 10-8. Applications of formula symbols.

Basic Machining Operations • 309

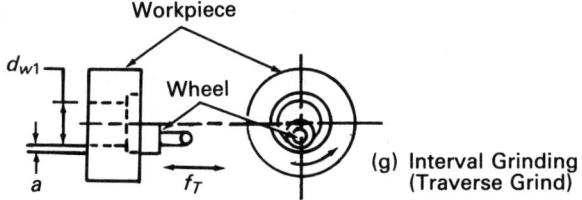

(g) Interval Grinding
(Traverse Grind)

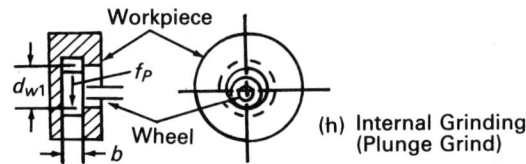

(h) Internal Grinding
(Plunge Grind)

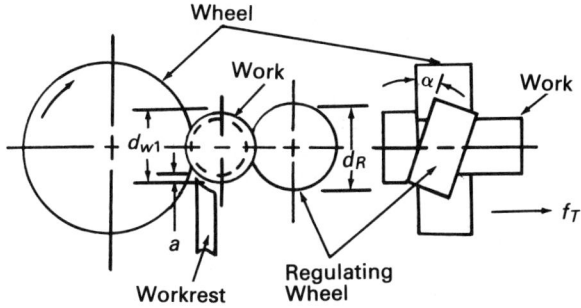

(i) Centerless Grinding
(Traverse Grind)

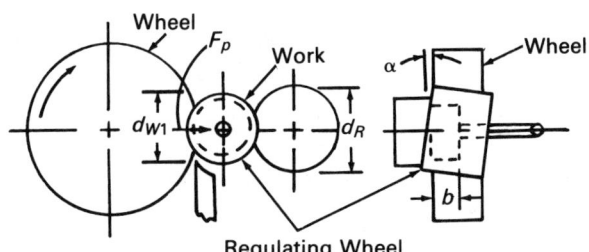

(j) Centerless Grinding
(Plunge Grind)

Fig. 10-8. (*continued*)

(reciprocating table) grinding and vertical spindle grinding with reciprocating or rotary tables.

An example of how to use these formulas would be in finding the work rpm and horsepower requirement at the motor during centerless grinding. Assume that the peripheral speed of the work

Table 10-5. Formulas for Grinding Operations

Unknown	Grinding Operation	Formula
Peripheral wheel speed (v_S)	All	$0.262 d_{S_1} n_S$
Peripheral work speed (v_W)	Horizontal spindle and vertical spindle with reciprocating table	v_W
	All others	$\dfrac{3.82 v_S}{d_{S_1}}$
Work rpm (n_T)	Vertical spindle with rotary table, center-type cylindrical, centerless, and internal grinding	$\dfrac{3.82 v_W}{d_{W_1}}$
Horsepower required at the motor (hp)	All	$\dfrac{PZ}{E}$
Traverse Grinding		
Table traverse feed rate (f_T)	Center-type cylindrical and internal grinding	$s n_W$
	Centerless grinding	$\pi d_R n_R \sin \alpha$
Metal removal rate (Z)	Horizontal spindle	$a b_C f_T$
	Vertical spindle reciprocating table	$a b_W f_T$
	Center-type cylindrical, centerless, and internal grinding	$\pi a f_T d_{W_1}$
Grinding ratio (G)	Horizontal spindle	$\dfrac{1.273 L b_W a_T}{b_S (d_{s1}^2 - d_{s2}^2)}$

(continued)

Table 10-5. (continued)

Unknown	Grinding Operation	Formula
	Vertical spindle reciprocating table	$\dfrac{1.273 L b_w a_T}{t(d_o^2 - d_c^2)}$
	Center-type cylindrical, centerless, and internal grinding	$\dfrac{L(d_{w1}^2 - d_{w2}^2)}{b_s(d_{s1}^2 - d_{s2}^2)}$
Plunge Grinding		
Rate of feed	Horizontal spindle and vertical spindle reciprocating table	f_T
	Vertical spindle with rotary table	sn_T
	Center-type cylindrical, centerless, and internal grinding	f_P
Metal removal rate (Z)	Horizontal spindle	abf_T
	Vertical spindle reciprocating table	$ab_w f_T$
	Vertical spindle rotary table	$\pi w_M f_P d_T$
	Center-type cylindrical, centerless, and internal grinding	$\pi b f_B d_{W_1}$
Grinding ratio (G)	Horizontal spindle	$\dfrac{1.273 L b a_T}{b_S(d_{s1}^2 - d_{s2}^2)}$
	Vertical spindle reciprocating and rotary tables	$\dfrac{1.273 L b_w a_T}{t(d_o^2 - d_I^2)}$
	Center-type cylindrical, centerless, and internal grinding	$\dfrac{b(d^2{}_{W1} - d^2{}_{w2})}{b_S(d_{s1}^2 - d_{s2}^2)}$

is 2500 fpm with the rough stock diameter being 3.750 in. Also, the unit horsepower is 20 hp/in.³/min, with the metal being removed at a rate of 0.720 in.³/min, and at an efficiency of 0.90 (90%). To find the solutions, the following formulas are used:

$$n_W = \frac{3.82 v_W}{d_{W_1}}$$
$$= \frac{(3.82)(2500)}{3.750}$$
$$= 2547 \text{ rpm}$$

$$\text{hp} = \frac{PZ}{E}$$
$$= (20)(0.720)/0.90$$
$$= 16$$

Many modern shapers and planers are designed with devices that will indicate the speed of travel of the table or arm. Older and small job-shop shapers, however, are not. Thus, it is advisable to know how these speeds can be calculated. The symbols used in these formulas are:

v = cutting speed
v_R = return speed
s = number of cutting strokes (usually per minute)
L = length of the cutting stroke
w = width of surface being shaped or planed
t = time
f = length of feed

It should be noted that cutting speeds are *estimates*, but are considered to be reasonably accurate. First, the formulas used to calculate planer cutting speed and cutting strokes per minute are:

$$v = sL$$
$$s = \frac{v}{L}$$

The approximate time required to plane a surface is determined by the following formula:

$$t = \frac{w}{f\,[L(1/v + 1/v_R) + 0.025]}$$

To find the cutting speed of a planer that is making 16 strokes per minute over 24 meters, the following procedures are followed:

$v = sL$
$= (16)(24)$
$= 384$ m/min (approximate)

Exercises

7. What are the recommended cutting speeds for drilling operations employing ⅛-, ¼-, ½-, and ¾-in. drills when used at spindle speeds of 300 and 500 rpm?

8. If a 2.22-in.-diameter milling cutter is rotating at 1800 rpm, what would the recommended cutting speed be?

9. What amount of horsepower is needed by a turret lathe motor to cut 5 in.³/min of 200 BHN stainless steel with a carbide cutting tool and an efficiency of 95%?

10. What amount of horsepower is needed for a drilling operation meeting the same specifications as those in problem 9?

11. What would be the peripheral speed of two aluminum oxide grinding wheels if they were turning at a rate of 1200 rpm and had diameters of 6 and 8 in.?

12. Calculate the amount of horsepower that would be required for centerless grinding where 4.12 in.³/min was removed with 98% efficiency, and the unit horsepower was 10 hp/in.³/min.

13. Find the feed rate for plunge grinding on a vertical spindle with a 2400-rpm rotary table where a feed of 0.020 in./rev is used.

14. What is the metal removal rate for a horizontal spindle grinding unit used in traverse grinding to remove 0.002 in. of material per pass if the cross-feed is set at 0.024 in. per pass and the table traverse feed rate is 3 in./min?

Power and Force Calculations in Machining

When machining operations are being executed, a number of different forces are exerted upon the machine and cutting tools. Most of the more important forces can be measured by use of instrumentation. Torque and thrust forces generated during drilling, milling, and turning are frequently determined by dynamometer reading. Hence the formulas presented in this section provide the procedures for calculating power and torque from dynamometer readouts.

Machining Power Requirements

Several methods are available for determining power requirements for machining. No one method will meet all situational needs. Therefore, one should be familiar with each technique. In all, there are three major methods used to find machining power requirements.

1. *Watt meters and ammeters.* Either a wattmeter or an ammeter will provide the simplest and most practical method used for measuring horsepower. Of the two, the wattmeter is perhaps the most accurate. Horsepower requirements at the cutter are determined by placing the meter at the spindle motor and calculating the difference between the idle horsepower and horsepower during cutting.

2. *Published power units.* A satisfactory estimate of horsepower requirements can be obtained from published tables or charts. Many times, these tables are used in conjunction with specific equations.

3. *Calculations.* The last method available is to calculate power needs by utilizing obtained measures in mathematical formulas. In addition, cutting force values can also be obtained by use of dynamometer readings and horsepower calculations.

Power-and-Force Formulas

The basic formulas used for calculating power-and-force requirements during machining use unit horsepower values (P). In Table 10-6 are the formulas used for turning, milling, and drilling operations.

An example of a power-and-force problem would be to find the horsepower required at the motor and the torque required at a 48-hp spindle during milling if the spindle had a 70% spindle drive

Table 10-6. Power and Force Formulas

Unknown	Operation	Formula
Unit power (P)	Turning, milling, and drilling	$\dfrac{HP_S}{Q}$
Horsepower at the spindle (HP_S)	Turning and milling	$\dfrac{F_c V_c}{33,000}$
	Drilling	$\dfrac{(T_S)(\text{rpm})}{63,030}$
Horsepower at the motor (HP_M)	Turning, milling, and drilling	$\dfrac{HP_S}{E}$
Rate of metal removal (Q)	Turning Milling Drilling	$12 df_R V_C$ $(wdf_T n)(\text{rpm})$ $(\pi D^2 {}_D f_R)(\text{rpm})$
Torque at the spindle (T_S)	Turning and milling Drilling	$\dfrac{63,030 HP_S}{\text{rpm}}$ $49,500 f_R D^2 {}_D P$

efficiency while it was rotating at a rate of 850 rpm. To find these values, the following procedures are used:

$$HP_M = \frac{HP_S}{E}$$
$$= \frac{48}{0.70}$$
$$= 68.57$$

$$T_S = \frac{63030 HP_S}{\text{rpm}}$$
$$= \frac{(63030)(48)}{850}$$
$$= 3559.34 \text{ in. lb}$$

Exercises

15. Find the horsepower required at the motor and the spindle torque for a turning operation with a cutting force of 150 lb,

cutting speed of 3.2 fpm, turning at 950 rpm, and having a spindle efficiency drive of 65%.

16. If the feed for the turning operation in problem 15 was 0.015 in./rev and the depth of cut was 0.063 in., find its unit power.

17. If the torque at a drilling machine's spindle is 550 in.-lb, find the amount of horsepower at that spindle if it were drilling at 758 rpm.

PART III

Industrial Management Formulas and Calculations

CHAPTER 11

Management Controls

- **Industrial Organization**
- **Span, Production, and Inventory Controls**
- **Quality Assurance, Manufacturing Controls, and Budget Cost Controls**

The appropriate use of management controls is critical to the successful operation of any industrial organization. The professional manager must be able to plan, organize, operate, and measure various elements of work to assure the most economical and highest quality products and/or services. To accomplish this, controls are calculated for personnel, production, inventory, quality control, scheduling, manufacturing, and budgeting functions. This chapter is a brief review of industrial organization and presents formulas used in management control functions.

Industrial Organization

Industrial managers are required to operate within a wide spectrum of areas. They not only must participate in the structuring of work and personnel assignments, but must also design systems and procedures that link the organization to business objectives. This process involves four basic elements:

1. *Planning.* This involves the setting of objectives and goals, the establishment of production standards, work performances, and profit margins.

2. *Organizing.* To meet business and production objectives, managers are required to divide work and procedures into manageable units and jobs. Such a process involves the grouping of personnel and operations into logical and orderly structures, assigning personnel to the structure, formulating systems, and establishing procedures for executing work.
3. *Operating.* The operational phase of management involves decision making, delegating, directing, and communicating. Also involved here are the interpretation and restructuring of work objectives.
4. *Measuring.* Measurement is central to management controls. Without accurate measures of performance and productivity, it would be extremely difficult to apply effective management controls. Measuring involves the recording, reporting, interpreting, and analyzing of data as they compare to established standards.

Span, Production, and Inventory Controls

Good organizational design and management employ simple and easy-to-implement controls. Three controls that lend themselves to this concept are span, production, and inventory. This section will present the basic formulas used for establishing these management controls.

Span Control

The *span-of-control theory* is used to determine the number of units and levels needed in an organization to manage efficiently and effectively and to assign responsibilities. This theory is based on the concept that individuals can manage and supervise a specific number of subordinates. The number of subordinates that can be managed will depend on the activity taking place (i.e., the more uniform and routine the job and lower the personnel level, the greater the span of control).

The general rule of thumb states that four to seven first-line managers (or supervisors) can effectively report to a manager, and fifteen workers can report to a first-line supervisor. Shown in Table 11–1 is the total number of employees that can be supervised for a managerial span of four, with a first-line span of fifteen.

Table 11-1. Managerial Spans of Control*

Level Number	Total Number of Levels				
	2	3	4	5	6
1	1	1	1	1	1
2	15	4	4	4	4
3		60	16	16	16
4			240	64	64
5				960	254
6					3810
Total employees	16	65	261	1045	4149

*Managerial span of four and first-line supervisor span of fifteen.

The span shown here is just one example. In today's businesses, the actual managerial span will vary from function to function. Hence, production functions will support higher spans than engineering functions. Ratios typically applied here are given in Table 11-2.

Two formulas are primarily used for span-of-control calculations. The first is for determining the number of levels of line supervisors required for a given situation:

$$L = \frac{\log p - \log f}{\log r} + 1$$

Here, L = the number of levels of line supervision responsible for a unit (including the top manager or executive), p = number of pri-

Table 11-2. Managerial Ratios

Function	Ratios of First-Line Supervisors to Manager
Research	8:1–10:1
Engineering	10:1–12:1
Manufacturing	15:1–20:1
Quality Control	12:1–14:1
Purchasing	10:1–12:1
Finance	10:1–12:1
Employee Relations	10:1–12:1

mary workers, $f=$ span of control for first-line supervision, and $r=$ span of control above the first-line supervision.

An example of how this formula would be used is in an industrial organization of 3000 primary operatives (workers), a span control of four for the first-line supervisor, and a span control of two for above the first-line supervisor. To determine the number of levels of line supervision required, the following calculations are made:

$$L = \frac{\text{Log } p - \text{Log } f}{\text{Log } r} + 1$$

$$= \frac{\text{Log } 3000 - \text{Log } 4}{\text{Log } 2} + 1$$

$$= \frac{3.4771213 - 0.6020599}{0.30103} + 1$$

$$= 10.5507466 \text{ or}$$

$$= 11 \text{ levels}$$

The second formula employing span-of-control theory is for determining the number of line supervision personnel needed for an organization. This formula is:

$$e = \frac{p(1 + 1/r + 1/r^2 + \cdots 1/r^{L-1})}{f}$$

where $e=$ the number of line supervision personnel, $p=$ number of primary workers, $r=$ span of control above first-line supervision, and $L=$ number of levels of line supervisors (including top management or executive in direct charge of the work unit).

A typical problem involves a service organization of 2800 primary workers, with three levels of line supervision, and a span control of five for above first-line supervision and eight for first-line supervision. To determine the number of line supervisors needed in this organization, the following calculations are made (the superscript $L-1$ will be $3-1$, or 2):

$$e = \frac{p(1 + 1/r + 1/r^2 + \cdots 1/r^{L-1})}{f}$$

$$= 2800 \left(\frac{1 + 1/5 + 1/5^2}{8} \right)$$

$$= 434 \text{ line supervision personnel}$$

Exercises

1. A company has 7000 employees directly involved with production, needs a span control of ten for the first-line supervision level, and is anticipating a span control of five for positions above the first-line supervisor. Calculate the levels of supervision required for this to be accomplished.
2. An industrial organization of 12,000 primary employees has a span control of fifteen for first-line supervision and a span control of four for above first-line supervision. Find the number of levels of line supervision that it should have.
3. A business of 7000 primary operatives has five levels of line supervision, a span control of five above first-line supervision, and a span control of ten for first-line supervision. Find the total number of line supervisors required.
4. A service industry of 1200 primary employees requires six levels of line supervision, a span control of four for above first-line supervision, and fifteen for first-line supervision. How many line supervisors are required?

Production Control

Both production planning and control are closely linked to the successful operation of a business. The exact nature of production planning and control will vary from one industry to the next, as well as between different-sized organizations. What is common among all organizations, however, is the *economics* of production.

The goal of all production engineers and managers is to produce a quality product at the lowest possible cost. Most of these calculations are based upon inventory and purchase costs. Hence, the two basic formulas presented here incorporate these factors.

The first formula to be considered for production control addresses the economical production of lot quantity—that is, production and inventory costs. The formula used here is:

$$Q = \sqrt{2rs/i}$$

where Q = the economical production in lot quantity units, r = annual production of units required, s = setup cost per order, and i = annual inventory cost per unit.

An example of how this formula would be used is to calculate the economical production lot quantity for the Masters Company, which has an annual production requirement of 5000 units of their product. The setup cost per order is $75.00 and the annual inventory cost per unit is $0.96. To find the number of lot quantities needed to be produced to make the venture economical, the following calculations are made:

$$\begin{aligned} Q &= \sqrt{2rs/i} \\ &= \sqrt{2(5000)(75)/0.96} \\ &= \sqrt{781250} \\ &= 883.883 \text{ or} \\ &= 884 \text{ lots} \end{aligned}$$

The second formula also deals with production lot quantities but takes into consideration quantity discounts. It addresses the possibility of purchasing inventory at given quantity discounts. The basic formula used here is:

$$C_A = C_N - C_P$$

where:

$$C_N = rp_N + sN_N + p_N n_N i \text{ and}$$
$$C_P = rp_D + sN_D + p_D n_D i$$

C_A = the difference in annual cost between the discount price and no discount price, C_N = the cost with no discount, and C_P = the discount price. Other variables considered here are:

p_N = purchased unit cost with no discount
p_D = discounted purchase unit cost
N_N = annual orders with no discount
N_D = annual orders with discount
n_N = number of units per order with no discount
n_D = number of units per order with discount

Given the Masters Company example, let's say that the unit cost of product A from the company's supplier has been $7.00. Now, the supplier is offering a unit price of $6.00 for 2500 units or more. The annual number of orders with no discount has been six. If a discount were used, Masters would have to order only twice annually. In addition, with no discount the Masters Company purchases 835 units per order; with the discount, they would purchase

2500 units per order. To find out the difference in annual costs between discount and no-discount procurements, the following calculations are made:

$$C_N = rp_N + sN_N + p_N n_N i$$
$$= (5000)(7) + (75)(6) + (7)(835)(0.96)$$
$$= \$41,061.20$$

$$C_P = rp_D + sN_D + p_D n_D i$$
$$= (5000)(6) + (75)(2) + (6)(2500)(0.96)$$
$$= \$44,550.00$$

$$C_A = C_N - C_P$$
$$= 41,061.20 - 44,550.00$$
$$= -\$3,488.80$$

As can be seen in this example, though there is a discount offered by the supplier, it would cost the Masters Company $3,488.80 annually to take advantage of the lower cost.

Exercises

5. The R. M. Gordon Company requires a minimum production rate of 600 units annually of mechniques. The set cost per order is $15.00 and the annual inventory cost per unit is $0.20. What is the economical production lot quantity?

6. Considering the R. M. Gordon Company in problem 5, the unit cost from H & K supplier was $5.00 per unit. H & K has now offered a 10% discount to all its customers for orders of 500 or more. Gordon officials have traditionally made four orders annually of 300 units per order. They note that with the discount they will only need to make one order per year of 600 units per order. Calculate the amount of savings or loss that would be realized if R. M. Gordon took advantage of the discount.

Inventory Control

The control of inventory is an important factor for any type of business operation. It is not uncommon to find businesses with inventories that exceed 25% of its total annual inventory at one time. This

326 • MANAGEMENT CONTROLS

can add up to substantial costs. These costs include the inventory purchase itself, handling and storage costs, insurance fees, ordering and expediting costs, review and inspection costs, plus losses due to obsolescence, damage, and deterioration.

Central to inventory control is the principle of *economical order-quantity*, which is based on the concept that for every unit, there is an optimum order quantity that results in the lowest possible total per unit. The more units ordered, the higher the annual carrying cost; however, the fixed cost of placing the order will be lowered. Thus, an optimum quantity order must be determined. This is graphically illustrated in Figure 11-1.

The optimum quantity per order is mathematically shown as:

$$\sqrt{2(u_A)(w_S)/w_U p_{CC}}$$

where u_A = annual unit usage, w_S = cost per setup or order, w_U = cost per unit, and p_{CC} = percent annual carrying cost. This formula is quite simple for internally manufactured parts, but becomes more complicated for purchased parts and material. This latter situation is by far more common.

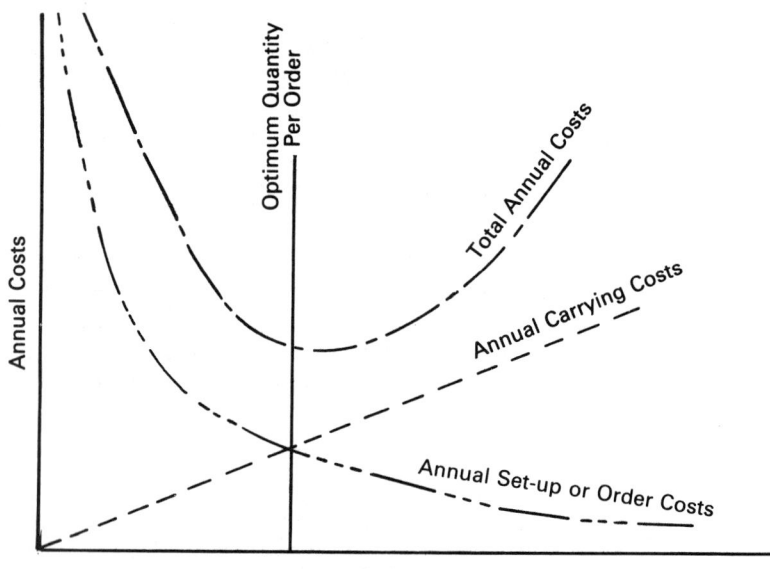

Fig. 11-1. **Optimum quantity order.**

There are five primary formulas used for inventory control. The first is known as the reorder point (ro_P) and is found by the formula:

$$ro_P = Lu_W$$

Here, L = lead time in weeks and u_W = average usage in weekly units. For example, if a business uses 345 units per week and the order must be placed six weeks ahead of time, then the reorder point will be determined by:

$$\begin{aligned} ro_P &= Lu_W \\ &= (6)(345) \\ &= 2070 \text{ units} \end{aligned}$$

The second formula is used to calculate average inventory during the lead time (a_H), measured in units, and is expressed as:

$$a_H = ro_P/2$$

A third formula is for calculating the average hold cost. The formula is:

$$W_H = a_W c_H$$

where w_H = the average weekly holding cost, a_W = average weekly inventory during the lead time in units, and c_H = holding cost per unit per week. An example of a typical problem is for a company that has an average weekly inventory of 240 units, with the holding cost per unit per week being $1.04. The average weekly holding cost, then, is calculated as follows:

$$\begin{aligned} w_H &= a_W c_H \\ &= (240)(1.04) \\ &= \$249.60 \end{aligned}$$

Another important factor involved in inventory control is cost attributed to out-of-stock items. The formula used for calculating average weekly cost of out-of-stock inventory is:

$$w_O = p_O c_O$$

In this formula, w_O = the cost of out-of-stock inventory on a weekly average, p_O = percent of time that the business is out of stock, and c_O = the average cost per order of out-of-stock inventory.

If ILS Company finds that it is out of stock 12.4% of the time and the average cost per order of out-of-stock inventory is $246.48, then the average out-of-stock costs would be:

$$w_O = p_O c_O$$
$$= (12.4)(246.48)$$
$$= \$30.56$$

The last formula to be considered concerns total weekly holding and out-of-stock-costs. These costs are calculated by applying the following formula:

$$w_T = w_H + w_O$$

where w_T = the total costs per week, w_H = average holding costs per week, and w_O = average cost per week of out-of-stock inventory.

Considering the ILS Company again, should its average weekly holding costs be $0.87 per unit, then the total cost per week for holding and out-of-stock items would be calculated as follows:

$$w_T = w_H + w_O$$
$$= 0.87 + 30.56$$
$$= \$31.43$$

Exercises

7. Veremex Corporation has a midwestern plant with an average lead time for orders of three weeks and an average usage of 80 units per week. Calculate the reorder time for this plant.

8. Find the average weekly inventory during the reorder point period for the plant in problem 7.

9. The West Company has an average weekly inventory of 120 units and a holding cost of $0.60 per unit. Find the average weekly holding costs.

10. If the West Company is found to be out of stock 14% of the time and has an average cost per order of out-of-stock inventory of $150.00, compute the average weekly cost of out-of-stock inventory.

11. The West Company is in the process of making an inventory cost analysis of its product. Based upon the information given and calculated in problems 9 and 10, compute the total holding and out-of-stock costs.

Quality Assurance, Manufacturing Controls, and Budget Cost Controls

Three factors that are critical to the success of a well-managed operation are quality assurance, manufacturing controls, and budget cost controls. The management of these controls often requires the use of specific and accurate data, which may be obtained by direct observation or mathematical calculations.

Quality Assurance

The concept of quality assurance can be successfully applied to both production and servicing operations. It is concerned with the delivery of a product and/or process with minimal defects and errors. To assure product and service quality, it is essential that both inspection and other quality control measures be used.

Inspection is the most basic method for controlling quality. The inspection process may require that a 100% check be made on all items, or that an acceptable sample of items be used. The highest level of inspection is the 100% check, but it is also the most expensive, since every item must be checked. However, a small sample, while being more economical, may be inadequate to assure a given level of quality. Therefore, a balance must be struck between the two. Presented in Figure 11-2 are the characteristics of a sampling curve that shows the probability of finding a defect or error using various sampling percentages.

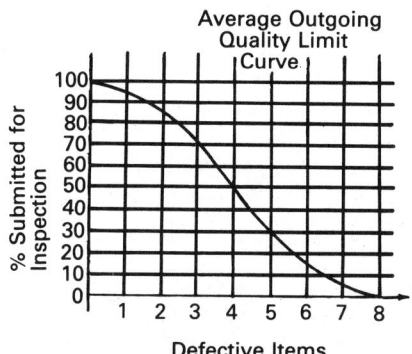

Fig. 11-2. Characteristics of sampling inspection and defective items generated for a hypothetical situation.

330 • MANAGEMENT CONTROLS

When a process is "under control," it is said to be in a state of statistical equilibrium. By this we mean that the end product can be characterized or described within a given probability (i.e., how many parts will probably be defective). Critical here is for the engineer, technologist, or technician to detect when there is a lack of control, and to make the appropriate corrections.

A tool that is frequently used for statistical analysis is the *control chart*. In most cases, the control chart is used for high-volume quality control. When the chart is used, control limits are established in terms of upper and lower limits. To establish these limits, the following formulas are used:

$$UCL = p' + 3\sqrt{p'(1-p')/n}$$
$$LCL = p' - 3\sqrt{p'(1-p')/n}.$$

In these formulas, UCL = the upper control limit, LCL = the lower control limit, and p' is a mathematical expression that is calculated by the formula:

$$p' = \frac{n_R}{nn_T}$$

where n_R = the total number of rejects detected in all lots inspected, n = average number of units in each inspected lot, and n_T = total number of lots inspected.

For example, a company has gathered data for 100 lots with an average of 300 units inspected per lot and a total number of 90 rejects for all inspected lots. To compute the upper and lower control limits, the following procedures are followed:

$$p' = \frac{n_R}{nn_T} = \frac{900}{(300)(100)} = 0.03$$

$$\begin{aligned}
UCL &= p' + 3\sqrt{p'(1-p')/n} \\
&= 0.03 + (3)(\sqrt{0.03(1-0.03)/300}) \\
&= 0.03 + 0.029\ 5465 \\
&= 0.059\ 5465
\end{aligned}$$

$$\begin{aligned}
LCL &= p' - 3\sqrt{p'(1-p')/n} \\
&= 0.03 - (3)(\sqrt{0.03(1-0.03)/300}) \\
&= 0.03 - 0.029\ 5465 \\
&= 0.000\ 4535
\end{aligned}$$

Tools frequently used in sampling inspection are sampling inspection tables, of which the Dodge-Romig tables are the most common (see: H. F. Dodge and H.G. Romig, *Sampling Inspection Tables* [New York: John Wiley and Sons, Inc.]). The Dodge-Romig tables were originated in the Bell Telephone Laboratories for use within the Bell Telephone System. They were designed to minimize inspection costs and sustain product quality. In all there are four sets of tables:

1. Single-sampling lot tolerance tables
2. Double-sampling lot tolerance tables
3. Single-sampling AOQL tables
4. Double-sampling AOQL tables

The initials $AOQL$ stand for "average outgoing quality limits." Table 11-3 is an example of a Dodge-Romig tolerance table. In this table, n = the sample size and c = the acceptance number of defects.

Table 11-3. Example of Partial Dodge-Romig Single-Sampling Lot Tolerance Tables

	Process Average (%)					
	0–0.05			0.06–0.05		
Lot Size	n	c	AOQL %	N	C	AOQL %
1–30	all	0	0	all	0	0
31–50	30	0	0.49	30	0	0.49
51–100	37	0	0.63	37	0	0.63
101–200	40	0	0.74	40	0	0.74
.						
.						
1001–2000	45	0	0.80	75	1	1.0
2001–3000	75	1	1.1	105	2	1.3
3001–4000	75	1	1.1	105	2	1.3
.						
.						
20001–50000	75	1	1.1	135	3	1.4
50001–100000	75	1	1.1	160	4	1.6

Exercises

12. A manufacturing company has decided to incorporate a control chart for a high-volume product. As a result, it must establish the upper and lower control limits for production. If it has recorded data for 40 lots with an average of 250 units per lot, what would the limits be if there were a total of 500 rejects?

13. Calculate the limits for the company in problem 12 if there were a total of 240 rejects.

Manufacturing Economics

In many firms, profits are not maximized because of insufficient analysis and the inability or unwillingness of management to change. Three common elements have been identified as causes for poor profits:

1. Too much effort is spent on increasing sales volume, maintaining market shares, and widening product offers rather than maximizing the present worth of long-run profits.
2. Technical specialists tend to strive for personal recognition and excessive levels of quality that are not cost-effective.
3. Managers tend to be overcautious, unimaginative, and afraid to make mistakes due to the lack of sufficient rewards.

Several techniques are available for analyzing the manufacturing economics of a firm. The first, and perhaps most important, is known as *break-even analysis*. Basically, this is an evaluation of the relationship between total cost and total revenue, and is frequently presented in chart form (Figure 11-3).

A formula that is frequently used to determine business profits is the business volume break-even point. The formula used here is:

$$BEP = \frac{c_F}{1 - c_V/s}$$

where BEP = the break-even point, c_F = fixed costs, c_V = variable costs, and s = sales. This formula is used to identify relationships between revenue and costs relative to production rates.

Consider a business whose sales volume for the year has been $5.5 million. The fixed costs were $0.5 million and variable costs were $1.12 million. To find its break-even point, the following procedures are used:

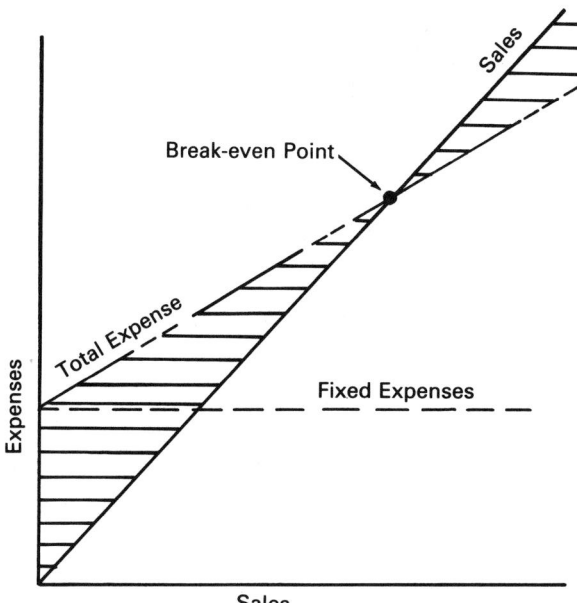

Fig. 11-3. Determining the break-even point.

$$BEP = \frac{c_F}{1 - c_V/s}$$
$$= \frac{500{,}000}{1 - 1{,}120{,}000/5{,}500{,}000}$$
$$= \$627{,}853.88$$

Another calculation essential to manufacturing economics is the contribution ratio. This ratio not only gives insight into the profits generated by businesses, but is also used to determine levels of profits. The basic formula used for finding the contribution ratio is:

$$CR = s - c_V$$

where CR = the contribution ratio and all other notations are the same as above. In addition, sales are calculated by:

$$s = c_F + c_V + p$$

The only new notation here is the profit (p) quantity, which is found with the following formula:

$$p = CR - c_F$$

An example of how these formulas are used would be in finding the contribution ratio and profits for a company whose fixed costs were $56,000 and which had a variable cost of 40% of sales, which totaled $435,000. The calculations made are as follows:

$$CR = s - c_V$$
$$= 435{,}000 - 174{,}000$$
$$= \$261{,}000$$

$$p = CR - c_F$$
$$= 261{,}000 - 56{,}000$$
$$= \$205{,}000$$

The last major factor frequently considered in manufacturing economics deals with replacement costs of units. A common method used here is the annual cost method, which calculates the annual cost of capital recovery (ACCR). To find ACCR, the following formula is used:

$$ACCR = (p - L)CRF + Li$$

where p = first cost, L = salvage value, CRF = capital recovery factor, and i = interest rate. The capital recovery factor can be obtained from standardized tables that present the factor for various interest rates over given time periods (n = life years). Table 11-4 is an abbreviated CRF table.

Consider a business that is considering purchasing for $150,000 a new piece of equipment that has an estimated life period of 20 years with a salvage value of $21,000. If the purchase rate can be obtained at a nominal interest rate of 12%, the annual cost of capital recovery would be calculated as follows:

$$ACCR = (p - L)CRF + Li$$
$$= (150{,}000 - 21{,}000)(0.14022) + (21{,}000)(0.12)$$
$$= \$20{,}608.38$$

Exercises

14. Find the break-even point for a business that has a sales volume of $1.25 million, fixed costs of $210,000, and variable costs of $850,000. If the firm can cut variable costs by 12%, what will its break-even point be then?

Table 11-4. Capital Recovery Factor of 1 at Continuously Compounded Nominal Interest Rates to 21 Years

Life Years (n)	Nominal Interest Rates (%)						
	6	8	10	12	14	16	18
1	1.06183	1.08329	1.10517	1.12750	1.15027	1.17351	1.19722
2	0.54684	0.56330	0.58019	0.59753	0.61533	0.63360	0.65234
3	0.37538	0.39034	0.40578	0.42172	0.43818	0.45515	0.47266
4	0.28981	0.30413	0.31901	0.33445	0.35046	0.36706	0.38425
5	0.23858	0.25263	0.26729	0.28258	0.29851	0.31509	0.33233
6	0.20454	0.21848	0.23310	0.24841	0.26443	0.28117	0.29863
7	0.18031	0.19424	0.20892	0.22435	0.24056	0.25754	0.27531
8	0.16221	0.17619	0.19099	0.20660	0.22305	0.24033	0.25845
9	0.14820	0.16227	0.17723	0.19306	0.20978	0.22738	0.24588
10	0.13705	0.15125	0.16638	0.18245	0.19946	0.21740	0.23627
11	0.12799	0.14232	0.15765	0.17397	0.19128	0.20957	0.22881
12	0.12048	0.13496	0.15050	0.16708	0.18470	0.20332	0.22293
13	0.11418	0.12882	0.14457	0.16146	0.17932	0.19828	0.21824
14	0.10881	0.12362	0.13959	0.15670	0.17491	0.19418	0.21447
15	0.10420	0.11918	0.13538	0.15275	0.17124	0.19082	0.21143
16	0.10020	0.11536	0.13178	0.14940	0.16818	0.18805	0.20895
17	0.09671	0.11204	0.12868	0.14655	0.16560	0.18575	0.20692
18	0.09363	0.10915	0.12600	0.14412	0.16342	0.18383	0.20526
19	0.09091	0.10660	0.12367	0.14202	0.16158	0.18223	0.20389
20	0.08849	0.10436	0.12163	0.14022	0.16000	0.18088	0.20276
21	0.08632	0.10237	0.11985	0.13865	0.15866	0.17975	0.20182

15. What is the contribution ratio for a firm with $250,000 of fixed costs, variable costs of $585,000, and total sales of $1,000,000?

16. Find the profit for the firm in problem 15.

17. Consider that a unit has a salvage value of $16,000 and, if overhauled, will have another five life years, at which time its salvage value will be $5000. The overhauling cost will be $40,000 at an interest rate of 10%. Calculate the actual cost of capital recovery for this unit.

18. If the firm in problem 17 purchases a new unit at an interest rate of 10%, it will cost $100,000 and have an estimated life period of 20 years, at which time its salvage value will be $15,000. Compute the alternate cost.

Budgetary Cost Control

The fundamental measure of a successful business is its profits. Business profits fall into three broad theoretical groups:

1. Rewards for taking risks and uncertainties
2. Imperfections in the economy that allow for monopolies
3. Rewards for successful innovations

Business budgeting is often a complex and difficult process that is designed to match prospective investments with financial resources and management abilities and is considered to be one of the most important processes for management decision making. This is because the level of profitability is dependent on the selection and implementation of investments. These investments not only include financial factors but also account for investments in personnel, equipment, materials, and procedures.

An important piece of information that is needed for the budgeting process is to find the profit-to-sales ratio (p/s), which is usually expressed in percentages. Here, profits are calculated as the difference between sales and expenses. Expenses are calculated by summing material (m) and labor and other costs (e). These figures can be expressed in the following expressions:

$$p = s - (m + e)$$

Consider a firm that must find its profit-to-sales ratio for budgetary decision making. If its sales totaled \$750,000, material costs were \$320,000, labor cost \$125,000, and other expenses were \$34,000, the following procedures would be used:

- $p = s - (m + e)$
  $= 750,000 - (320,000 + 125,000 + 34,000)$
  $= \$271,000$
  $\dfrac{p}{s} = \dfrac{271,000}{750,000}$
  $= 36.133\%$

Exercises

19. Compute the profits and profit-to-sales ratio for Iota Corporation if its sales for the first quarter of the year were \$64,000,

material costs were $12,400, labor costs were $18,600, and rental and warehousing costs were $4600.

20. The Mech Company had annual sales amounting to $2,400,000. Its expenses included: labor = $800,000, material costs = $700,000, and other expenses = $125,000. What are the company's profits and profit-to-sales ratio?

CHAPTER 12

Economics in Machining

- **Machining Costs**
- **Cost and Production Rate Equations**
- **Formulas for Optimization**

The profitability of many companies can be enhanced by improving the economics in machining. Tool and manufacturing engineering is one of the major functions of production industries that give the necessary services for producing goods and enabling the organization to operate smoothly. Machining operations, then, can be either production and/or maintenance oriented.

When machining a part, all engineering specifications must be satisfied: surface finish, dimensional accuracy, and surface integrity. The economics of machining is directly related to the given specifications and the nature of the material to be machined. For example, aluminum alloys and plain carbon-steel can be turned more economically than tantalum alloys or nickel-base high-temperature alloys. This chapter will discuss and present those factors and formulas that are used to describe the economics of machining.

Machining Costs

Machining costs are significantly affected by the nature of the materials to be machined. Table 12-1 shows the comparative costs of lathe turning different materials. (The machining costs factor is a

Table 12-1. Relative Lathe Turning Costs for Various Alloys (Aluminum Alloy as a Base 1)

Material	Lathe Turning Cost Basis
Aluminum alloy	1
Plain carbon-steel	3
Stainless steel (martensitic)	4.5
Low-alloy steel	5.5
Stainless steel	6.25
PH stainless steel	8
Titanium alloy	8.75–13
Low-alloy steel	11.25
Molybdenum alloy	11.25
Iron-base high-temperature alloy	13.5
Cobalt-base high-temperature alloy	15.25
Nickel-base high-temperature alloy	19.75–39
Tantalum alloy	22.25

comparison of costs rather than exact dollar amount. For example, using aluminum alloy as a base 1, plain carbon-steel will be approximately three times as expensive to machine.) Here, the relative cost of producing a part can vary greatly from one alloy to the next.

The cost of machining is also influenced by the type of surface finish specified. As in the case of material type, the cost of machining will vary according to the type of operation and finish required. The cost increase will naturally be influenced by the need for additional finishing processes or operations such as grinding, buffing, or honing. Table 12-2 is a comparison of finishing costs.

In addition, the overall costs associated with machining are also affected by the level of tolerance required. The relationship between surface finish and tolerance is close, since the close control of tolerance will usually require finer surface finishes.

Total machining costs is made up of a number of factors. These include:

340 • ECONOMICS IN MACHINING

Table 12-2. Machining Cost Increase for Surface Finish (With Cast, Sawed, or Other Rough Finishing as Basis)

Machining Cost increase (%)	Operation	Finish (micro inches)
0	Cast, sawed, etc.	2000
22	Rough turning	1000
31	Rough turning	500
38	Rough turning	250
62	Semifinish turning	125
100	Finish turning	63
240	Grinding	32
400	Honing	16

1. Machine tool operation
2. Cutter purchase or manufacture
3. Cutter reconditioning and maintenance
4. Idle time
5. Tool changing
6. Machining

Idle costs will often remain constant as changes in cutting speeds occur. Machining costs, however, will change with increases and decreases in speed, while tooling reconditioning and maintenance will increase as machining speed increases. The total cost of machining will be derived as the total sum of all these factors.

Cost and Production Rate Equations

Before it is possible to present the various cost and production rate equations used for machining processes, it is first necessary to identify the factors used, their symbols, and applications. It is important here to note that not all factors are applicable to all operations. Therefore, one must not only be aware of formulas and symbols but also which symbols have relevancy to individual processes. A list of symbols, with their meanings and applications, is presented in Table 12-3.

Operation Time and Production Rate

Two important measures used in describing the economics of machining operations are operation time per piece and production rate. Both measures are used as input in determining machining and production costs. Presented in this section are formulas used for turning, milling, drilling, reaming, tapping, and center drilling or chamfering operations.

OPERATION TIME PER PIECE

In all cases, operation time per piece is calculated by finding the sum of four basic factors:

1. Feeding time
2. Rapid traverse time
3. Cutting index time
4. Dull-tool replacement time

To calculate the operation time per piece for turning operations, the following formulas are used (note that the formulas presented here are the sum of four algebraic quantities; in each formula, they correspond to the four factors just identified):

$$t_M = \frac{D(L+e)}{3.82 f_R v} + \frac{R}{r} + t_I + \frac{DLt_D}{3.82 f_R v T}$$

To find the operation time per piece for milling operations, the following formula should be used:

$$t_M = \frac{D(L+e)}{3.82 Z f_R v} + \frac{R}{r} + t_I + \frac{Lt_D}{T_T}$$

The formula used for calculating operation time per piece for both drilling and reaming operations is:

$$t_M = \frac{D(L+e)}{3.82 f_R v} + \frac{R}{r} + t_I + \frac{Lt_D}{T_T}$$

Tapping time per piece can be determined by using the following equation:

$$t_M = \frac{mD(L+e)}{191 v} + \frac{R}{r} + t_I + \frac{Ltd}{T_T}$$

342 • ECONOMICS IN MACHINING

Table 12-3. Cost and Productive Rate Equation Factors

Symbol	Process Application	Meaning
C	Turning Milling Drilling and reaming Tapping Center drilling	Cost for machining workpiece in dollars/workpiece
C_C	Turning Milling	Cost of each insert or inserted blade in dollars/insert
C_P	Turning Milling Drilling and reaming Tapping Center drilling	The purchase cost of tooling or cutter in dollars/cutter
C_W	Turning Milling	Cost of grinding wheel for the resharpening of the cutting tool in dollars/cutting tool
d	Turning Milling	Depth of cut
D	Turning Milling Drilling and reaming Tapping Center drilling	Diameter of workpiece
e	Turning Milling Drilling and reaming Tapping Center drilling	The extra travel at feed rate
f_R	Turning Drilling and reaming Center drilling	Feed per revolution
f_T	Milling	Feed per tooth
G	Turning Milling Drilling and reaming Tapping Center drilling	Tool recondition cost in dollars/minute—calculated as the sum of labor plus overhead
k_1	Turning Milling Drilling and reaming	Number of times that the cutting tool is resharpened before it is discarded

Table 12-3. (continued)

Symbol	Process Application	Meaning
	Tapping Center drilling	
k_2	Turning Milling	Number of times the cutting tool is resharpened before inserts or blades are inserted or reset.
k_3	Turning Milling	Number of times the cutting tool is resharpened before blades or inserts are discarded
L	Turning Milling Drilling and reaming Tapping Center drilling	Length of workpiece or sum of length of all holes in respective operation
m	Tapping	Threads per inch
M	Turning Milling Drilling and reaming Tapping Center drilling	Cost in dollars/minute in machine operation—calculated as sum of labor and overhead
n	Turning Milling Drilling and reaming Tapping	Tool life exponent (in Taylor's equation)
N_L	Turning Milling Drilling and reaming Trapping Center drilling	Total number of workpieces per lot
P	Turning Milling Drilling and reaming Tapping Center drilling	Production rate per hour in units or workpiece/hr
r	Turning Milling Drilling and reaming Tapping Center drilling	Rapid traverse rate

(continued)

344 • ECONOMICS IN MACHINING

Table 12-3. (*continued*)

Symbol	Process Application	Meaning
R	Turning Milling Drilling and reaming Tapping Center drilling	Total rapid traverse distance for the cutting tool on a given part
S	Turning	The reference cutting speed for a standard tool life of 1 min in ft/min
S_T	Milling Drilling and reaming Tapping	The reference cutting speed for standard tool life of 1 min. in ft/min
t_B	Turning Milling	Time to replace or reset cutting tool
t_D	Turning Milling Drilling and reaming Tapping Center drilling	Time to replace dull cutting tool in tool changer storage unit
t_I	Turning Milling Drilling and reaming Tapping Center drilling	Indexing time from one cutter to the next between operations
t_L	Turning Milling Drilling and reaming Tapping Center drilling	Loading and unloading time of workpiece
t_M	Turning Milling Drilling and reaming Tapping Center drilling	Average time to complete one operation. The operating time per piece
t_O	Turning Milling Drilling and reaming Tapping Center drilling	Setup time for machine tool for given operation

Table 12-3. (continued)

Symbol	Process Application	Meaning
t_P	Turning Milling Drilling and reaming Tapping Center drilling	Time to preset tooling away from machine area
t_S	Turning Milling Drilling and reaming Tapping Center drilling	Cutting tool resharpening time
T	Turning	Tool life, expressed in minutes to dull a sharpened tool
T_H	Center drilling	Number of holes per resharpening
T_T	Milling Drilling and reaming Tapping	Tool life, expressed in inches of travel of work or tool to dull cutting tool
u_C	Center drilling	Number of holes center drilled or chamfered in the workpiece
v	Turning Milling Drilling and reaming Tapping Center drilling	Cutting speed
w	Milling	Width of cut
Z	Milling Tapping	Number of teeth in cutting tool

The final formula presented here is for finding the operation time per piece for both drilling and chamfering operations:

$$t_M = \frac{D(L+e)}{3.82 f_R v} + \frac{R}{r} + t_I + \frac{u_C t_D}{T_H}$$

For example, these formulas are used to determine the operation time per piece for turning a workpiece that has a diameter of 4.50 in. and is 14.75 in. long. The feed rate is 0.2 in./rev, the cutting speed is 112.4 fpm, and there is a calculated extra travel of 0.750 in./revolution. The total rapid traverse distance for the tool is 1.75 in. with a rapid traverse rate of 4 in./min, and time required to index between cutters is 2.5 min. If the tool life is 270 min and the time needed to replace the dull cutting tool is 12 min, then the operation time per piece should be calculated as follows:

$$\begin{aligned}t_M &= \frac{D(L+e)}{3.82 f_R v} + \frac{R}{r} + t_I + \frac{DLt_D}{3.82 f_R v T}\\ &= \frac{4.5(14.75 + 0.750)}{(3.82)(0.2)(112.4)} + \left(\frac{1.75}{4}\right) + 2.5 + \frac{(4.5)(14.75)(12)}{(3.82)(0.2)(112.4)(270)}\\ &= 0.5800 + 0.4375 + 2.5 + 0.0344\\ &= 3.5519 \text{ min or}\\ &= 3 \text{ min } 33.114 \text{ sec}\end{aligned}$$

PRODUCTION RATE

The formula used for calculating production rate is the same for all operations. This formula is:

$$P = \frac{60}{\Sigma t_M + t_L + (t_O/N_L)}$$

This formula is a proportion of one hour (in minutes) to the sum of times to load and complete an operation on the workpiece, and machine tool setup time per total workpieces. An example of how this formula is used is to find the production rate for a lot containing 20 workpieces if the times needed to complete individual operations are 2.2, 3.4, and 5.7 min, respectively, with 4.5 min for loading and unloading the workpiece, and the setup time for the machine tool is 1.75 min. The calculations made here are:

$$\begin{aligned}P &= \frac{60}{\Sigma t_M + t_L + (t_O/N_L)}\\ &= \frac{60}{11.3 + 4.5 + (1.75/20)}\\ &= \frac{60}{15.8875}\\ &= 3.78 \text{ workpieces per hour}\end{aligned}$$

Exercises

1. The feeding and tool replacement times for a product are 2.75 and 7.35 min, respectively. Assuming that these times will be the same for all machining operations, calculate the operation time per piece for each of the data collected on the following machining processes:

Operation	Rapid Traverse Total Distance (min)	Rapid Traverse Rate (in./mm)	Time to Index for Cutters (min)
Turning	0.75	0.012	1.00
Milling	0.95	0.003	1.75
Chamfering	1.12	2.185	3.50

2. What would be the feeding time for a milled workpiece with the following specifications: 18.750 in. long, 0.750 in. of excess cutter travel, using a cutter 0.500 in. in diameter and with 18 teeth, a cutter speed of 12 in./min, and a feed rate per tooth of 0.0021 in.

3. Calculate the dull-tool replacement time for the product in problem 2 if it took 12 min to replace the dull cutter in the storage unit. Assume that the cutter has a total of six flutes and tool life of 2400 in.

4. Using the information given in problems 1 through 3, calculate the operation time per piece for the milled product.

5. Find the production rate for the following product that has 14 units per lot. (*Hint:* Use the average times for loading and unloading time, and machine tool setup time):

Operation Times	Loading and Unloading Times (all times in min)	Machine Tool Setup Times
1.25	3.5	4.0
2.2	3.5	4.0
2.01	3.0	5.1
1.75	0.0	0.5
4.55	3.2	4.6

Machining Costs

Machining cost estimates are attempts to determine expenses involved in machining operations or manufacturing. The actual factors used to determine this amount will differ from one situation to the next. Generalized machining cost estimates employ as part of their calculations the operation time per piece multiplied by a given dollar-per-minute amount, plus other cost factors. The dollar-per-minute factor will vary from business to business, and is influenced by a number of factors such as wages, material costs, indirect costs, and utility cost.

The first part of machining cost calculations is the cost of operating time per piece. The second half of the formulas will involve the totals of indirect costs. These indirect costs account for the following factors involved with machining:

1. Tool depreciation cost
2. Tool resharpening cost
3. Rebrazing or blade reset cost
4. Insert or blade cost
5. Grinding wheel cost

Generalized machining costs for both turning and milling take into account all these factors. The formulas used for turning and milling are presented in order as follows:

$$C = Mt_M + \left[\frac{DL}{3.82 f_R vT}\right]\left[\frac{C_P}{K_1 + 1} + Gt_s + \frac{Gt_B}{k_2} + \frac{C_C}{k_3} + C_W + Gt_P\right]$$

$$C = Mt_M + \left[\frac{L}{ZT_T}\right]\left[\frac{C_P}{k_2 + 1} + Gt_S + \frac{Gt_B}{k_2} + \frac{ZC_C}{k_3} + C_W + Gt_P\right]$$

The formula used to calculate machining costs for drilling, reaming, and tapping operations is the same in each case. This formula is:

$$C = Mt_M + \left[\frac{L}{T_T}\right]\left[\frac{C_P}{K_1 + 1} + Gt_S + Gt_P\right]$$

Costs for center drilling and chamfering operations employ the following formula:

$$C = M_{TM} + \left[\frac{u_C}{T_H}\right]\left[\frac{C_P}{K_1 + 1} + Gt_S + Gt_P\right]$$

Handling and setup costs are calculated by multiplying the dollar-per-minute factor by the sum of the loading and unloading time plus setup time. This formula is:

$$C = Mt_L + \frac{t_O}{N_L}$$

Assume that the operation time per piece for a chamfered workpiece is 2.5 min on average. Assume further that there are 8 holes to be chamfered, that the cutting tool is able to chamfer 880 holes before it requires resharpening and can be resharpened 50 times before discarding, and that it requires 3.25 min to sharpen the tool and 12.5 min to preset it away from the machine. If the purchase cost of the cutting tool is $45.00 and labor and overhead for machine operation and reconditioning are $1.12 and $0.90 per min respectively, the machining cost for this operation would be found by the following calculations:

$$C = Mt_M + \left[\frac{u_C}{T_H}\right]\left[\frac{C_P}{K_1+1} + Gt_S + Gt_P\right]$$
$$= (1.12)(2.5) + \left[\frac{8}{880}\right]\left[\frac{45}{50+1} + (0.90)(3.25) + (0.90)(12.5)\right]$$
$$= 2.8 + 0.13689$$
$$= \$2.93689 \text{ or } \$2.94$$

Exercises

6. Calculate the cost of a tapping operation with an operation time per piece of 2.45 min, and with the following specifications: labor and overhead for machine operation is $54.00/hr and for the machine tool maintenance department is $68.00/hr, length of the tap is 7.54 in., tool life is calculated at 48 in. of travel, cost of the tap is $24.00, the tap usually requires 5 min per tool for sharpening and 1 min for presetting, and can be resharpened 20 times before it must be totally replaced.

7. Find the handling and setup costs for the following product: labor and overhead involved is $28.00/hr, all lots have 18 units, the time required to unload and load a unit is 2 min, and the unit setup time on the processing machine is 24 min.

Formulas for Optimization

Optimized cutting speeds and tool life data are important for the economical efficiency of an operation. As might be expected, the term *optimized* pertains to the best or most effective conditions that can be expected given a set of circumstances. The data generated by these formulas establish machining and maintenance goals, *which assume that optimum cutting conditions are beyond available data.* Therefore, it must be realized that optimum cutting speeds and tool life are extremely difficult to maintain over the span of production.

Optimization formulas are usually broken down into two broad areas: minimum-cost formulas and maximum-production formulas. The production engineer always attempts to achieve minimum cost and maximum productivity.

The procedures required to calculate all data for these formulas are beyond the context of this book. What will be presented here, however, is a listing of those formulas as a reference source only. Use of factors derived from Taylor's formula should be obtained from engineering reference materials, or by the formula itself.

The most common relationship found for Taylor's equation is:

$$S = Vt^n$$

The exponent n is dependent upon the characteristics of the workpiece, and usually falls within the ranges given in Table 12-4. When the V is plotted as a function of T on log-log paper, a straight line results (see Figure 12-1), the characteristics of which depend upon a given set of conditions that vary from one situation to the next. The value of n can then be found by using the formula:

$$n = \frac{\log V_2 - \log V_1}{\log T_2 - \log T_1}$$

Table 12-4. General Range of Values for *n* in Taylor's Equation

Type of Cutting Tool	*n*
High-speed steel	0.08–0.12
Carbide	0.13–0.25
Ceramic	0.40–0.55

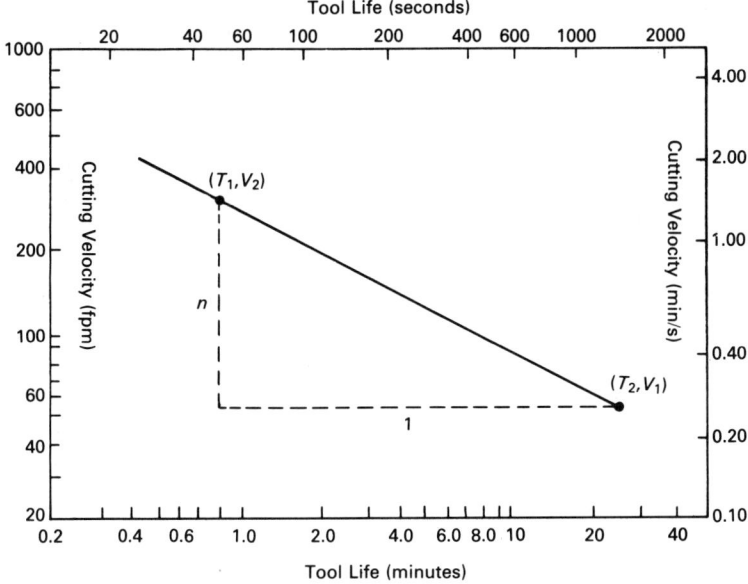

Fig. 12-1. Effect of cutting speed on tool life of cutting tool.

Optimized Cutting Speeds

The formulas used for calculating optimized cutting speeds at minimum cost ($v_{\text{MIN COST}}$) at maximum productivity ($v_{\text{MAX PROD}}$) are presented in this section. Note that the $v_{\text{MIN COST}}$ is specified in terms of dollars per unit and $v_{\text{MAX PROD}}$ is in feet-per-minute cutting speed.

TURNING

The formulas used for turning operations are:

$$v_{\text{MIN COST}} = S\left[\frac{nM(L+e)}{L(1-n)(Mt_D + \dfrac{C_P}{k_1+1} + Gt_S + \dfrac{Gt_B}{k_2} + \dfrac{C_C}{k_3} + C_W + Gt_P)}\right]^n$$

$$v_{\text{MAX PROD}} = S\left[\frac{n(L+e)}{Lt_D(1-n)}\right]^n$$

MILLING

The two formulas used for calculations in milling operations are:

$$v_{\text{MIN COST}} = [S_T]^{1/(n+1)} \left[\frac{nMD(L+e)}{3.82 f_T L (Mt_D + \dfrac{C_P}{k_1+1} + Gt_S + \dfrac{Gt_B}{k_2} + \dfrac{ZC_C}{k_3} + C_W + Gt_P)} \right]^{n/(n+1)}$$

$$v_{\text{MAX PROD}} = [S_T]^{1/(n+1)} \left[\frac{nD(L+e)}{3.82 f_T L t_D} \right]^{n/(n+1)}$$

DRILLING AND REAMING

The formulas used for drilling and reaming operations are:

$$v_{\text{MIN COST}} = [S_T]^{1/(n+1)} \left[\frac{mMD(L+e)}{3.82 f_R L (Mt_D + \dfrac{C_P}{k_1+1} + Gt_S + Gt_P)} \right]^{n/(n+1)}$$

$$v_{\text{MAX PROD}} = [S_T]^{1/(n+1)} \left[\frac{nD(L+e)}{3.82 f_R L t_D} \right]^{n/(n+1)}$$

TAPPING

The two basic formulas used for optimized speeds for tapping are:

$$v_{\text{MIN COST}} = [S_T]^{1/(n+1)} \left[\frac{mnMD(L+e)}{1.19 L (Mt_D + \dfrac{C_P}{k_1+1} + Gt_S + Gt_P)} \right]^{n/(n+1)}$$

$$v_{\text{MAX PROD}} = [S_T]^{1/(n+1)} \left[\frac{mnD(L+e)}{1.19 L t_D} \right]^{n/(n+1)}$$

Optimized Tool Life

Measures used to express optimized tool life are expressed in terms of time (T and T_T). There are two general formulas used here—for turning, and for milling, drilling, reaming, and tapping.

TURNING

The optimized-tool-life formulas generally used for turning operations are:

$$T_{\text{MIN COST}} = \left(\frac{S}{v_{\text{MIN COST}}}\right)^{1/n}$$

$$T_{\text{MAX PROD}} = \left(\frac{S}{v_{\text{MAX PROD}}}\right)^{1/n}$$

MILLING, DRILLING, REAMING, AND TAPPING

The two formulas applicable to other machining operations are presented as follows:

$$T_{T_{\text{MIN COST}}} = \left(\frac{S_T}{v_{\text{MAX PROD}}}\right)^{1/n}$$

$$T_{T_{\text{MIN COST}}} = \left(\frac{S_T}{v_{\text{MAX PROD}}}\right)^{1/n}$$

CHAPTER 13

Facility and Human Resources Management

- **Physical Facilities**
- **Human Resources Management**

There are two areas of management that are important to the successful operation of a business: physical facilities and human resources. Physical facility management pertains to the physical plant and its operation, and human resources management encompasses all functions dealing with personnel and staffing. This chapter will present formulas and calculations used in both facility and human resources management practices.

Physical Facilities

The modernization of today's businesses and industries results in more complex and precise planning, maintenance, and management. These facilities will have a number of large and sophisticated mechanical devices, computers, buildings, and systems that have to be integrated for achieving a single goal or purpose. Managers, then, must have a wide range of knowledge as to how these facilities are located, operated, and maintained.

The problem of locating a facility always arises whenever a business starts up or anticipates expansion. When selecting a site, there are generally two alternatives: to locate or expand at present locations or to locate at different locations. The best location will allow the company to produce and distribute, or service, its products with the highest margin of profit. There are six basic factors that must be considered when making these decisions:

1. Availability and transporting of raw materials or components.
2. Labor supply in the area.
3. Marketing.
4. Services available to the business.
5. Climatic conditions
6. Local and state laws, codes, and ordinances.

Plant Layout

An important phase of facility planning is plant layout. This procedure usually consists of a number of distinct problems. The first is to determine the amount of floor space required for a particular department. To accomplish this, a simple formula is available and applicable to most situations:

$$D = N_{WC} W_{CA}$$

Here, D = the floor area requirements for a given department, N_{WC} = the number of work centers or workstations, and W_{CA} = the total area requirement for the work center. Furthermore:

$$W_{CA} = P + S + A + B$$

where P = area required for the prime center, such as machinery, storage racks or tanks, tools, and cutters; S = supporting areas, which include the area needed by personnel to work, any space needed for incoming or outgoing materials, and operating areas; A = aisle areas between equipment and machinery (this includes both main and service aisles); B = facility structural areas, such as spaces occupied by columns, posts, and beams.

Consider a servicing business that has 54 work centers that are made up of two primary workstations of 64 and 82 ft^2. If the supporting and aisle areas require 7.5% and 32%, respectively, of the major area and the structural areas are fixed at 1.28 and 0.82 ft^2, the amount of total floor area required for the business is:

$$W_{CA} = P + S + A + B$$
$$= (64 + 82) + (4.8 + 6.15) + (20.48 + 26.24) + (1.28 + .82)$$
$$= 146 + 10.95 + 46.72 + 2.1$$
$$= 205.77 \text{ ft}^2$$

$$D = N_{WC} W_{CA}$$
$$= (54)(205.77)$$
$$= 11,111.58 \text{ ft}^2$$

When designing the general construction outline, or "footprint," of the building, a number of shapes and configurations can be used. The most basic shape is the rectangle. Critical here is to calculate the cost per running length of the building's outside wall. The formula used to find the optimized shape is:

$$B_{WL} = 2L + 2W$$

where B_{WL} = the total length of the outside building wall, L = longest building wall, and W = shortest building wall. It is assumed that the area of the building will be equal to the product of L and W.

An example of how this formula would be used is to find the total linear length of the outside building wall for three proposed designs totalling 160,000 ft². These designs are:

$$800 \times 200 = 160,000 \text{ ft}^2$$
$$400 \times 400 = 160,000 \text{ ft}^2$$
$$500 \times 320 = 160,000 \text{ ft}^2$$

$$B_{WL} = 2L + 2W$$
$$= (2)(800) + (2)(200)$$
$$= 2000 \text{ ft}$$

$$= (2)(400) + (2)(400)$$
$$= 1600 \text{ ft}$$

$$= (2)(500) + (2)(320)$$
$$= 1640 \text{ ft}$$

With these findings, it will be possible to estimate the cost for each building based on the linear length of its outside walls. Of course, other factors will effect the cost of the building, but this will give planners a good idea of building costs and provide information upon which other factors can be considered.

The last formula to be considered here is used to determine

space utilization. Space utilization pertains to the percentage of total space being used for production, storage, or other activities. Such utilization is based on occupied floor space and building height or usable vertical space. There are two basic formulas used here:

$$V = \frac{100A}{P}$$
$$R = SV$$

In these formulas, V = the percentage of vertical space utilization, A = the present (average) storage height, P = the potential (average) storage height, R = the floor space requirements, and S = the actual floor area occupied.

Consider a company that needs to determine the level of utilization of warehouse space. If the existing floor space being occupied is 260,000 ft^2, and it is stacking units an average of 16 of a potential 22 ft, the vertical space utilization effectiveness and floor space requirements are found as follows:

$$V = \frac{100A}{P}$$
$$= \frac{(100)(16)}{22}$$
$$= 72.73\%$$
$$R = SV$$
$$= (260,000)(0.7273)$$
$$= 189,098 \text{ ft}^2$$

Exercises

1. The V. A. Pace Company had recently decided to expand its present facilities and must determine the floor space requirements for its components department, which has 150 work centers. If each prime center area needs 60 ft^2, the supporting area is 5% of the prime center area, aisle area required is 21 ft^2, and the building structural area is 0.60 ft^2, calculate the total floor area needed.

2. A newly formed business has concluded that it needs a total of 360,000 ft^2 for materials handling, warehousing, and storage. The following combination dimensions for a building footprint

are under consideration: 600 ft × 600 ft, 900 ft × 400 ft, and 1200 ft × 300 ft. Calculate the total linear length of outside building wall for each design.
3. The Bells Corporation found that it is stacking its raw materials an average height of 12 ft, while the building has a stacking capacity of 14 ft. What is the vertical space utilization effectiveness?
4. What is the actual floor space requirements for the building in problem 3 if the floor area occupied is 183,000 ft^2?

Plant Services

There are two major plant service areas where calculations are commonly made: electricity, heating, water, and air. Each is critical to the operation and management of industrial processes. Considered here are formulas are basic to most operations.

ELECTRICITY

The basic notations used in electrical formulas are:

I = current measured in amperes
E = electromotive force (emf) in volts
R = resistance in units of ohms
W = electrical power in watts
HP = electrical horsepower.

The first formula to be considered is known as *Ohm's law*, which illustrates the mathematical relationship between resistance, current, and voltage. These relationships are:

$$E = IR$$
$$I = \frac{E}{R}$$
$$R = \frac{E}{I}$$

An example of how these relationships are used is to determine the current flow within a conductor where the emf is 220 volts and the resistance is 12 ohms. The solution to this problem would be:

$$I = \frac{E}{R}$$
$$= \frac{220}{12}$$
$$= 18.333 \text{ amperes}$$

Circuitry designs are critical in determining electrical characteristics. The first is the series connection, where current passes through a series of units in a circuit. The basic formula used here is:

$$E = I\Sigma R_S$$

where ΣR_S is the sum total resistance of the connections in ohms (i.e., $R_1 + R_2 + R_3 + \cdots R_S$). Consider an electrical circuit where four resistors are connected in series with values of 12, 8, 6, and 24 ohms, respectively. If 110 volts of emf are fed into the circuit, the current flow through the circuit would be calculated as:

$$I = \frac{E}{\Sigma R_S}$$
$$= \frac{110}{50}$$
$$= 2.2 \text{ amperes}$$

The second circuit design is the parallel connection. Here, current divides between units connected in a circuit. The basic formula used here is:

$$E = I\Sigma R_P$$

where ΣR_P = the total resistance in units of ohms represented as:

$$\frac{1}{\Sigma R_P} = \frac{1}{R_1} + \frac{1}{R_2} + \frac{1}{R_3} + \cdots \frac{1}{R_P}$$

A formula used for calculating electrical power in terms of wattage is:

$$W = EI$$

In most cases, wattage will be expressed in units of kilowatts, or kw, which is 1000 watts, or $w \times 10^3$. An example of such a problem is to find the electrical power required for a circuit that requires

880 volts and 1250 amperes of current. The following calculations are made:

$$W = EI$$
$$= (880)(1250)$$
$$= 1,100,000 \text{ or}$$
$$= 1100 \text{ kw}$$

The last formula to be considered is for determining electrical horsepower. The formula is:

$$HP = \frac{EI}{746} \text{ or}$$
$$HP = \frac{W}{746}$$

Consider a generator that must maintain 440 volts of electromotive force with a current of 120 amperes. To compute the electrical horsepower generated, the following procedures are used:

$$W = EI$$
$$= (440)(120)$$
$$= 52.8 \text{ kw}$$

$$HP = \frac{W}{746}$$
$$= \frac{52800}{746}$$
$$= 70.78 \text{ electrical horsepower}$$

HEAT, WATER, AND AIR

There are two primary calculations made in the areas of heat, water, and air services. The first is for heating and cooling, which incorporates the use of the *British thermal unit*, or BTU. Basically, 1 BTU is the amount of heat that is needed to raise the temperature of 1 lb of water 1°F. The formula used here is:

$$\text{BTU} = (sh)(W)(t_D)$$

where sh = the specific heat of the substance being heated or cooled, W = the weight of the substance in pounds, and t_D = the temperature difference (original temperature − final temperature) of state of substance. It should be noted that the specific heat of a substance is the amount of BTUs required to raise the temperature of 1 lb of that substance by 1°F.

A typical problem would be to compute the number of BTUs needed to raise the temperature of 150 lb of water 130°F (the specific heat of water is 1.0). The procedures used here are:

$$\text{BTU} = (sh)(W)(t_D)$$
$$= (1.0)(150)(130)$$
$$= 19,500 \text{ BTU}$$

The second calculation made for heat, water, and air services is *fuel consumption*. The formula used here is:

$$C = \frac{k}{e}$$

where C = the annual consumption given in fuel units, k = a constant computed in BTUs, and e = the effective fuel valued in BTUs/hr. Furthermore:

$$k = \frac{24hd}{65 - t_O}$$

where h = heating requirements in BTUs/hr, t_O = outside temperature design of the plant, d = degree-days (difference in temperature for any one day between 65°F and the mean or average temperature for that day).

To find the fuel consumption for a building using natural gas with an effective fuel value of 78,000 BTU s/hr., which is designed for an outside temperature of $-25°F$ when the heating requirements are 200,000 BTUs/hr., and is located in a geographic region where the average annual degree-days are 6500, the following procedures are used:

$$k = \frac{24hd}{65 - t_0}$$
$$= \frac{(24)(200,000)(6500)}{65 - (-25)}$$
$$= \frac{31,200,000,000}{90}$$
$$= 346,666,666.7 \text{ BTUs}$$

$$C = \frac{k}{e}$$
$$= \frac{346,666,666.7}{78,000}$$
$$= 4,444.44 \text{ BTUs/hr.}$$

Exercises

5. Find the resistance of a circuit with 110 volts and 1.1 amperes of current flow.

6. A 220-volt circuit has three resistors connected in series with values of 3, 5, and 8.5 ohms. What is the current that will flow through this circuit?

7. Conductors of 6 and 4.75 ohms are connected in parallel to a 2-volt battery. What is the total resistance and the current flow in this circuit?

8. If a 60-watt bulb is connected to an emf of 115 volts, what will be the current required to operate the bulb? Compute the electrical horsepower being generated.

9. An industrial generator maintains an electromotive force of 440 volts with a current usage of 95 amperes. How much electrical horsepower is being generated?

10. What is the amount of heat given off in cooling 50 lbs of copper, with a specific heat of 0.093, from 250° to 72°F?

11. How much heat is required to raise the temperature of 50 lb water, with a specific heat of 1.0, a total of 120°F?

12. An office building is designed for an outside temperature of −75°F, is located in a city where the average annual degree-days are 5800, and has a heating requirement of 150,000 BTUs/hr. Compute the annual consumption of fuel units for fuel oil with an effective fuel value of 100,000 BTUs/gal.

13. If the building in problem 12 were to convert over to total electricity with an effective fuel value of 3412 BTUs/kw.hr., what would the annual consumption of fuel units be?

Human Resources Management

Of all areas of responsibility assigned to management, one is unique: its task to integrate operations and systems with human beings. Managers must be able to bring together information and knowledge about people, their behavior, capabilities, motivations, and competencies to mesh effectively with industrial organizations. The field of human resources development evolved to address the needs of business and industrial organizations for effective and efficient human productivity.

To adequately discuss and present all aspects of human resources management would be beyond the scope and purpose of this book. This section will present formulas and computations most frequently used in the field of human resources management.

Work Measurement

There are two major concepts involved in work measurement. The first is known as *work standards*; the second is *manpower planning*. Work standards pertain to the measurement of work output that is expected from personnel relative to working conditions. Manpower planning, on the other hand, addresses the number of individuals required to work to meet production goals.

The first formula to be considered is for establishing the basic work standard (w_S). This is found by the summation of the standard basic time values (t_V) and base time allowances (t_B):

$$w_S = t_V + t_B$$

Standard basic time values pertain to functions that have to be performed to complete an operation. They must be recognized and distinguishable as a logical step or task in an operation. Time values assigned to these tasks must be reasonable and fair in terms of expected personnel performance.

Base time allowances are specified in terms of minutes per time period or shift. These allowances include such factors as worker fatigue, work breaks, lunch or dinner, and personal needs. This proportion, then, is usually expressed as a percentage figure and is calculated by using the following formula:

$$t_B = \left(\frac{b_T}{s - b_T}\right) 100$$

where s = shift time in minutes and b_T = the total sum of individual allowances in minutes/person/shift.

364 • FACILITY AND HUMAN RESOURCES MANAGEMENT

Assume that a company is analyzing its operation for the purpose of establishing a work standard. All personnel functions for this study are the same, and have been identified as follows:

Task	Time (min)
1. Layout	8.5
2. Bend hems	2.0
3. Bend one side to right angle	1.0
4. Fit fingers and bend	1.5
5. Bend end	1.0
Total	13.5 min/operation

Allowances	Time (min/shift)
1. Coffee breaks	30
2. Personal	20
3. Other	5
Total	55 min/shift

From this analysis, the standard basic time value for the operation is 13.5 min. To determine the percentage of basic time allowances and work standard, the following procedures are used:

$$t_B = \left(\frac{b_T}{s - b_T}\right) 100$$

$$= \left(\frac{55}{480 - 55}\right) 100$$

$$= 12.94\%$$

$$w_S = t_V + t_B$$

$$= t_V + (\% t_B)(t_V)$$

$$= 13.5 + (0.1294)(13.5)$$

$$= 13.5 + 1.7469$$

$$= 14.7469 \text{ min/operation or}$$

$$= 4.069 \text{ operations/hr}$$

The second work measurement formula to be presented deals with manpower planning. The process of identifying correct numbers of personnel required is critical to the cost-effectiveness and efficiency of a well-run organization. If too few workers are hired, then orders and requests cannot be met. If too many employees are on the payroll, then the cost per unit of production increases and reduces the level of profits.

The basic measure used in manpower planning is the *equivalent standard manpower requirement* (M_{ESR}), which is a proportion of productive standard hours (t_{PS}) to available productive hours (t_A). This relationship is shown in the formula:

$$M_{ESR} = \frac{t_{PS}}{t_A}$$

An example of how this formula is used might be for a company that must estimate the equivalent standard manpower requirements for the coming fiscal year given the following information:

1. Projected requirement of 42,000 productive standard hours for a basically repetitive operation.
2. Productive requirements will be fairly constant from month to month.
3. The average employee is estimated to have 2200 available productive hours after vacation time and other nonwork activities.

To solve this problem, the following calculations are made:

$$\begin{aligned} M_{ESR} &= \frac{t_{PS}}{t_A} \\ &= \frac{42000}{2200} \\ &= 19.1 \text{ workers.} \end{aligned}$$

Exercises

14. A company works 8-hour shifts and has two scheduled breaks of 20 minutes each, estimates that an additional 30 minutes per shift are required for personal needs, and figures that an-

other 5 minutes per shift is for minor interruptions and distractions. Calculate the work standard for the following two operations:

Task Number	Time (minutes)
Operation A	
1	1.0
2	3.5
3	4.0
4	10.0
5	2.0
Operation B	
1	3.0
2	15.0
3	24.0
4	4.5

15. A firm is attempting to compute the base time allowance percentage for a job that has a standard basic time value of 13 minutes per order. If it is estimated that 1 hour per 8-hour shift is spent for nonproductive activities, what is the percent base time allowance?

16. The ILS Corporation is estimating 560,000 productive standard hours for its plants in the forthcoming year. Most operations are repetitive, and production requirements are basically the same from month to month and plant to plant. According to its records, it is estimated that the average employee has 1920 available productive hours. What is the equivalent standard manpower requirements for the forthcoming year?

Personnel Evaluation

Most organizations have taken the position that personnel evaluation is necessary and a prerequisite to corporate profitability. The process of evaluating job performance is not totally objective, since many judgments must be made in assessing how well one does one's job. Some measures, or criteria, used to evaluate performance are easy to obtain, such as meeting sales quotas or total production output.

Performances dealing with attitudes and interaction with others are much more difficult to measure objectively.

Personnel specialists attempt to develop methods and procedures that will keep all evaluation as objective as possible. Furthermore, the analysis and interpretation of these data must also be accomplished in an objective manner. Presented in this section will be a discussion of calculations made in job evaluation and merit rating situations.

JOB EVALUATION

One of the most important tools used in job evaluation is the *evaluation instrument*. This tool is used by the evaluator to rate or grade job performance. These instruments should be developed by a group of trained professionals who are familiar with the tasks performed on a given job. Afterwards, the instrument can be refined with input from supervisors, foremen, and employees.

One of the most common job evaluation tools used is the job rating plan shown in Table 13-1. This is used to rate the performance of an individual for a specific job title.

It should be noted that this mathematical table is only an example and should not be adopted for any one situation. In fact, each company should design its own modified version of the job rating plan. With this rating scale, it will then be possible to calculate the point range per grade (p_{RG}) by using the following formula:

$$p_{RG} = \frac{p_{MAX} - p_{MIN}}{n}$$

Here, p_{MAX} = the maximum number of points it is possible to obtain in a job rating plan, p_{MIN} = the minimum number of points possible in the job rating plan, and n = number of grades in a job.

An example of how this formula is used is for a company that used the job rating plan in Table 13-1. If the company is thinking of slotting this job title into eight job grades using the results of the rating plan, the point ranges for each of these eight grades would be calculated as follows:

$$p_{RG} = \frac{p_{MAX} - p_{MIN}}{n}$$
$$= \frac{740 - 490}{8}$$
$$= 31.25 \text{ or } 31$$

Table 13-1. Job Rating Plan

Evaluation Factors	Rating (points)						Max. Pts. (%)
	1	2	3	4	5	6	
Technical Skills							
Education	30	28	25	24	22	20	
Training	60	62	65	66	68	70	
Experience	90	95	100	105	115	150	
Subtotal	180	185	190	195	205	240	32
Effort							
Mental	20	23	25	27	28	30	
Physical	70	75	80	85	88	90	
Subtotal	90	98	105	112	116	120	16
Responsibility							
Machinery	70	75	80	85	95	105	
Materials	20	25	30	35	40	45	
Servicing	30	35	40	45	48	50	
Supervision	0	5	10	12	15	30	
Subtotal	120	140	160	177	198	230	31
Work Conditions							
Environment	60	63	65	66	68	70	
Stress	30	35	40	45	50	60	
Hazards	10	12	15	17	19	20	
Subtotal	100	110	120	128	137	150	20
TOTAL	490	533	575	590	612	740	100

Job Grade	Point Range
1	490–521
2	522–553
3	554–585
4	586–617
5	618–649
6	650–681
7	682–713
8	714–740

Based on this table, if a person were to receive a job evaluation of 619 total points, his or her job grade would be 5.

MERIT RATING

When determining merit ratings, the point range per grade formula can again be used. Consider the job merit rating plan presented in Table 13-2.

To find the point range for merit ratings of A, B, and C (A being the highest rating), the following calculations are made:

$$p_{RG} = \frac{p_{MAX} - p_{MIN}}{n}$$
$$= \frac{435 - 325}{3}$$
$$= 36.67 \text{ or } 37$$

Grade	Merit Rating
A	401–435
B	363–400
C	325–362

Table 13-2. Merit Rating Plan

Factors	Rating (points)				Max Pts. (%)
	1	2	3	4	
Dependability	60	65	70	75	17
Quality of work	60	65	70	75	17
Quantity of work	60	65	70	75	17
Cooperation	50	55	60	65	15
Attitudes	40	45	50	60	14
Initiative	30	35	40	45	10
Punctuality	25	30	35	40	9
Total	325	360	395	435	100

Exercises

17. Given the information in Table 13-1, determine the job rating ranges for a company that wants to slot the job into 12 grades.

18. Using the information in Table 13-2, calculate the merit rating ranges for 5 grade levels.

Answers to Exercises

Chapter 1

1. 64
2. 65,536
3. 16,807
4. 3^2
5. $2^3/3^3$
6. 5^8
7. 0.000 003 6134
8. 4,740,000,000
9. 4
10. 4
11. 2
12. 4
13. 0
14. -1
15. -6
16. 2
17. $9.9014-10$
18. 3.0959
19. 1.3703
20. $7.0057-10$
21. 3.851
22. 19,140
23. 0.000 000 3352
24. 0.7300
25. 1.627
26. 114.6
27. -22.30
28. 5.407
29. $5xy$
30. $6x12b$
31. $5ay^2+3$
32. $\dfrac{x}{2y}$
33. $4z$
34. a^4
35. c^{AB}
36. 9
37. 1
38. $3x+5y+z$
39. $3a$
40. $2x+4y$
41. $2a^2+4a-6$
42. $16a^3-25a^2+21a-9$
43. $3z-3$

ANSWERS TO EXERCISES

Chapter 2

1. 33.0625 m² and 8.1305 m
2. 98 in.² and 9.8994 in.
3. 25.375 in.² and 8.051 in.
4. 15.75 cm
5. 192 ft.²
6. 56 in.² and 11.490 in.
7. (a) 499.55 yd.²
   (b) 875 mm²
8. (a) 906.5 in.²
   (b) 348.98 cm²
9. 147 dm²
10. 1190 m²
11. 210.438 in.², 2660.352 in.², and 9986.712 in.²
    9 in., 32 in., and 62 in.
12. 497.31 m² and 14.149 m
13. 1,012.97 cm²
14. 18.850 in., 25.133 in., and 36.568 in.
15. 12.56 m
16. 93.83 m²
17. 232.532° or 232°31′55.2″
18. 38.197° or 38°11′49.2″
19. 13.35 in.²
20. 0.00352 in.² and 1.317 ft²
21. 320.442 in.² and 17,492.373 cm²
22. (a) 154 in.²
    (b) 440 m2
23. 27 in.³, 13,824 in.³, and 6581.256 in.³
24. 14 ft.
25. 2539.95 cm³
26. 278,092.8 mm³
27. 18.651 in.³
28. 60.013 ft³
29. 163.245 cm³
30. 11,808.33 mm³ and 2,834,000 grams
31. 21,714 in.³ and 3,619.11 in.²
32. 3.654 m³ and 2.077 m
33. 3,335,332 ft.³
34. 1017.875 in.² and 3053.625 in.³
35. 9.94 m³
36. 961.327 in.² and 3562.56 in.³
37. 1959.39 cm³ and 12.97 cm

38. 0.0576 ft³, 902.016 m³, and 3,875.37 yd³
39. 90,075.67 cm³
40. 2.77 ft³ and 26.56 ft²
41. 10,096.60 mm² and 78,248.66 mm³

Chapter 3

1. 9.22 in.
2. 4.27 m
3. $31°19'43''$
4. $A = 22°32'45'', B = 65°27'15''$
5. $A = 71°36'5'', B = 18°23'55''$
6. $B = 62°49', a = 19.33$ cm, $b = 30.24$ cm
7. $A = 30°42'46'', B = 59°17'14'', c = 23.50$ cm
8. $a = 22.41$ in.
9. 148.671 m
10. $38°27'13''$
11. $\angle CHG = 36°52'12'', \angle HCG = 53°7'48''$
12. 1.47 mi
13. $76°12'30''$
14. 9.526 mm
15. 0.117 in.
16. $B = 68°, a = 3.54$ cm, $b = 3.69$ cm
17. $C = 32°50', B = 123°10', b = 12.35$ m
18. $A = 38°27', B = 116°6', b = 8.88$ ft.
19. $C = 90°, B = 60°, b = 4.24$ km
20. $B = 73°19', C = 39°41', a = 2.88$ yds.
21. $C = 50°6', D = 62°54', d = 2.90$ m
22. $A = 126°47', B = 29°13', a = 4.92$ in.
23. $F = 26°23', G = 36°20', H = 117°17'$
24. $Y = 99°19', x = 35.09$ in., $z = 24.47$ in.
25. $M = 43°56', N = 121°34', n = 26.78$ mm
26. $P = 25°13', Q = 106°36', R = 48°11'$
27. $H = 67°51', f = 2.91$ in., $g = 5.78$ in.
28. $Q = 49°10'20'', R = 63°18', p = 48.80$ in.
29. $A = 13°49'30'', B = 133°51'40'', c = 6.77$ cm
30. $X = 22°39'40'', Z = 34°08', b = 1.17$ km
31. $(\tan A + \tan B)/(1 - \tan A \tan B)$
32. $(\tan P - \tan Q)/(1 + \tan P \tan Q)$
33. $\sin X \cos Y - \cos X \sin Y$
34. $\cos(37° - 22°)$
35. 0.845
36. $-\cos 78°, \cot 56°, \tan 84°, \cos 47°$

ANSWERS TO EXERCISES

37. 0.06699, 0.05731, 0.40075
38. 21.5
39. 1,423 cm^2
40. 127.28 m

Chapter 4

1. (a) 65, (b) 780 kg, (c) 8.7 lb, (d) 40 g
2. 800 kg, and 1.875 m to the right of force A
3. (a) 53 kg, (b) 17.3 kips
4. (a) 45.7 tons, (b) 12.3, (c) 20.5
5. (a) -20 m-Kgm, (b) -832 cm-lb.
6. $F_1 = 195.959$ kips, $F_2 = 267.685$ kips
7. $F_1 = 335.877$ kg, $F_2 = 335.877$ kg
8. 73.540 lb
9. 65.637 kg
10. 962.5 m-kg, 19.89 yd-t
11. 92.30 lb
12. 52.716 kg, at 40°56′3″
13. $F_R = 281.647$ kg, $O_X = 60°11'36''$, $O_Y = 41°47'17''$, $O_Z = 63°39'8''$
14. $F_R = 532.756$, $O_{XR} = 54°54'32''$, $O_{YR} = 41°48'39''$, $O_{ZR} = 70°16'11''$
15. 64.705 kg, 12.94 kg
16. 209.595 kg, 41.919 kg
17. 0.6735 t
18. 0.175 t
19. 1.474 t
20. 66.988 lb
21. 26.625 lb
22. 24.095 kips
23. 0.7092 mm
24. 117.333 g
25. 53.333 yd
26. 1 : 2
27. 128.75 lb
28. 1.7167 in.

Chapter 5

1. 1998 kg
2. 166.67 kg
3. 1/2 of the velocity of the 25-lb force applied

Answers to Exercises • 375

4. 25.981 kips
5. 115.47 lb
6. 85 lb
7. 61.167 kg
8. 308.99 kg
9. 62.51 lb
10. $x = 5.26$, $y = 0.574$, and $z = -1.22$
11. $c = 11.04$ mm, $d = 8.96$ mm, and $e = 16.52$ mm
12. 27.37 in.
13. $x = 1.117$ m and $y = 1.117$ m
14. $a = 27$ cm
15. (a) $a = 6.74$ cm, (b) $a = 12.36$ cm
16. $a = 4$m
17. (a) $a = 1.46$ in., (b) $a = 1.42$ m
18. $I_X = 106\ 667$ m^4, $I_Y = 26\ 667$ m^4
19. $I_X = 4.5$ cm^4, $I_Y = 18$ cm^4
20. $I_X = 0.0425$ m^4, $I_Y = 0.232$ m^4
21. $I_X = I_Y = 15.34$ in^4
22. $I_X = I_Y = 31\ 400$ m^4
23. 17.23 cm^2
24. 245.9 yd^4
25. 82.4 in.^{4}
26. $I_X = 214\ 000$ ft^4, $I_Y = 67\ 600$ ft^4
27. 19,900 m^4
28. $I_X = 480$ in.^{4}, $I_Y = 240$ in.^{4}, $I_Z = 624$ in.^{4}
29. $I_X = 9,686,590$ cm^4, $I_Y = 1,154,995$ cm^4, $I_Z = 9,686,590$ cm^4
30. $I_X = I_Y = I_Z = 145,760$ m^4
31. 30,600 m or 30.6 km
32. 13 minutes 51 seconds
33. 0.785 rad/s
34. 25°
35. 94.25 ft.
36. 9.4 s
37. 1944 m/hr, and 19,440 mi
38. 12.5 (mi/hr)/s or 18.3 ft/s^2
39. 6.4 s
40. 1.64 s and 3.28 s
41. 32 m/s and 1 s
42. 1260 m
43. 8.74 rad
44. 5236 rad/s/s
45. 34.656 rad
46. 3.415 s

376 • ANSWERS TO EXERCISES

47. −11.728 rad/s/s
48. −7.158 rad/s/s
49. 1040 j
50. 1040 j
51. 75 200 j
52. 3.03 hp
53. 50,000 ft-lbs
54. 75,000 ft-lb
55. 451.688 kg
56. 918.336 lb
57. 286.57 ft/s

Chapter 6

1. 114,273,345.7 MPa
2. 568 psi
3. 10,800 psi
4. $A\text{–}A = 43.53$ MPa, $B\text{–}B = 14.53$ MPa
5. 480 N
6. 1945.88 psi
7. 99.949 Pa
8. 0.002 569 44
9. 0.044
10. 0.027 78
11. 1.7685 MPa and 256,500 psi
12. 0.120 467 MPa and 1747.2 psi
13. 0.000 88
14. 0.002 919
15. 152.2898 N
16. 0.639 mm
17. 1.26 mm
18. 11.987 in.
19. 0.44352 in.
20. 0.000 004 899
21. 1,680.68 psi
22. 47 977 969.78 N-mm
23. 2.48 in.
24. 0.000 692 68
25. 2,625 psi
26. 58.5 MPa
27. 12,700 psi
28. 2°11′54″

29. 0°0'1"
30. 2°30'9"
31. 1.50 in.
32. 0.390 in.
33. 80
34. 27.8
35. 179,000 N
36. 173,000 lb
37. 74,600 N

Chapter 7

1. 160 lbs, 98 N, 14 700 dyn
2. 0.80 g/cm^3, 800.0 kg/m^3, 49.9 slugs/cu. ft.
3. 60 lbs/cu.ft., 0.0714 dyn/cm^3, 72 N/m^3
4. 27.0 cSt
5. 22.1 cP
6. 9 800 000 dyn/cm^2
7. 103 000 dyn/cm^2
8. 1 037 000 dyn/cm^2
9. 0.857 inch
10. 747 psi
11. 50.6 ft/s
12. 9799.2
13. turbulent
14. laminar
15. pressure drop = −212 psi
    head loss = −313 ft/100 ft
16. pressure drop = 1500 psi
    head loss = 3880 ft/100 ft
17. 29.4 psig
18. 13,500 cu.ft.
19. 155.794 psia, 141.098 psig
20. 14.04 cu.ft.

Chapter 8

1. 54°
2. 72°
3. 3.750 in.
4. 0, 0.015 625, 0.062 500, and 0.140 625

5. 0.500, 0.875, 1.250
6. 1.804 688, 1.929 688, and 1.992 188
7. 16 in.
8. 1.421 in.
9. Diametral pitch = 0.667 cm
   Addendum = 1.499 cm
   Dedendum = 1.801 cm
   Circular pitch = 4.710 cm
   Tooth thickness = 2.355 cm
   Whole depth = 3.300 cm
   Working depth = 2.999 cm
   Clearance standard = 0.302 cm
10. Addendum = 0.036 39 in.
    Dedendum = 0.085 17 in.
    Circular pitch = 0.174 53 in.
    Outside diameter = 7.069 96 in.
    Whole depth = 0.123 56 in.
    Working depth = 0.111 11 in.
11. Addendum = 0.074 72 in.
    Dedendum = 0.046 84 in.
    Circular pitch = 0.174 53 in.
    Outside diameter = 4.143 65 in.
    Whole depth = 0.123 56 in.
    Working depth = 0.111 11 in.
12. 52.3182 in
13. Whole depth = 12.6074 in.
    Working depth = 11.4588 in.
14. 76°27′28″
15. Addendum = 0.0702 in.
    Tooth thickness = 0.1102 in.
    Whole depth = 0.1514 in.

Chapter 9

1. pitch = 0.125 in., 0.7143 in., 0.3846 in., 0.02632 in., flat = 0.0156 in., 0.0893 in., 0.0481 in., 0.00329 in.
2. d = 0.87875 in., 0.03222 in., 0.17203 in., t = 1.0825 in., 0.03969 in., 0.21174 in.
3. 3,750 psi and 10,781 psi
4. 26.15 lb
5. 880.56 lbs

Answers to Exercises • 379

6. 66,240 psi
7. 0.074 in.
8. 0.125 in.
9. 33.941 in.
10. Diametral pitch = 12 in. for all,
    stub pitch = 24 in. for all,
    circular pitch = 0.2618 in. for all, and
    minimum effective space width: $30° = 0.1309$ in., $37.5° = 0.1392$ in., and $45° = 0.1476$ in.
11. 140.625 in.-lb, 250.313 in.-lb, 337.5 in.-lb, 646.875 in.-lb.
12. Bearing force = 2500 lb
    modulus $M = 54,004$.
13. Open not ground = 0.0532 in.,
    open plain ground = 0.0539 in.,
    closed not ground = 0.0517 in., and
    closed ground = 0.0524 in.
14. 1.1175 in.
15. 0.000 081 25 psi
16. Clearance modulus = 0.0012 in.
17. 0.355
18. 6.283 in.
19. 4974 ft/min
20. 1.1398, 1,139,800 revolutions

Chapter 10

1. 150 fpm, 10 fpm, 50 fpm, 70 fpm
2. 0.250 in.: 2293 rpm, 153 rpm, 763 rpm, 306 rpm;
   5.54 mm: 8619 rpm, 575 rpm, 2873 rpm, 4022 rpm
3. 225 fpm, 516 rpm
4. 0.354 m/min
5. 22.44 m/min
6. 73.63 fpm
7. 300 rpm: 9.83 fpm, 19.65 fpm, 39.3 fpm, 58.59 fpm;
   500 rpm: 16.38 fpm, 32.75 fpm, 65.5 fpm, 98.25 fpm
8. 1046.95 hp
9. 68.42 hp
10. 5.79 hp
11. 1886.4 fpm, 2515.2 fpm
12. 42.04 hp
13. 48 in./min or 4 fpm.

14. 0.00014 in.^{3}/min
15. 0.02 hp, 0.965 in.-lb
16. 0.4
17. 6.6 hp

Chapter 11

1. 5.07 or 5
2. 5.82 or 6
3. 3.15 or 3
4. 2.02 or 2
5. 300
6. $105
7. 240
8. 120
9. $72
10. $21
11. $93
12. $UCL = 0.0914$, $LCL = 0.0086$
13. $UCL = 0.1310$, $LCL = -0.0300$
14. $656,250 $522,908.32
15. $415,000
16. $165,000
17. $13,954
18. $11,484
19. $28,400   0.44375
20. $775,000   0.32292

Chapter 12

1. Turning: 73.6 min/workpiece
   milling: 328.52 min/workpiece
   chamfering: 14.11 min/workpiece
2. 5.63 in./min.
3. 0.016 min
4. 321.186 min/workpiece
5. 14.39 workpieces/hr
6. $3.45/workpiece
7. $1.56/workpiece

Chapter 13

1. 12,690 ft^2
2. 2400 ft, 2600 ft, 3000 ft.
3. 86%
4. 157,380 ft^2
5. 100 ohms
6. 13.33 amp
7. 2.65 ohms
8. 0.52 amp, 0.080 hp
9. 56.03 hp
10. 827.7 Btus
11. 6000 Btus
12. 1491.43 gal
13. 43,711.27 kw-hr
14. 55.5 min/operation, 81.5 min/operation
15. 14.3%
16. 291.61 workers
17. $p_{RC} = 20.83$ or 21
18. $p_{RC} = 22$

Appendixes

APPENDIX A

Decimal Equivalents of Common Fractions

8ths	16ths	32ths	64ths	Exact Decimal Values
			1	0.015625
		1	2	0.03125
			3	0.046875
	1	2	4	0.0625
			5	0.078125
		3	6	0.09375
			7	0.109375
1	2	4	8	0.125
			9	0.140625
		5	10	0.15625
			11	0.171875
	3	6	12	0.1875
			13	0.203125
		7	14	0.21875
			15	0.234375
2	4	8	16	0.25
			17	0.265625
		9	18	0.28125
			19	0.296875
	5	10	20	0.3125

8ths	16ths	32ths	64ths	Exact Decimal Values
			21	0.328125
		11	22	0.34375
			23	0.359375
3	6	12	24	0.375
			25	0.390625
		13	26	0.40625
			27	0.421875
	7	14	28	0.4375
			29	0.453125
		15	30	0.46875
			31	0.484375
4	8	16	32	0.50
			33	0.515625
		17	34	0.53125
			35	0.546875
	9	18	36	0.5625
			37	0.578125
		19	38	0.59375
			39	0.609375
5	10	20	40	0.625
			41	0.640625
		21	42	0.65625
			43	0.671875
	11	22	44	0.6875
			45	0.703125
		23	46	0.71875
			47	0.734375
6	12	24	48	0.75
			49	0.765625
		25	50	0.78125
			51	0.796875
	13	26	52	0.8125
			53	0.828125
		27	54	0.84375
			55	0.859375
7	14	28	56	0.875
			57	0.890625
		29	58	0.90625
			59	0.921875
	15	30	60	0.9375

8ths	16ths	32ths	64ths	Exact Decimal Values
			61	0.953125
		31	62	0.96875
			63	0.984375
8	16	32	64	1

APPENDIX B

Functions of Whole Numbers from 1 to 100

Number (N)	Square (N^2)	Cube (N^3)	Square Root ($\sqrt{N}$)	Cube Root ($\sqrt[3]{N}$)	Reciprocal $\left(\dfrac{1}{N}\right)$
1	1	1	1.0000	1.0000	1.000000000
2	4	8	1.4142	1.2599	0.500000000
3	9	27	1.7321	1.4422	0.333333333
4	16	64	2.0000	1.5874	0.250000000
5	25	125	2.2361	1.7100	0.200000000
6	36	216	2.4495	1.8171	0.166666667
7	49	343	2.6458	1.9129	0.142857143
8	64	512	2.8284	2.0000	0.125000000
9	81	729	3.0000	2.0801	0.111111111
10	100	1,000	3.1623	2.1544	0.100000000
11	121	1,331	3.3166	2.2240	0.090909091
12	144	1,728	3.4641	2.2894	0.083333333
13	169	2,197	3.6056	2.3513	0.076923077
14	196	2,744	3.7417	2.4101	0.071428571
15	225	3,375	3.8730	2.4662	0.066666667
16	256	4,096	4.0000	2.5198	0.062500000
17	289	4,913	4.1231	2.5713	0.058823529
18	324	5,832	4.2426	2.6207	0.055555556
19	361	6,859	4.3589	2.6684	0.052631579
20	400	8,000	4.4721	2.7144	0.050000000

Number (N)	Square (N^2)	Cube (N^3)	Square Root ($\sqrt{N}$)	Cube Root ($\sqrt[3]{N}$)	Reciprocal ($\frac{1}{N}$)
21	441	9,261	4.5826	2.7589	0.047619048
22	484	10,648	4.6904	2.8020	0.045454545
23	529	12,167	4.7958	2.8439	0.043478261
24	576	13,824	4.8990	2.8845	0.041666667
25	625	15,625	5.0000	2.9240	0.040000000
26	676	17,576	5.0990	2.9625	0.038461538
27	729	19,683	5.1962	3.0000	0.037037037
28	784	21,952	5.2915	3.0366	0.035714286
29	841	24,389	5.3852	3.0723	0.034482759
30	900	27,000	5.4772	3.1072	0.033333333
31	961	29,791	5.5678	3.1414	0.032258065
32	1,024	32,768	5.6569	3.1748	0.031250000
33	1,089	35,937	5.7446	3.2075	0.030303030
34	1,156	39,304	5.8310	3.2396	0.029411765
35	1,225	42,875	5.9161	3.2711	0.028571429
36	1,296	46,656	6.0000	3.3019	0.027777778
37	1,369	50,653	6.0828	3.3322	0.027027027
38	1,444	54,872	6.1644	3.3620	0.026315789
39	1,521	59,319	6.2450	3.3812	0.025641026
40	1,600	64,000	6.3246	3.4200	0.025000000
41	1,681	68,921	6.4031	3.4482	0.024390244
42	1,764	74,088	6.4807	3.4760	0.023809524
43	1,849	79,507	6.5574	3.5034	0.023255814
44	1,936	85,184	6.6332	3.5303	0.022727273
45	2,025	91,125	6.7082	3.5569	0.022222222
46	2,116	97,336	6.7823	3.5830	0.021739130
47	2,209	103,823	6.8557	3.6088	0.021276600
48	2,304	110,592	6.9282	3.6342	0.020833333
49	2,401	117,649	7.0000	3.6593	0.020408163
50	2,500	125,000	7.0711	3.6840	0.020000000
51	2,601	132,651	7.1414	3.7084	0.019607843
52	2,704	140,608	7.2111	3.7325	0.019230769
53	2,809	148,877	7.2801	3.7563	0.018867925
54	2,916	157,464	7.3485	3.7798	0.018518519
55	3,025	166,375	7.4162	3.8030	0.018181818
56	3,136	175,616	7.4833	3.8259	0.017857143
57	3,249	185,193	7.5498	3.8485	0.017543860
58	3,364	195,112	7.6158	3.8709	0.017241379
59	3,481	205,379	7.6811	3.8930	0.016949153
60	3,600	216,000	7.7460	3.9149	0.016666667

Number (N)	Square (N^2)	Cube (N^3)	Square Root ($\sqrt{N}$)	Cube Root ($\sqrt[3]{N}$)	Reciprocal ($\frac{1}{N}$)
61	3,721	226,981	7.8102	3.9365	0.016393443
62	3,844	238,328	7.8740	3.9579	0.016129032
63	3,969	250,047	7.9391	3.9791	0.015873016
64	4,096	262,144	8.0000	3.0000	0.015625000
65	4,225	274,625	8.0623	4.0207	0.015384615
66	4,356	287,496	8.1240	4.0412	0.015151515
67	4,489	300,763	8.1854	4.0615	0.014925373
68	4,624	314,432	8.2462	4.0817	0.014705882
69	4,761	328,509	8.3066	4.1016	0.014492754
70	4,900	343,000	8.3666	4.1213	0.014285714
71	5,041	357,911	8.4261	4.1408	0.014084510
72	5,182	373,248	8.4853	4.1602	0.013888889
73	5,329	389,017	8.5440	4.1793	0.013698630
74	5,476	405,224	8.6023	4.1983	0.013513514
75	5,625	421,875	8.6603	4.2172	0.013333333
76	5,776	438,976	8.7178	4.2358	0.013157895
77	5,929	456,533	8.7750	4.2543	0.012987013
78	6,084	474,552	8.8318	4.2727	0.012820513
79	6,241	493,039	8.8882	4.2908	0.012658228
80	6,400	512,000	8.9443	4.3089	0.012500000
81	6,561	531,441	9.0000	4.3267	0.012345679
82	6,724	551,368	9.0554	4.3445	0.012195122
83	6,889	571,787	9.1104	4.3621	0.012048193
84	7,056	592,704	9.1652	4.3795	0.011904762
85	7,225	614,125	9.2195	4.3968	0.011764706
86	7,396	636,056	9.2736	4.4140	0.011627907
87	7,569	658,503	9.3274	4.4310	0.011494253
88	7,744	681,472	9.3808	4.4480	0.011363636
89	7,921	704,969	9.4340	4.4647	0.011235955
90	8,100	729,000	9.4860	4.4814	0.011111111
91	8,281	753,571	9.5394	4.4979	0.010989011
92	8,464	778,688	9.5917	4.5144	0.010869565
93	8,649	804,357	9.6437	4.4307	0.010752688
94	8,836	830,584	9.6954	4.5468	0.010638298
95	9,025	857,375	9.7468	4.5629	0.010526316
96	9,216	884,736	9.7980	4.5789	0.010416667
97	9,409	912,673	9.8489	4.5947	0.010209278
98	9,604	941,192	9.8995	4.6104	0.010204082
99	9,801	970,299	9.9499	4.6261	0.010101010
100	10,000	1,000,000	10.0000	4.6416	0.010000000

APPENDIX C

Segments of Circles for Radius = 1 Mathematical Tables (English or Metric Units)

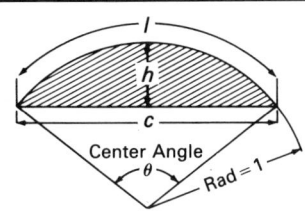

Length of arc, height of segment, length of chord, and area of segment for angles from 1 to 180 degrees and radius = 1. For other radii, multiply the values of *l*, *h*, and *c* in the table by the given radius *r*, and the values for areas, by r^2, the square of the radius.

The values in the tables can be used for English or metric units.

Center Angle θ (Degrees)	*l*	*h*	*c*	Area of Segment A	Center Angle θ (Degrees)	*l*	*h*	*c*	Area of Segment A
1	0.01745	0.00004	0.01745	0.00000	11	0.19199	0.00460	0.19169	0.00059
2	0.03491	0.00015	0.03490	0.00000	12	0.20944	0.00548	0.20906	0.00076
3	0.05236	0.00034	0.05235	0.00001	13	0.22689	0.00643	0.22641	0.00097
4	0.06981	0.00061	0.06980	0.00003	14	0.24435	0.00745	0.24374	0.00121
5	0.08727	0.00095	0.08724	0.00006	15	0.26180	0.00856	0.26105	0.00149
6	0.10472	0.00137	0.10467	0.00010	16	0.27925	0.00973	0.27835	0.00181
7	0.12217	0.00187	0.12210	0.00015	17	0.29671	0.01098	0.29562	0.00217
8	0.13963	0.00244	0.13951	0.00023	18	0.31416	0.01231	0.31287	0.00257
9	0.15708	0.00308	0.15692	0.00032	19	0.33161	0.01371	0.33010	0.00302
10	0.17453	0.00381	0.17431	0.00044	20	0.34907	0.01519	0.34730	0.00352

Center Angle θ (Degrees)	l	h	c	Area of Segment A	Center Angle θ (Degrees)	l	h	c	Area of Segment A
21	0.36652	0.01675	0.36447	0.00408	68	1.187	0.1710	1.118	0.12982
22	0.38397	0.01837	0.38162	0.00468	69	1.204	0.1759	1.133	0.13535
23	0.40143	0.02008	0.39874	0.00535	70	1.222	0.1808	1.147	0.14102
24	0.41888	0.02185	0.41582	0.00607	71	1.239	0.1859	1.161	0.14683
25	0.43633	0.02370	0.43288	0.00686	72	1.257	0.1910	1.176	0.15279
26	0.45379	0.02563	0.44990	0.00771	73	1.274	0.1961	1.190	0.15889
27	0.47124	0.02763	0.46689	0.00862	74	1.292	0.2014	1.204	0.16514
28	0.48869	0.02970	0.48384	0.00961	75	1.309	0.2066	1.218	0.17154
29	0.50615	0.03185	0.50076	0.01067	76	1.326	0.2120	1.231	0.17808
30	0.52360	0.03407	0.51764	0.01180	77	1.344	0.2174	1.245	0.18477
31	0.54105	0.03637	0.53448	0.01301	78	1.361	0.2229	1.259	0.19160
32	0.55851	0.03874	0.55127	0.01429	79	1.379	0.2284	1.272	0.19859
33	0.57596	0.04118	0.56803	0.01566	80	1.396	0.2340	1.286	0.20573
34	0.59341	0.04370	0.58474	0.01711	81	1.414	0.2396	1.299	0.21301
35	0.61087	0.04628	0.60141	0.01864	82	1.431	0.2453	1.312	0.22045
36	0.62832	0.04894	0.61803	0.02027	83	1.449	0.2510	1.325	0.22804
37	0.64577	0.05168	0.63461	0.02198	84	1.466	0.2569	1.338	0.23578
38	0.66323	0.05448	0.65114	0.02378	85	1.484	0.2627	1.351	0.24367
39	0.68068	0.05736	0.66761	0.02568	86	1.501	0.2686	1.364	0.25171
40	0.69813	0.06031	0.68404	0.02767	87	1.518	0.2746	1.377	0.25990
41	0.71558	0.06333	0.70041	0.02976	88	1.536	0.2807	1.389	0.26825
42	0.73304	0.06642	0.71674	0.03195	89	1.553	0.2867	1.402	0.27675
43	0.75049	0.06958	0.73300	0.03425	90	1.571	0.2929	1.414	0.28540
44	0.76794	0.07282	0.74921	0.03664	91	1.588	0.2991	1.427	0.2942
45	0.78540	0.07612	0.76537	0.03915	92	1.606	0.3053	1.439	0.3032
46	0.803	0.0795	0.781	0.04176	93	1.623	0.3116	1.451	0.3123
47	0.820	0.0829	0.797	0.04448	94	1.641	0.3180	1.463	0.3215
48	0.838	0.0865	0.813	0.04731	95	1.658	0.3244	1.475	0.3309
49	0.855	0.0900	0.829	0.05025	96	1.676	0.3309	1.486	0.3405
50	0.873	0.0937	0.845	0.05331	97	1.693	0.3374	1.498	0.3502
51	0.890	0.0974	0.861	0.05649	98	1.710	0.3439	1.509	0.3601
52	0.908	0.1012	0.877	0.05978	99	1.728	0.3506	1.521	0.3701
53	0.925	0.1051	0.892	0.06319	100	1.745	0.3572	1.532	0.3803
54	0.942	0.1090	0.908	0.06673	101	1.763	0.3639	1.543	0.3906
55	0.960	0.1130	0.923	0.07039	102	1.780	0.3707	1.554	0.4010
56	0.977	0.1171	0.939	0.07417	103	1.798	0.3775	1.565	0.4117
57	0.995	0.1212	0.954	0.07808	104	1.815	0.3843	1.576	0.4224
58	1.012	0.1254	0.970	0.08212	105	1.833	0.3912	1.587	0.4333
59	1.030	0.1296	0.985	0.08629	106	1.850	0.3982	1.597	0.4444
60	1.047	0.1340	1.000	0.09059	107	1.868	0.4052	1.608	0.4556
61	1.065	0.1384	1.015	0.09502	108	1.885	0.4122	1.618	0.4669
62	1.082	0.1428	1.030	0.09958	109	1.902	0.4193	1.628	0.4784
63	1.100	0.1474	1.045	0.10428	110	1.920	0.4264	1.638	0.4901
64	1.117	0.1520	1.060	0.10911	111	1.937	0.4336	1.648	0.5019
65	1.134	0.1566	1.075	0.11408	112	1.955	0.4408	1.658	0.5138
66	1.152	0.1613	1.089	0.11919	113	1.972	0.4481	1.668	0.5259
67	1.169	0.1661	1.104	0.12443	114	1.990	0.4554	1.677	0.5381

Center Angle θ (Degrees)	l	h	c	Area of Segment A	Center Angle θ (Degrees)	l	h	c	Area of Segment A
115	2.007	0.4627	1.687	0.5504	148	2.583	0.7244	1.923	1.0266
116	2.025	0.4701	1.696	0.5629	149	2.601	0.7328	1.927	1.0428
117	2.042	0.4775	1.705	0.5755	150	2.618	0.7412	1.932	1.0590
118	2.059	0.4850	1.714	0.5883	151	2.635	0.7496	1.936	1.0753
119	2.077	0.4925	1.723	0.6012	152	2.653	0.7581	1.941	1.0917
120	2.094	0.5000	1.732	0.6142	153	2.670	0.7666	1.945	1.1082
121	2.112	0.5076	1.741	0.6273	154	2.688	0.7750	1.949	1.1247
122	2.129	0.5152	1.749	0.6406	155	2.705	0.7836	1.953	1.1413
123	2.147	0.5228	1.758	0.6540	156	2.723	0.7921	1.956	1.1580
124	2.164	0.5305	1.766	0.6676	157	2.740	0.8006	1.960	1.1747
125	2.182	0.5383	1.774	0.6813	158	2.758	0.8092	1.963	1.1915
126	2.199	0.5460	1.782	0.6950	159	2.775	0.8178	1.967	1.2084
127	2.217	0.5538	1.790	0.7090	160	2.793	0.8264	1.970	1.2253
128	2.234	0.5616	1.798	0.7230	161	2.810	0.8350	1.973	1.2422
129	2.251	0.5695	1.805	0.7372	162	2.827	0.8436	1.975	1.2592
130	2.269	0.5774	1.813	0.7514	163	2.845	0.8522	1.978	1.2763
131	2.286	0.5853	1.820	0.7658	164	2.862	0.8608	1.981	1.2934
132	2.304	0.5933	1.827	0.7803	165	2.880	0.8695	1.983	1.3105
133	2.321	0.6013	1.834	0.7950	166	2.897	0.8781	1.985	1.3277
134	2.339	0.6093	1.841	0.8097	167	2.915	0.8868	1.987	1.3449
135	2.356	0.6173	1.848	0.8245	168	2.932	0.8955	1.989	1.3621
136	2.374	0.6254	1.854	0.8395	169	2.950	0.9042	1.991	1.3794
137	2.391	0.6335	1.861	0.8546	170	2.967	0.9128	1.992	1.3967
138	2.409	0.6416	1.867	0.8697	171	2.985	0.9215	1.994	1.4140
139	2.426	0.6498	1.873	0.8850	172	3.002	0.9302	1.995	1.4314
140	2.443	0.6580	1.879	0.9003	173	3.019	0.9390	1.996	1.4488
141	2.461	0.6662	1.885	0.9158	174	3.037	0.9477	1.997	1.4662
142	2.478	0.6744	1.891	0.9314	175	3.054	0.9564	1.998	1.4836
143	2.496	0.6827	1.897	0.9470	176	3.072	0.9651	1.999	1.5010
144	2.513	0.6910	1.902	0.9627	177	3.089	0.9738	1.999	1.5184
145	2.531	0.6993	1.907	0.9786	178	3.107	0.9825	2.000	1.5359
146	2.548	0.7076	1.913	0.9945	179	3.124	0.9913	2.000	1.5533
147	2.566	0.7160	1.918	1.0105	180	3.142	1.0000	2.000	1.5708

APPENDIX D

Metric Conversion and Equivalency Tables

Table D-1. Inches to Millimeters
(Based on 1 in. = 25.4 millimeters, exactly)

Inches	0.000	0.001	0.002	0.003	0.004	0.005	0.006	0.007	0.008	0.009
					Millimeters					
0.000	...	0.0254	0.0508	0.0762	0.1016	0.1270	0.1524	0.1778	0.2032	0.2286
0.010	0.2540	0.2794	0.3048	0.3302	0.3556	0.3810	0.4064	0.4318	0.4572	0.4826
0.020	0.5080	0.5334	0.5588	0.5842	0.6096	0.6350	0.6604	0.6858	0.7112	0.7366
0.030	0.7620	0.7874	0.8128	0.8382	0.8636	0.8890	0.9144	0.9398	0.9652	0.9906
0.040	1.0160	1.0414	1.0668	1.0922	1.1176	1.1430	1.1684	1.1938	1.2192	1.2446
0.050	1.2700	1.2954	1.3208	1.3462	1.3716	1.3970	1.4224	1.4478	1.4732	1.4986
0.060	1.5240	1.5494	1.5748	1.6002	1.6256	1.6510	1.6764	1.7018	1.7272	1.7526
0.070	1.7780	1.8034	1.8288	1.8542	1.8796	1.9050	1.9304	1.9558	1.9812	2.0066
0.080	2.0320	2.0574	2.0828	2.1082	2.1336	2.1590	2.1844	2.2098	2.2352	2.2606
0.090	2.2860	2.3114	2.3368	2.3622	2.3876	2.4130	2.4384	2.4638	2.4892	2.5146
0.100	2.5400	2.5654	2.5908	2.6162	2.6416	2.6670	2.6924	2.7178	2.7432	2.7686
0.110	2.7940	2.8194	2.8448	2.8702	2.8956	2.9210	2.9464	2.9718	2.9972	3.0226
0.120	3.0480	3.0734	3.0988	3.1242	3.1496	3.1750	3.2004	3.2258	3.2512	3.2766
0.130	3.3020	3.3274	3.3528	3.3782	3.4036	3.4290	3.4544	3.4798	3.5052	3.5306
0.140	3.5560	3.5814	3.6068	3.6322	3.6576	3.6830	3.7084	3.7338	3.7592	3.7846
0.150	3.8100	3.8354	3.8608	3.8862	3.9116	3.9370	3.9624	3.9878	4.0132	4.0386
0.160	4.0640	4.0894	4.1148	4.1402	4.1656	4.1910	4.2164	4.2418	4.2672	4.2926
0.170	4.3180	4.3434	4.3688	4.3942	4.4196	4.4450	4.4704	4.4958	4.5212	4.5466
0.180	4.5720	4.5974	4.6228	4.6482	4.6736	4.6990	4.7244	4.7498	4.7752	4.8006
0.190	4.8260	4.8514	4.8768	4.9022	4.9276	4.9530	4.9784	5.0038	5.0292	5.0546

Table D-1. Inches to Millimeters (continued)

Inches	0.000	0.001	0.002	0.003	0.004	0.005	0.006	0.007	0.008	0.009
	Millimeters									
0.200	5.0800	5.1054	5.1308	5.1562	5.1816	5.2070	5.2324	5.2578	5.2832	5.3086
0.210	5.3340	5.3594	5.3848	5.4102	5.4356	5.4610	5.4864	5.5118	5.5372	5.5626
0.220	5.5880	5.6134	5.6388	5.6642	5.6896	5.7150	5.7404	5.7658	5.7912	5.8166
0.230	5.8420	5.8674	5.8928	5.9182	5.9436	5.9690	5.9944	6.0198	6.0452	6.0706
0.240	6.0960	6.1214	6.1468	6.1722	6.1976	6.2230	6.2484	6.2738	6.2992	6.3246
0.250	6.3500	6.3754	6.4008	6.4262	6.4516	6.4770	6.5024	6.5278	6.5532	6.5786
0.260	6.6040	6.6294	6.6548	6.6802	6.7056	6.7310	6.7564	6.7818	6.8072	6.8326
0.270	6.8580	6.8834	6.9088	6.9342	6.9596	6.9850	7.0104	7.0358	7.0612	7.0866
0.280	7.1120	7.1374	7.1628	7.1882	7.2136	7.2390	7.2644	7.2898	7.3152	7.3406
0.290	7.3660	7.3914	7.4168	7.4422	7.4676	7.4930	7.5184	7.5438	7.5692	7.5946
0.300	7.6200	7.6454	7.6708	7.6962	7.7216	7.7470	7.7724	7.7978	7.8232	7.8486
0.310	7.8740	7.8994	7.9248	7.9502	7.9756	8.0010	8.0264	8.0518	8.0772	8.1026
0.320	8.1280	8.1534	8.1788	8.2042	8.2296	8.2550	8.2804	8.3058	8.3312	8.3566
0.330	8.3820	8.4074	8.4328	8.4582	8.4836	8.5090	8.5344	8.5598	8.5852	8.6106
0.340	8.6360	8.6614	8.6868	8.7122	8.7376	8.7630	8.7884	8.8138	8.8392	8.8646
0.350	8.8900	8.9154	8.9408	8.9662	8.9916	9.0170	9.0424	9.0678	9.0932	9.1186
0.360	9.1440	9.1694	9.1948	9.2202	9.2456	9.2710	9.2964	9.3218	9.3472	9.3726
0.370	9.3980	9.4234	9.4488	9.4742	9.4996	9.5250	9.5504	9.5758	9.6012	9.6266
0.380	9.6520	9.6774	9.7028	9.7282	9.7536	9.7790	9.8044	9.8298	9.8552	9.8806
0.390	9.9060	9.9314	9.9568	9.9822	10.0076	10.0330	10.0584	10.0838	10.1092	10.1346
0.400	10.1600	10.1854	10.2108	10.2362	10.2616	10.2870	10.3124	10.3378	10.3632	10.3886
0.410	10.4140	10.4394	10.4648	10.4902	10.5156	10.5410	10.5664	10.5918	10.6172	10.6426
0.420	10.6680	10.6934	10.7188	10.7442	10.7696	10.7950	10.8204	10.8458	10.8712	10.8966
0.430	10.9220	10.9474	10.9728	10.9982	11.0236	11.0490	11.0744	11.0998	11.1252	11.1506
0.440	11.1760	11.2014	11.2268	11.2522	11.2776	11.3030	11.3284	11.3538	11.3792	11.4046
0.450	11.4300	11.4554	11.4808	11.5062	11.5316	11.5570	11.5824	11.6078	11.6332	11.6586
0.460	11.6840	11.7094	11.7348	11.7602	11.7856	11.8110	11.8364	11.8618	11.8872	11.9126
0.470	11.9380	11.9634	11.9888	12.0142	12.0396	12.0650	12.0904	12.1158	12.1412	12.1666
0.480	12.1920	12.2174	12.2428	12.2682	12.2936	12.3190	12.3444	12.3698	12.3952	12.4206
0.490	12.4460	12.4714	12.4968	12.5222	12.5476	12.5730	12.5984	12.6238	12.6492	12.6746
0.500	12.700	12.7254	12.7508	12.7762	12.8016	12.8270	12.8524	12.8778	12.9032	12.9286
0.510	12.9540	12.9794	13.0048	13.0302	13.0556	13.0810	13.1064	13.1318	13.1572	13.1826
0.520	13.2080	13.2334	13.2588	13.2842	13.3096	13.3350	13.3604	13.3858	13.4112	13.4366
0.530	13.4620	13.4874	13.5128	13.5382	13.5636	13.5890	13.6144	13.6398	13.6652	13.6906
0.540	13.7160	13.7414	13.7668	13.7922	13.8176	13.8430	13.8684	13.8938	13.9192	13.9446
0.550	13.9700	13.9954	14.0208	14.0462	14.0716	14.0970	14.1224	14.1478	14.1732	14.1986
0.560	14.2240	14.2494	14.2748	14.3002	14.3256	14.3510	14.3764	14.4018	14.4272	14.4526
0.570	14.4780	14.5034	14.5288	14.5542	14.5796	14.6050	14.6304	14.6558	14.6812	14.7066
0.580	14.7320	14.7574	14.7828	14.8082	14.8336	14.8590	14.8844	14.9098	14.9352	14.9606
0.590	14.9860	15.0114	15.0368	15.0622	15.0876	15.1130	15.1384	15.1638	15.1892	15.2146
0.600	15.2400	15.2654	15.2908	15.3162	15.3416	15.3670	15.3924	15.4178	15.4432	15.4686
0.610	15.4940	15.5194	15.5448	15.5702	15.5956	15.6210	15.6464	15.6718	15.6972	15.7226
0.620	15.7480	15.7734	15.7988	15.8242	15.8496	15.8750	15.9004	15.9258	15.9512	15.9766

Table D-1. Inches to Millimeters (continued)

Inches	0.000	0.001	0.002	0.003	0.004	0.005	0.006	0.007	0.008	0.009
					Millimeters					
0.630	16.0020	16.0274	16.0528	16.0782	16.1036	16.1290	16.1544	16.1798	16.2052	16.2306
0.640	16.2560	16.2814	16.3068	16.3322	16.3576	16.3830	16.4084	16.4338	16.4592	16.4846
0.650	16.5100	16.5354	16.5608	16.5862	16.6116	16.6370	16.6624	16.6878	16.7132	16.7386
0.660	16.7640	16.7894	16.8148	16.8402	16.8656	16.8910	16.9164	16.9418	16.9672	16.9926
0.670	17.0180	17.0434	17.0688	17.0942	17.1196	17.1450	17.1704	17.1958	17.2212	17.2466
0.680	17.2720	17.2974	17.3228	17.3482	17.3736	17.3990	17.4244	17.4498	17.4752	17.5006
0.690	17.5260	17.5514	17.5768	17.6022	17.6276	17.6530	17.6784	17.7038	17.7292	17.7546
0.700	17.7800	17.8054	17.8308	17.8562	17.8816	17.9070	17.9324	17.9578	17.9832	18.0086
0.710	18.0340	18.0594	18.0848	18.1102	18.1356	18.1610	18.1864	18.2118	18.2372	18.2626
0.720	18.2880	18.3134	18.3388	18.3642	18.3896	18.4150	18.4404	18.4658	18.4912	18.5166
0.730	18.5420	18.5674	18.5928	18.6182	18.6436	18.6690	18.6944	18.7198	18.7452	18.7706
0.740	18.7960	18.8214	18.8468	18.8722	18.8976	18.9230	18.9484	18.9738	18.9992	19.0246
0.750	19.0500	19.0754	19.1008	19.1262	19.1516	19.1770	19.2024	19.2278	19.2532	19.2786
0.760	19.3040	19.3294	19.3548	19.3802	19.4056	19.4310	19.4564	19.4818	19.5072	19.5326
0.770	19.5580	19.5834	19.6088	19.6342	19.6596	19.6850	19.7104	19.7358	19.7612	19.7866
0.780	19.8120	19.8374	19.8628	19.8882	19.9136	19.9390	19.9644	19.9898	20.0152	20.0406
0.790	20.0660	20.0914	20.1168	20.1422	20.1676	20.1930	20.2184	20.2438	20.2692	20.2946
0.800	20.3200	20.3454	20.3708	20.3962	20.4216	20.4470	20.4724	20.4978	20.5232	20.5486
0.810	20.5740	20.5994	20.6248	20.6502	20.6756	20.7010	20.7264	20.7518	20.7772	20.8026
0.820	20.8280	20.8534	20.8788	20.9042	20.9296	20.9550	20.9804	21.0058	21.0312	21.0566
0.830	21.0820	21.1074	21.1328	21.1582	21.1836	21.2090	21.2344	21.2598	21.2852	21.3106
0.840	21.3360	21.3614	21.3868	21.4122	21.4376	21.4630	21.4884	21.5138	21.5392	21.5646
0.850	21.5900	21.6154	21.6408	21.6662	21.6916	21.7170	21.7424	21.7678	21.7932	21.8186
0.860	21.8440	21.8694	21.8948	21.9202	21.9456	21.9710	21.9964	22.0218	22.0472	22.0726
0.870	22.0980	22.1234	22.1488	22.1742	22.1996	22.2250	22.2504	22.2758	22.3012	22.3266
0.880	22.3520	22.3774	22.4028	22.4282	22.4536	22.4790	22.5044	22.5298	22.5552	22.5806
0.890	22.6060	22.6314	22.6568	22.6822	22.7076	22.7330	22.7584	22.7838	22.8092	22.8346
0.900	22.8600	22.8854	22.9108	22.9362	22.9616	22.9870	23.0124	23.0378	23.0632	23.0886
0.910	23.1140	23.1394	23.1648	23.1902	23.2156	23.2410	23.2664	23.2918	23.3172	23.3426
0.920	23.3680	23.3934	23.4188	23.4442	23.4696	23.4950	23.5204	23.5458	23.5712	23.5966
0.930	23.6220	23.6474	23.6728	23.6982	23.7236	23.7490	23.7744	23.7998	23.8252	23.8506
0.940	23.8760	23.9014	23.9268	23.9522	23.9776	24.0030	24.0284	24.0538	24.0792	24.1046
0.950	24.1300	24.1554	24.1808	24.2062	24.2316	24.2570	24.2824	24.3078	24.3332	24.3586
0.960	24.3840	24.4094	24.4348	24.4602	24.4856	24.5110	24.5364	24.5618	24.5872	24.6126
0.970	24.6380	24.6634	24.6888	24.7142	24.7396	24.7650	24.7904	24.8158	24.8412	24.8666
0.980	24.8920	24.9174	24.9428	24.9682	24.9936	25.0190	25.0444	25.0698	25.0952	25.1206
0.990	25.1460	25.1714	25.1968	25.2222	25.2476	25.2730	25.2984	25.3238	25.3492	25.3746
1.000	25.4000	...	...	...	...	...	...	...	...	...

Table D-1. Inches to Millimeters—*Continued*

\multicolumn{12}{c	}{Inches to Millimeters}

in.	mm	in.	mm	in.	mm	in.	mm	in.	mm	in.	mm
10	254.00000	1	25.40000	.1	2.54000	.01	.25400	.001	.02540	.0001	.00254
20	508.00000	2	50.80000	.2	5.08000	.02	.50800	.002	.05080	.0002	.00508
30	762.00000	3	76.20000	.3	7.62000	.03	.76200	.003	.07620	.0003	.00762
40	1,016.00000	4	101.60000	.4	10.16000	.04	1.01600	.004	.10160	.0004	.01016
50	1,270.00000	5	127.00000	.5	12.70000	.05	1.27000	.005	.12700	.0005	.01270
60	1,524.00000	6	152.40000	.6	15.24000	.06	1.52400	.006	.15240	.0006	.01524
70	1,778.00000	7	177.80000	.7	17.78000	.07	1.77800	.007	.17780	.0007	.01778
80	2,032.00000	8	203.20000	.8	20.32000	.08	2.03200	.008	.20320	.0008	.02032
90	2,286.00000	9	228.60000	.9	22.86000	.09	2.28600	.009	.22860	.0009	.02286
100	2,540.00000	10	254.00000	1.0	25.40000	.10	2.54000	.010	.25400	.0010	.02540

Table D-2. Millimeters to Inches
(Based on 1 in. = 25.4 millimeters, exactly)

Milli-meters	0	1	2	3	4	5	6	7	8	9
					Inches					
0	...	0.03937	0.07874	0.11811	0.15748	0.19685	0.23622	0.27559	0.31496	0.35433
10	0.39370	0.43307	0.47244	0.51181	0.55118	0.59055	0.62992	0.66929	0.70866	0.74803
20	0.78740	0.82677	0.86614	0.90551	0.94488	0.98425	1.02362	1.06299	1.10236	1.14173
30	1.18110	1.22047	1.25984	1.29921	1.33858	1.37795	1.41732	1.45669	1.49606	1.53543
40	1.57480	1.61417	1.65354	1.69291	1.73228	1.77165	1.81102	1.85039	1.88976	1.92913
50	1.96850	2.00787	2.04724	2.08661	2.12598	2.16535	2.20472	2.24409	2.28346	2.32283
60	2.36220	2.40157	2.44094	2.48031	2.51969	2.55906	2.59843	2.63780	2.67717	2.71654
70	2.75591	2.79528	2.83465	2.87402	2.91339	2.95276	2.99213	3.03150	3.07087	3.11024
80	3.14961	3.18898	3.22835	3.26772	3.30709	3.34646	3.38583	3.42520	3.46457	3.50394
90	3.54331	3.58268	3.62205	3.66142	3.70079	3.74016	3.77953	3.81890	3.85827	3.89764
100	3.93701	3.97638	4.01575	4.05512	4.09449	4.13386	4.17323	4.21260	4.25197	4.29134
110	4.33071	4.37008	4.40945	4.44882	4.48819	4.52756	4.56693	4.60630	4.64567	4.68504
120	4.72441	4.76378	4.80315	4.84252	4.88189	4.92126	4.96063	5.00000	5.03937	5.07874
130	5.11811	5.15748	5.19685	5.23622	5.27559	5.31496	5.35433	5.39370	5.43307	5.47244
140	5.51181	5.55118	5.59055	5.62992	5.66929	5.70866	5.74803	5.78740	5.82677	5.86614
150	5.90551	5.94488	5.98425	6.02362	6.06299	6.10236	6.14173	6.18110	6.22047	6.25984
160	6.29921	6.33858	6.37795	6.41732	6.45669	6.49606	6.53543	6.57480	6.61417	6.65354
170	6.69291	6.73228	6.77165	6.81102	6.85039	6.88976	6.92913	6.96850	7.00787	7.04724
180	7.08661	7.12598	7.16535	7.20472	7.24409	7.28346	7.32283	7.36220	7.40157	7.44094
190	7.48031	7.51969	7.55906	7.59843	7.63780	7.67717	7.71654	7.75591	7.79528	7.83465
200	7.87402	7.91339	7.95276	7.99213	8.03150	8.07087	8.11024	8.14961	8.18898	8.22835
210	8.26772	8.30709	8.34646	8.38583	8.42520	8.46457	8.50394	8.54331	8.58268	8.62205
220	8.66142	8.70079	8.74016	8.77953	8.81890	8.85827	8.89764	8.93701	8.97638	9.01575
230	9.05512	9.09449	9.13386	9.17323	9.21260	9.25197	9.29134	9.33071	9.37008	9.40945
240	9.44882	9.48819	9.52756	9.56693	9.60630	9.64567	9.68504	9.72441	9.76378	9.80315
250	9.84252	9.88189	9.92126	9.96063	10.0000	10.0394	10.0787	10.1181	10.1575	10.1969
260	10.2362	10.2756	10.3150	10.3543	10.3937	10.4331	10.4724	10.5118	10.5512	10.5906
270	10.6299	10.6693	10.7087	10.7480	10.7874	10.8268	10.8661	10.9055	10.9449	10.9843
280	11.0236	11.0630	11.1024	11.1417	11.1811	11.2205	11.2598	11.2992	11.3386	11.3780
290	11.4173	11.4567	11.4961	11.5354	11.5748	11.6142	11.6535	11.6929	11.7323	11.7717
300	11.8110	11.8504	11.8898	11.9291	11.9685	12.0079	12.0472	12.0866	12.1260	12.1654
310	12.2047	12.2441	12.2835	12.3228	12.3622	12.4016	12.4409	12.4803	12.5197	12.5591
320	12.5984	12.6378	12.6772	12.7165	12.7559	12.7953	12.8346	12.8740	12.9134	12.9528
330	12.9921	13.0315	13.0709	13.1102	13.1496	13.1890	13.2283	13.2677	13.3071	13.3465
340	13.3858	13.4252	13.4646	13.5039	13.5433	13.5827	13.6220	13.6614	13.7008	13.7402
350	13.7795	13.8189	13.8583	13.8976	13.9370	13.9764	14.0157	14.0551	14.0945	14.1339
360	14.1732	14.2126	14.2520	14.2913	14.3307	14.3701	14.4094	14.4488	14.4882	14.5276
370	14.5669	14.6063	14.6457	14.6850	14.7244	14.7638	14.8031	14.8425	14.8819	14.9213
380	14.9606	15.0000	15.0394	15.0787	15.1181	15.1575	15.1969	15.2362	15.2756	15.3150
390	15.3543	15.3937	15.4331	15.4724	15.5118	15.5512	15.5906	15.6299	15.6693	15.7087

Table D-2. Millimeters to Inches (continued)

Millimeters	0	1	2	3	4	5	6	7	8	9
					Inches					
400	15.7480	15.7874	15.8268	15.8661	15.9055	15.9449	15.9843	16.0236	16.0630	16.1024
410	16.1417	16.1811	16.2205	16.2598	16.2992	16.3386	16.3780	16.4173	16.4567	16.4961
420	16.5354	16.5748	16.6142	16.6535	16.6929	16.7323	16.7717	16.8110	16.8504	16.8898
430	16.9291	16.9685	17.0079	17.0472	17.0866	17.1260	17.1654	17.2047	17.2441	17.2835
440	17.3228	17.3622	17.4016	17.4409	17.4803	17.5197	17.5591	17.5984	17.6378	17.6772
450	17.7165	17.7559	17.7953	17.8346	17.8740	17.9134	17.9528	17.9921	18.0315	18.0709
460	18.1102	18.1496	18.1890	18.2283	18.2677	18.3071	18.3465	18.3858	18.4252	18.4646
470	18.5039	18.5433	18.5827	18.6220	18.6614	18.7008	18.7402	18.7795	18.8189	18.8583
480	18.8976	18.9370	18.9764	19.0157	19.0551	19.0945	19.1339	19.1732	19.2126	19.2520
490	19.2913	19.3307	19.3701	19.4094	19.4488	19.4882	19.5276	19.5669	19.6063	19.6457
500	19.6850	19.7244	19.7638	19.8031	19.8425	19.8819	19.9213	19.9606	20.0000	20.0394
510	20.0787	20.1181	20.1575	20.1969	20.2362	20.2756	20.3150	20.3543	20.3937	20.4331
520	20.4724	20.5118	20.5512	20.5906	20.6299	20.6693	20.7087	20.7480	20.7874	20.8268
530	20.8661	20.9055	20.9449	20.9843	21.0236	21.0630	21.1024	21.1417	21.1811	21.2205
540	21.2598	21.2992	21.3386	21.3780	21.4173	21.4567	21.4961	21.5354	21.5748	21.6142
550	21.6535	21.6929	21.7323	21.7717	21.8110	21.8504	21.8898	21.9291	21.9685	22.0079
560	22.0472	22.0866	22.1260	22.1654	22.2047	22.2441	22.2835	22.3228	22.3622	22.4016
570	22.4409	22.4803	22.5197	22.5591	22.5984	22.6378	22.6772	22.7165	22.7559	22.7953
580	22.8346	22.8740	22.9134	22.9528	22.9921	23.0315	23.0709	23.1102	23.1496	23.1890
590	23.2283	23.2677	23.3071	23.3465	23.3858	23.4252	23.4646	23.5039	23.5433	23.5827
600	23.6220	23.6614	23.7008	23.7402	23.7795	23.8189	23.8583	23.8976	23.9370	23.9764
610	24.0157	24.0551	24.0945	24.1339	24.1732	24.2126	24.2520	24.2913	24.3307	24.3701
620	24.4094	24.4488	24.4882	24.5276	24.5669	24.6063	24.6457	24.6850	24.7244	24.7638
630	24.8031	24.8425	24.8819	24.9213	24.9606	25.0000	25.0394	25.0787	25.1181	25.1575
640	25.1969	25.2362	25.2756	25.3150	25.3543	25.3937	25.4331	25.4724	25.5118	25.5512
650	25.5906	25.6299	25.6693	25.7087	25.7480	25.7874	25.8268	25.8661	25.9055	25.9449
660	25.9843	26.0236	26.0630	26.1024	26.1417	26.1811	26.2205	26.2598	26.2992	26.3386
670	26.3780	26.4173	26.4567	26.4961	26.5354	26.5748	26.6142	26.6535	26.6929	26.7323
680	26.7717	26.8110	26.8504	26.8898	26.9291	26.9685	27.0079	27.0472	27.0866	27.1260
690	27.1654	27.2047	27.2441	27.2835	27.3228	27.3622	27.4016	27.4409	27.4803	27.5197
700	27.5591	27.5984	27.6378	27.6772	27.7165	27.7559	27.7953	27.8346	27.8740	27.9134
710	27.9528	27.9921	28.0315	28.0709	28.1102	28.1496	28.1890	28.2283	28.2677	28.3071
720	28.3465	28.3858	28.4252	28.4646	28.5039	28.5433	28.5827	28.6220	28.6614	28.7008
730	28.7402	28.7795	28.8189	28.8583	28.8976	28.9370	28.9764	29.0157	29.0551	29.0945
740	29.1339	29.1732	29.2126	29.2520	29.2913	29.3307	29.3701	29.4094	29.4488	29.4882
750	29.5276	29.5669	29.6063	29.6457	29.6850	29.7244	29.7638	29.8031	29.8425	29.8819
760	29.9213	29.9606	30.0000	30.0394	30.0787	30.1181	30.1575	30.1969	30.2362	30.2756
770	30.3150	30.3543	30.3937	30.4331	30.4724	30.5118	30.5512	30.5906	30.6299	30.6693
780	30.7087	30.7480	30.7874	30.8268	30.8661	30.9055	30.9449	30.9843	31.0236	31.0630
790	31.1024	31.1417	31.1811	31.2205	31.2598	31.2992	31.3386	31.3780	31.4173	31.4567
800	31.4961	31.5354	31.5748	31.6142	31.6535	31.6929	31.7323	31.7717	31.8110	31.8504

Table D-2. Millimeters to Inches (continued)

Milli-meters	0	1	2	3	4	5	6	7	8	9
					Inches					
810	31.8898	31.9291	31.9685	32.0079	32.0472	32.0866	32.1260	32.1654	32.2047	32.2441
820	32.2835	32.3228	32.3622	32.4016	32.4409	32.4803	32.5197	32.5591	32.5984	32.6378
830	32.6772	32.7165	32.7559	32.7953	32.8346	32.8740	32.9134	32.9528	32.9921	33.0315
840	33.0709	33.1102	33.1496	33.1890	33.2283	33.2677	33.3071	33.3465	33.3858	33.4252
850	33.4646	33.5039	33.5433	33.5827	33.6220	33.6614	33.7008	33.7402	33.7795	33.8189
860	33.8583	33.8976	33.9370	33.9764	34.0157	34.0551	34.0945	34.1339	34.1732	34.2126
870	34.2520	34.2913	34.3307	34.3701	34.4094	34.4488	34.4882	34.5276	34.5669	34.6063
880	34.6457	34.6850	34.7244	34.7638	34.8031	34.8425	34.8819	34.9213	34.9606	35.0000
890	35.0394	35.0787	35.1181	35.1575	35.1969	35.2362	35.2756	35.3150	35.3543	35.3937
900	35.4331	35.4724	35.5118	35.5512	35.5906	35.6299	35.6693	35.7087	35.7480	35.7874
910	35.8268	35.8661	35.9055	35.9449	35.9843	36.0236	36.0630	36.1024	36.1417	36.1811
920	36.2205	36.2598	36.2992	36.3386	36.3780	36.4173	36.4567	36.4961	36.5354	36.5748
930	36.6142	36.6535	36.6929	36.7323	36.7717	36.8110	36.8504	36.8898	36.9291	36.9685
940	37.0079	37.0472	37.0866	37.1260	37.1654	37.2047	37.2441	37.2835	37.3228	37.3622
950	37.4016	37.4409	37.4803	37.5197	37.5591	37.5984	37.6378	37.6772	37.7165	37.7559
960	37.7953	37.8346	37.8740	37.9134	37.9528	37.9921	38.0315	38.0709	38.1102	38.1496
970	38.1890	38.2283	38.2677	38.3071	38.3465	38.3858	38.4252	38.4646	38.5039	38.5433
980	38.5827	38.6220	38.6614	38.7008	38.7402	38.7795	38.8189	38.8583	38.8976	38.9370
990	38.9764	39.0157	39.0551	39.0945	39.1339	39.1732	39.2126	39.2520	39.2913	39.3307
1000	39.3701	...	...	...	...	...	...	...	...	...

Millimeters to Inches

mm	in.	mm	in.	mm	in.	mm	in.	mm	in.	mm	in.
100	3.93701	10	.39370	1	.03937	.1	.00394	.01	.00039	.001	.00004
200	7.87402	20	.78740	2	.07874	.2	.00787	.02	.00079	.002	.00008
300	11.81102	30	1.18110	3	.11811	.3	.01181	.03	.00118	.003	.00012
400	15.74803	40	1.57480	4	.15748	.4	.01575	.04	.00157	.004	.00016
500	19.68504	50	1.96850	5	.19685	.5	.01969	.05	.00197	.005	.00020
600	23.62205	60	2.36220	6	.23622	.6	.02362	.06	.00236	.006	.00024
700	27.55906	70	2.75591	7	.27559	.7	.02756	.07	.00276	.007	.00028
800	31.49606	80	3.14961	8	.31496	.8	.03150	.08	.00315	.008	.00031
900	35.43307	90	3.54331	9	.35433	.9	.03543	.09	.00354	.009	.00035
1,000	39.37008	100	3.93701	10	.39370	1.0	.03937	.10	.00394	.010	.00039

Table D-3. Fractional Inch to Millimeters and Feet to Millimeters Conversion Tables (Based on 1 in. = 2.54 millimeters, exactly)

\<td colspan=8\> Fractional Inch to Millimeters							
in.	mm	in.	mm	in.	mm	in.	mm
1/64	0.397	17/64	6.747	33/64	13.097	49/64	19.447
1/32	0.794	9/32	7.144	17/32	13.494	25/32	19.844
3/64	1.191	19/64	7.541	35/64	13.891	51/64	20.241
1/16	1.588	5/16	7.938	9/16	14.288	13/16	20.638
5/64	1.984	21/64	8.334	37/64	14.684	53/64	21.034
3/32	2.381	11/32	8.731	19/32	15.081	27/32	21.431
7/64	2.778	23/64	9.128	39/64	15.478	55/64	21.828
1/8	3.175	3/8	9.525	5/8	15.875	7/8	22.225
9/64	3.572	25/64	9.922	41/64	16.272	57/64	22.622
5/32	3.969	13/32	10.319	21/32	16.669	29/32	23.019
11/64	4.366	27/64	10.716	43/64	17.066	59/64	23.416
3/16	4.762	7/16	11.112	11/16	17.462	15/16	23.812
13/64	5.159	29/64	11.509	45/64	17.859	61/64	24.209
7/32	5.556	15/32	11.906	23/32	18.256	31/32	24.606
15/64	5.953	31/64	12.303	47/64	18.653	63/64	25.003
1/4	6.350	1/2	12.700	3/4	19.050	1	25.400

Inches to Millimeters

in.	mm	in.	mm	in.	mm	in.	mm	in.	mm
1	25.4	3	76.2	5	127.0	7	177.8	9	228.6
2	50.8	4	101.6	6	152.4	8	203.2	10	254.0

(continued: 11 | 279.4 ; 12 | 304.8)

Feet to Millimeters

ft	mm	ft	mm	ft	mm	ft	mm	ft	mm
100	30,480	10	3,048	1	304.8	0.1	30.48	0.01	3.048
200	60,960	20	6,096	2	609.6	0.2	60.96	0.02	6.096
300	91,440	30	9,144	3	914.4	0.3	91.44	0.03	9.144
400	121,920	40	12,192	4	1,219.2	0.4	121.92	0.04	12.192
500	152,400	50	15,240	5	1,524.0	0.5	152.40	0.05	15.240
600	182,880	60	18,288	6	1,828.8	0.6	182.88	0.06	18.288
700	213,360	70	21,336	7	2,133.6	0.7	213.36	0.07	21.336
800	243,840	80	24,384	8	2,438.4	0.8	243.84	0.08	24.384
900	274,320	90	27,432	9	2,743.2	0.9	274.32	0.09	27.432
1,000	304,800	100	30,480	10	3,048.0	1.0	304.80	0.10	30.480

Table D-4. Microinches to Micrometers (microns)
(Based on 1 microinch = 0.0254 micrometers, exactly)

Micro-inches	0	1	2	3	4	5	6	7	8	9
	\multicolumn{10}{c}{Micrometers (microns)}									
0	.....	0.025	0.051	0.076	0.102	0.127	0.152	0.178	0.203	0.229
10	0.254	0.279	0.305	0.330	0.356	0.381	0.406	0.432	0.457	0.483
20	0.508	0.533	0.559	0.584	0.610	0.635	0.660	0.686	0.711	0.737
30	0.762	0.787	0.813	0.838	0.864	0.889	0.914	0.940	0.965	0.991
40	1.016	1.041	1.067	1.092	1.118	1.143	1.168	1.194	1.219	1.245
50	1.270	1.295	1.321	1.346	1.372	1.397	1.422	1.448	1.473	1.499
60	1.524	1.549	1.575	1.600	1.626	1.651	1.676	1.702	1.727	1.753
70	1.778	1.803	1.829	1.854	1.880	1.905	1.930	1.956	1.981	2.007
80	2.032	2.057	2.083	2.108	2.134	2.159	2.184	2.210	2.235	2.261
90	2.286	2.311	2.337	2.362	2.388	2.413	2.438	2.464	2.489	2.515
100	2.540	2.565	2.591	2.616	2.642	2.667	2.692	2.718	2.743	2.769
110	2.794	2.819	2.845	2.870	2.896	2.921	2.946	2.972	2.997	3.023
120	3.048	3.073	3.099	3.124	3.150	3.175	3.200	3.226	3.251	3.277
130	3.302	3.327	3.353	3.378	3.404	3.429	3.454	3.480	3.505	3.531
140	3.556	3.581	3.607	3.632	3.658	3.683	3.708	3.734	3.759	3.785
150	3.810	3.835	3.861	3.886	3.912	3.937	3.962	3.988	4.013	4.039
160	4.064	4.089	4.115	4.140	4.166	4.191	4.216	4.242	4.267	4.293
170	4.318	4.343	4.369	4.394	4.420	4.445	4.470	4.496	4.521	4.547
180	4.572	4.597	4.623	4.648	4.674	4.699	4.724	4.750	4.775	4.801
190	4.826	4.851	4.877	4.902	4.928	4.953	4.978	5.004	5.029	5.055
200	5.080	5.105	5.131	5.156	5.182	5.207	5.232	5.258	5.283	5.309
210	5.334	5.359	5.385	5.410	5.436	5.461	5.486	5.512	5.537	5.563
220	5.588	5.613	5.639	5.664	5.690	5.715	5.740	5.766	5.791	5.817
230	5.842	5.867	5.893	5.918	5.944	5.969	5.994	6.020	6.045	6.071
240	6.096	6.121	6.147	6.172	6.198	6.223	6.248	6.274	6.299	6.325
250	6.350	6.375	6.401	6.426	6.452	6.477	6.502	6.528	6.553	6.579
260	6.604	6.629	6.655	6.680	6.706	6.731	6.756	6.782	6.807	6.833
270	6.858	6.883	6.909	6.934	6.960	6.985	7.010	7.036	7.061	7.087
280	7.112	7.137	7.163	7.188	7.214	7.239	7.264	7.290	7.315	7.341
290	7.366	7.391	7.417	7.442	7.468	7.493	7.518	7.544	7.569	7.595

μin.	μm	μin.	μm	μin.	μm	μin.	μm	μin.	μm
300	7.620	900	22.860	1500	38.100	2100	53.340	2700	68.580
600	15.240	1200	30.480	1800	45.720	2400	60.960	3000	76.200

**Table D-5. Micrometers (microns) to Microinches
(Based on 1 microinch = 0.0254 micrometers, exactly)**

Micrometers (microns)	0	0.01	0.02	0.03	0.04	0.05	0.06	0.07	0.08	0.09
	Microinches									
0	....	0.4	0.8	1.2	1.6	2.0	2.4	2.8	3.1	3.5
0.10	3.9	4.3	4.7	5.1	5.5	5.9	6.3	6.7	7.1	7.5
0.20	7.9	8.3	8.7	9.1	9.4	9.8	10.2	10.6	11.0	11.4
0.30	11.8	12.2	12.6	13.0	13.4	13.8	14.2	14.6	15.0	15.4
0.40	15.7	16.1	16.5	16.9	17.3	17.7	18.1	18.5	18.9	19.3
0.50	19.7	20.1	20.5	20.9	21.3	21.7	22.0	22.4	22.8	23.2
0.60	23.6	24.0	24.4	24.8	25.2	25.6	26.0	26.4	26.8	27.2
0.70	27.6	28.0	28.3	28.7	29.1	29.5	29.9	30.3	30.7	31.1
0.80	31.5	31.9	32.3	32.7	33.1	33.5	33.9	34.3	34.6	35.0
0.90	35.4	35.8	36.2	36.6	37.0	37.4	37.8	38.2	38.6	39.0
1.00	39.4	39.8	40.2	40.6	40.9	41.3	41.7	42.1	42.5	42.9
1.10	43.3	43.7	44.1	44.5	44.9	45.3	45.7	46.1	46.5	46.9
1.20	47.2	47.6	48.0	48.4	48.8	49.2	49.6	50.0	50.4	50.8
1.30	51.2	51.6	52.0	52.4	52.8	53.1	53.5	53.9	54.3	54.7
1.40	55.1	55.5	55.9	56.3	56.7	57.1	57.5	57.9	58.3	58.7
1.50	59.1	59.4	59.8	60.2	60.6	61.0	61.4	61.8	62.2	62.6
1.60	63.0	63.4	63.8	64.2	64.6	65.0	65.4	65.7	66.1	66.5
1.70	66.9	67.3	67.7	68.1	68.5	68.9	69.3	69.7	70.1	70.5
1.80	70.9	71.3	71.7	72.0	72.4	72.8	73.2	73.6	74.0	74.4
1.90	74.8	75.2	75.6	76.0	76.4	76.8	77.2	77.6	78.0	78.3
2.00	78.7	79.1	79.5	79.9	80.3	80.7	81.1	81.5	81.9	82.3
2.10	82.7	83.1	83.5	83.9	84.3	84.6	85.0	85.4	85.8	86.2
2.20	86.6	87.0	87.4	87.8	88.2	88.6	89.0	89.4	89.8	90.2
2.30	90.6	90.9	91.3	91.7	92.1	92.5	92.9	93.3	93.7	94.1
2.40	94.5	94.9	95.3	95.7	96.1	96.5	96.9	97.2	97.6	98.0
2.50	98.4	98.8	99.2	99.6	100.0	100.4	100.8	101.2	101.6	102.0
2.60	102.4	102.8	103.1	103.5	103.9	104.3	104.7	105.1	105.5	105.9
2.70	106.3	106.7	107.1	107.5	107.9	108.3	108.7	109.1	109.4	109.8
2.80	110.2	110.6	111.0	111.4	111.8	112.2	112.6	113.0	113.4	113.8
2.90	114.2	114.6	115.0	115.4	115.7	116.1	116.5	116.9	117.3	117.7
3.00	118.1	118.5	118.9	119.3	119.7	120.1	120.5	120.9	121.3	121.7
3.10	122.0	122.4	122.8	123.2	123.6	124.0	124.4	124.8	125.2	125.6
3.20	126.0	126.4	126.8	127.2	127.6	128.0	128.3	128.7	129.1	129.5
3.30	129.9	130.3	130.7	131.1	131.5	131.9	132.3	132.7	133.1	133.5
3.40	133.9	134.3	134.6	135.0	135.4	135.8	136.2	136.6	137.0	137.4
3.50	137.8	138.2	138.6	139.0	139.4	139.8	140.2	140.6	140.9	141.3
3.60	141.7	142.1	142.5	142.9	143.3	143.7	144.1	144.5	144.9	145.3
3.70	145.7	146.1	146.5	146.9	147.2	147.6	148.0	148.4	148.8	149.2
3.80	149.6	150.0	150.4	150.8	151.2	151.6	152.0	152.4	152.8	153.1
3.90	153.5	153.9	154.3	154.7	155.1	155.5	155.9	156.3	156.7	157.1
4.00	157.5	157.9	158.3	158.7	159.1	159.4	159.8	160.2	160.6	161.0
4.10	161.4	161.8	162.2	162.6	163.0	163.4	163.8	164.2	164.6	165.0
4.20	165.4	165.7	166.1	166.5	166.9	167.3	167.7	168.1	168.5	168.9

Table D-5. Micrometers (microns) to Microinches (continued)

Micrometers (microns)	0	0.01	0.02	0.03	0.04	0.05	0.06	0.07	0.08	0.09
					Microinches					
4.30	169.3	169.7	170.1	170.5	170.9	171.3	171.7	172.0	172.4	172.8
4.40	173.2	173.6	174.0	174.4	174.8	175.2	175.6	176.0	176.4	176.8
4.50	177.2	177.6	178.0	178.3	178.7	179.1	179.5	179.9	180.3	180.7
4.60	181.1	181.5	181.9	182.3	182.7	183.1	183.5	183.9	184.3	184.6
4.70	185.0	185.4	185.8	186.2	186.6	187.0	187.4	187.8	188.2	188.6
4.80	189.0	189.4	189.8	190.2	190.6	190.9	191.3	191.7	192.1	192.5
4.90	192.9	193.3	193.7	194.1	194.5	194.9	195.3	195.7	196.1	196.5
5.00	196.9	197.2	197.6	198.0	198.4	198.8	199.2	199.6	200.0	200.4

μm	μin.	μm	μin.	μm	μin.	μm	μin.	μm	μin.
10	393.7	20	787.4	30	1,181.1	40	1,574.8	50	1,968.5
15	590.6	25	984.3	35	1,378.0	45	1,771.7	55	2,165.4

Table D-6. Area: Square Feet to Square Meters (Correct to the nearest 0.01 m^2)

ft^2	0	10	20	30	40	50	60	70	80	90
					m^2					
0	—	0.93	1.86	2.79	3.72	4.65	5.57	6.50	7.43	8.36
100	9.29	10.22	11.15	12.08	13.01	13.94	14.86	15.79	16.72	17.65
200	18.58	19.51	20.44	21.37	22.30	23.23	24.15	25.08	26.01	26.94
300	27.87	28.80	29.73	30.66	31.59	32.52	33.45	34.37	35.30	36.23
400	37.16	38.09	39.02	39.95	40.88	41.81	42.74	43.66	44.59	45.52
500	46.45	47.38	48.31	49.24	50.17	51.10	52.03	52.95	53.88	54.81
600	55.74	56.67	57.60	58.53	59.46	60.39	61.32	62.25	63.17	64.10
700	65.03	65.96	66.89	67.82	68.75	69.68	70.61	71.54	72.46	73.39
800	74.32	72.25	76.18	77.11	78.04	78.97	79.90	80.83	81.76	82.68
900	83.61	84.54	85.47	86.40	87.33	88.26	89.19	90.12	91.05	91.97
1000	92.90	—	—	—	—	—	—	—	—	—

AUXILIARY TABLE

ft^2	1	2	3	4	5	6	7	8	9
m^2	0.09	0.19	0.28	0.37	0.46	0.56	0.65	0.74	0.84

Table D-7. Area: Square Yards to Square Meters
(Correct to the nearest 0.01 m²)

yd²	0	10	20	30	40	50	60	70	80	90
					m²					
0	—	8.36	16.72	25.08	33.45	41.81	50.17	58.53	66.89	75.25
100	83.61	91.97	100.34	108.70	117.06	125.42	133.78	142.14	150.50	158.86
200	167.23	175.59	183.95	192.31	200.67	209.03	217.39	225.75	234.12	242.48
300	250.84	259.20	267.56	275.92	284.28	292.65	301.01	309.37	317.73	326.09
400	334.45	342.81	351.17	359.54	367.90	376.26	384.62	392.98	401.34	409.70
500	418.06	426.43	434.79	443.15	451.51	459.87	468.23	476.59	484.95	493.32
600	501.68	510.04	518.40	526.76	535.12	543.48	551.84	560.21	568.57	576.93
700	585.29	593.65	602.01	610.37	618.73	627.10	635.46	643.82	652.18	660.54
800	668.90	677.26	685.62	693.99	702.35	710.71	719.07	727.43	735.79	744.15
900	753.52	760.88	769.37	777.60	785.96	794.32	802.68	811.04	819.41	827.77
1000	836.13	—	—	—	—	—	—	—	—	—

AUXILIARY TABLE

yd²	1	2	3	4	5	6	7	8	9
m²	0.84	1.67	2.51	3.34	4.18	5.02	5.85	6.69	7.53

Table D-8. Volume: Cubic Inches to Cubic Centimeters or Milliliters (Correct to the nearest cm^3)

in^3	0	10	20	30	40	50	60	70	80	90
					cm^3					
0	—	164	328	492	655	819	983	1 147	1 311	1 475
100	1 639	1 803	1 966	2 130	2 294	2 458	2 622	2 786	2 950	3 114
200	3 277	3 441	3 605	3 769	3 933	4 097	4 261	4 425	4 588	4 752
300	4 916	5 080	5 244	5 408	5 572	5 735	5 899	6 063	6 227	6 391
400	6 555	6 719	6 883	7 046	7 210	7 374	7 538	7 702	7 866	8 030
500	8 194	8 357	8 521	8 685	8 849	9 013	9 177	9 341	9 505	9 668
600	9 832	9 996	10 160	10 324	10 488	10 652	10 816	10 979	11 143	11 307
700	11 471	11 635	11 799	11 963	12 126	12 290	12 454	12 618	12 782	12 946
800	13 110	13 274	13 437	13 601	13 765	13 929	14 093	14 257	14 421	14 585
900	14 748	14 912	15 076	15 240	15 404	15 568	15 732	15 896	16 059	16 223
1000	16 387	—	—	—	—	—	—	—	—	—

AUXILIARY TABLE

in^3	0	1	2	3	4	5	6	7	8	9
ml.	—	16	33	49	66	82	98	115	131	147

Table D-9. Volume: Cubic Feet to Cubic Meters
(Correct to the nearest 0.01 m³)

ft³	0	10	20	30	40	50	60	70	80	90
					m^3					
0	—	0.28	0.57	0.85	1.13	1.42	1.70	1.98	2.27	2.55
100	2.83	3.11	3.40	3.68	3.96	4.25	4.53	4.81	5.10	5.38
200	5.66	5.95	6.23	6.51	6.80	7.08	7.36	7.65	7.93	8.21
300	8.50	8.78	9.06	9.34	9.63	9.91	10.19	10.48	10.76	11.04
400	11.33	11.61	11.89	12.18	12.46	12.74	13.03	13.31	13.59	13.88
500	14.16	14.44	14.72	15.01	15.29	15.57	15.86	16.14	16.42	16.71
600	16.99	17.27	17.56	17.84	18.12	18.41	18.69	18.97	19.26	19.54
700	19.82	20.11	20.39	20.67	20.95	21.24	21.52	21.80	22.09	22.37
800	22.65	22.94	23.22	23.50	23.79	24.07	24.35	24.64	24.92	25.20
900	25.49	25.77	26.05	26.33	26.62	26.90	27.18	27.47	27.75	28.03
1000	28.32	—	—	—	—	—	—	—	—	—

AUXILIARY TABLE

ft³	0	1	2	3	4	5	6	7	8	9
m^3	—	0.03	0.06	0.08	0.11	0.14	0.17	0.20	0.23	0.25

Table D-10. Volume: Cubic Yards to Cubic Meters
(Correct to the nearest 0.01 m³)

yd³	0	1	2	3	4	5	6	7	8	9
					m³					
	—	0.76	1.53	2.29	3.06	3.82	4.59	5.35	6.12	6.88
10	7.65	8.41	9.17	9.94	10.70	11.47	12.23	13.00	13.76	14.53
20	15.29	16.06	16.82	17.58	18.35	19.11	19.88	20.64	21.41	22.17
30	22.94	23.70	24.47	25.23	25.99	26.76	27.52	28.29	29.05	29.82
40	30.58	31.35	32.11	32.88	33.64	34.41	35.17	35.93	36.70	37.46
50	38.23	38.99	39.76	40.52	41.29	42.05	42.82	43.58	44.34	45.11
60	45.87	46.64	47.40	48.17	48.93	49.70	50.46	51.23	51.99	52.75
70	53.52	54.28	55.05	55.81	56.58	57.34	58.11	58.87	59.64	60.40
80	61.16	61.93	62.69	63.46	64.22	64.99	65.75	66.52	67.28	68.05
90	68.81	69.57	70.34	71.10	71.87	72.63	73.40	74.16	74.93	75.69
100	75.46	—	—	—	—	—	—	—	—	—

Table D-11. Capacity: Gallons to Liters
(Correct to the nearest 10 ml)

gal	0	1	2	3	4	5	6	7	8	9
					liters					
0	—	4.55	9.09	13.64	18.18	22.73	27.28	31.82	36.37	40.91
10	45.46	50.01	54.55	59.10	63.65	68.19	72.74	77.28	81.83	86.38
20	90.92	95.47	100.01	104.56	109.11	113.65	118.20	122.74	127.29	131.84
30	136.38	140.93	145.48	150.02	154.57	159.11	163.66	168.21	172.75	177.30
40	181.84	186.39	190.94	195.48	200.03	204.57	209.12	213.67	218.21	222.76
50	227.31	231.85	236.40	240.94	245.49	250.04	254.58	259.13	263.67	268.22
60	273.77	277.31	281.86	286.40	290.95	295.50	300.04	304.59	309.13	313.68
70	318.23	322.77	327.32	331.87	336.41	340.96	345.50	350.05	354.60	359.14
80	363.69	368.23	372.78	377.33	381.87	386.42	390.96	395.51	400.06	404.60
90	409.15	413.69	418.24	422.79	427.33	431.88	436.43	440.97	445.52	450.06
100	454.61	—	—	—	—	—	—	—	—	—

AUXILIARY TABLE

Pint	1	2	3	4	5	6	7
liter	0.568	1.136	1.705	2.273	2.841	3.410	3.978

Appendixes • 409

Table D-12. Mass: Pounds to Kilograms (Correct to the nearest gram or 0.001 kg)

Pounds	0	1	2	3	4	5	6	7	8	9
					Kilograms					
0	—	0.454	0.907	1.361	1.814	2.268	2.722	3.175	3.629	4.082
10	4.536	4.990	5.443	5.897	6.350	6.804	7.257	7.711	8.165	8.618
20	9.072	9.525	9.979	10.433	10.886	11.340	11.793	12.247	12.701	13.154
30	13.608	14.061	14.515	14.969	15.422	15.876	16.329	16.783	17.237	17.690
40	18.144	18.597	19.051	19.505	19.958	20.412	20.865	21.319	21.772	22.226
50	22.680	23.133	23.587	24.040	24.494	24.948	25.401	25.855	26.308	26.762
60	27.216	27.669	28.123	28.576	29.030	29.484	29.937	30.391	30.844	31.298
70	31.752	32.205	32.659	33.112	33.566	24.019	34.473	34.927	35.380	35.834
80	36.287	36.741	37.195	37.648	38.102	38.555	39.009	39.463	39.916	40.370
90	40.823	41.277	41.731	42.184	42.638	43.091	43.549	43.999	44.452	44.906
100	45.359	—	—	—	—	—	—	—	—	—

Table D-13. Mass: Ton to Tonne

Tons	0	1	2	3	4	5	6	7	8	9
0	—	1.0160	2.0321	3.0481	4.0642	5.0802	6.0963	7.1123	8.1284	9.1444
10	10.1605	11.1765	12.1926	13.2086	14.2247	15.2407	16.2568	17.2728	18.2888	19.3049
20	20.3209	21.3370	22.3530	23.3691	24.3851	25.4012	26.4172	27.4333	28.4493	29.4654
30	30.4814	31.4975	32.5135	33.5295	34.5456	35.5616	36.5777	37.5937	38.6098	39.6258
40	40.6419	41.6579	42.6740	43.6900	44.7061	45.7221	46.7382	47.7542	48.7703	49.7865
50	50.8023	51.8184	52.8344	53.8505	54.8665	55.8826	56.8986	57.9147	58.9307	59.9468
60	60.9628	61.9789	62.9949	64.0110	65.0270	66.0430	67.0591	68.0751	69.0912	70.1072
70	71.1233	72.1393	73.1554	74.1714	75.1875	76.2035	77.2196	78.2356	79.2517	80.2677
80	81.2838	82.2998	83.3158	84.3319	85.3479	86.3640	87.3800	88.3961	89.4121	90.4282
90	91.4442	92.4603	93.4763	94.4924	95.5084	96.5245	97.5405	98.5566	99.5726	100.589
100	101.605	—	—	—	—	—	—	—	—	—

Table D-14. Mass: CWT—QTR to Tonne

qtr	0	1	2	3	qtr	0	1	2	3
cwt		tonne			cwt		tonne		
0	—	0.0127	0.0254	0.0381	10	0.5080	0.5207	0.5334	0.5461
1	0.0508	0.0635	0.0762	0.0889	11	0.3588	0.5715	0.5842	0.5969
2	0.1016	0.1143	0.1270	0.1397	12	0.6096	0.6223	0.6350	0.6477
3	0.1524	0.1651	0.1778	0.1905	13	0.6604	0.6731	0.6858	0.6985
4	0.2032	0.2159	0.2286	0.2413	14	0.7112	0.7239	0.7366	0.7493
5	0.2540	0.2667	0.2794	0.2921	15	0.7620	0.7747	0.7874	0.8001
6	0.3048	0.3175	0.3302	0.3429	16	0.8128	0.8255	0.8382	0.8509
7	0.3556	0.3683	0.3810	0.3937	17	0.8636	0.8763	0.8890	0.9017
8	0.4064	0.4191	0.4318	0.4445	18	0.9144	0.9271	0.9398	0.9525
9	0.4572	0.4699	0.4826	0.4953	19	0.9652	0.9779	0.9906	1.0033

Table D-15. Pressure: lb/in.^{2}f to 10^5N/m^2 (bar)

lb/in.^{2}f	0	1	2	3	4	5	6	7	8	9
0	—	0.069	0.138	0.207	0.276	0.345	0.414	0.483	0.552	0.621
10	0.689	0.758	0.827	0.896	0.965	1.034	1.103	1.172	1.241	1.310
20	1.379	1.448	1.517	1.586	1.654	1.724	1.793	1.862	1.931	1.999
30	2.068	2.137	2.206	2.275	2.344	2.413	2.482	2.551	2.620	2.689
40	2.758	2.827	2.895	2.965	3.034	3.103	3.172	3.241	3.309	3.378
50	3.447	3.516	3.585	3.654	3.723	3.792	3.861	3.930	3.999	4.068
60	4.137	4.206	4.275	4.344	4.413	4.442	4.551	4.619	4.688	4.757
70	4.826	4.895	4.964	5.033	5.102	5.171	5.240	5.309	5.378	5.447
80	5.516	5.585	5.654	5.723	5.792	5.861	5.929	5.998	6.067	6.136
90	6.205	6.274	6.343	6.412	6.481	6.550	6.619	6.688	6.757	6.826
100	6.895	—	—	—	—	—	—	—	—	—

AUXILIARY TABLE

lb/in.^{2}f	150	200	250	300	350	400	450	500	550	600
10^5N/m^2 (bar)	10.342	13.79	17.24	20.68	24.13	27.58	31.03	34.47	37.92	41.37
lb/in.^{2}f	650	700	750	800	850	900	950	1 000	—	—
10^5N/m^2 (bar)	44.82	48.26	51.71	55.16	58.61	62.05	65.50	68.95	—	—

To convert from lb/in.^{2}f to kN/m^2 multiply above factors by 100; thus for 15 lb/in.^{2}f = 1.034 bar = 103.4 kN/m^2

Table D-16. Stress: Tons/in.^{2}/F to MN/m^2 (N/mm^2)

ton/in.^{2}f	0	1	2	3	4	5	6	7	8	9
0	—	15.44	30.89	46.33	61.78	77.22	92.67	108.11	123.55	139.00
10	154.44	169.89	185.33	200.78	216.22	231.66	247.11	262.55	278.00	293.44
20	308.89	324.33	339.77	355.22	370.66	386.11	401.55	417.00	432.44	447.88
30	463.33	478.77	494.22	509.66	525.11	540.55	555.99	571.44	586.88	602.33
40	617.77	633.22	648.66	664.10	679.55	694.99	710.44	725.88	741.32	756.77
50	772.21	787.66	803.10	818.55	833.99	849.43	864.88	880.32	895.77	911.21
60	926.66	942.10	957.55	972.99	988.43	1003.88	1019.32	1034.77	1050.21	1065.65
70	1081.10	1096.54	1111.99	1127.43	1142.87	1158.32	1173.77	1189.21	1204.65	1220.10
80	1235.54	1250.98	1266.43	1281.87	1297.32	1312.76	1328.21	1343.65	1359.09	1374.54
90	1389.98	1405.43	1420.87	1436.32	1451.76	1467.21	1482.65	1498.09	1513.54	1528.98
100	1544.43	—	—	—	—	—	—	—	—	—

Table D-17. Stress: kgf/mm^2 to MN/m^2 (N/mm^2)

kgf/mm^2	0	1	2	3	4	5	6	7	8	9
0	0.00	9.81	19.61	29.42	39.23	40.03	58.84	68.65	78.45	88.26
10	98.07	107.87	117.68	127.49	137.29	147.10	156.91	166.71	176.52	186.33
20	196.13	205.94	215.75	225.55	235.36	245.17	254.97	264.78	274.59	284.39
30	294.20	304.01	313.81	323.62	333.43	343.23	353.04	362.85	372.65	382.46
40	392.27	402.07	411.88	421.69	431.49	441.30	451.11	460.91	470.72	480.53
50	490.33	500.14	509.95	519.75	529.56	539.37	549.17	558.98	568.79	578.59
60	588.40	598.21	608.01	617.82	627.63	637.43	647.24	657.05	666.85	676.66
70	686.47	696.27	706.08	715.89	725.69	735.50	745.31	755.11	764.92	774.73
80	784.53	794.34	804.15	813.95	823.76	833.57	843.37	853.18	862.99	872.79
90	882.60	892.41	902.21	912.02	921.83	931.63	941.44	951.25	961.05	970.86
100	980.67	—	—	—	—	—	—	—	—	—

AUXILIARY TABLE

kgf/mm^2	150	200	250	300	350	400	450	500	550	600
MN/m^2	1 471.00	1 961.33	2 451.66	2 942.00	3 432.33	3 922.66	4 412.99	4 903.33	5 393.66	5 883.99

Table D-18. SI Base-Units

Measure	Unit Name	Symbol	Technical Definition
Length	meter	m	The distance traveled by light in vacuum in 1/299,792,458 second, or 3.335641×10^{-9} seconds.
Mass	kilogram	kg	The quantity of matter of a standard prototype that is kept by the Bureau Internationale des Poids et Mesures at Sèvres, France.
Time	second	s	The amount of time of 9,192,631,770 periods of radiation that correspond to the transition between two superfine levels of an atom of cesium-133.
Electric Current	ampere	A	The constant current over given conductors that would produce a force equal to 2×10^{-7} N/m length.
Thermodynamic Temperature	degree Kelvin	K	1/273.16 of the thermodynamic temperature of the triple point of water.
Amount of a Substance	mole	mol	An amount of substance that has as many elementary entities as there are atoms.
Luminous Intensity	candela	cd	The amount of luminous intensity of a surface of 1/600,000 m^2 of a black body at a given temperature and pressure.

Table D-19. Derived SI Quantities

Quantity	Unit	Symbol	Definition
Force	newton	N	kgm/s^2
Work, Energy, Quantity of Heat	joule	J	NM
Power	watt	W	J/s
Electric Charge	coulomb	C	As
Electric Potential	volt	V	W/A
Electric Capacitance	farad	F	C/V
Electric Resistance	ohm	Ω	V/A
Magnetic Flux	weber	Wb	Vs
Inductance	henry	H	Vs/A
Luminous Flux	lumen	lm	$cd.sr$
Illumination	lux	lx	lm/m^2

Table D-20. Complex SI Units

Quantity	Unit	Notation
Area	square meter	m^2
Volume	cubic meter	m^3
Frequency	hertz	Hz (cycle/s)
Density (Mass Density)	kilogram/cubic meter	kg/m^3
Velocity	meter/second	m/s
Angular Velocity	radian/second	rad/s
Acceleration	meter/second squared	m/s^2
Angular Acceleration	radian/second squared	rad/s^2
Pressure	pascal	Pa (N/m^2)
Surface Tension	newton/meter	N/m
Dynamic Viscosity	newton second/meter squared	Ns/m^2
Kinematic Viscosity and Diffusion Coefficient	meter squared/second	m^2/s
Thermal Conductivity	watt/meter degree Kelvin	W/(m°K)
Electric Field Strength	volt/meter	V/m
Magnetic Flux Density	telsa	T (Wb/m^2)
Magnetic Field Strength	ampere/meter	A/m
Luminance	candela/meter squared	cd/m^2

Table D-21. Prefixes for Unit Multiples and Submultiples

Prefix	Symbol	Unit Factor Multiple
tera	T	10^{12}
giga	G	10^9
mega	M	10^6
kilo	k	10^3
hecto	h	10^2
deka	da	10
deci	d	10^{-1}
centi	c	10^{-2}
milli	m	10^{-3}
micro	μ	10^{-6}
nano	n	10^{-9}
pico	p	10^{-12}
femto	f	10^{-15}
atto	a	10^{-18}

Index

Absolute pressure, 200, 222
Absolute temperature, 222, 223
Absolute unit system, 85
Absolute viscosity, 203
Acceleration, 85, 152
Acute triangles, area of, 21
Addendum, 244
Addendum angle, 250
Addition, vector, 89
Air, density of, 226
Air services, 360–361
Algebra, 10–14
Algebraic expressions, 10–13
Algebraic operations, 13–14
Algebraic symbols, 10
Alphabetical exponents, 12
Ammeters, 314
Angle(s)
    complementary, 75–76
    multiple, 76–77
    negative, 76
    of repose, 113
    of twist, 190–191
Angular contact ball bearings, 286
Angular motion, 152–153
Angular velocities, conversion of, 160, 161
    to radians per second, 160
Annual cost method, 334
Answers to exercises, 371–381
Apothem of triangle, 79
Arc
    center angle of, 29
    center of gravity of, 134

Area
    composite, moment of inertia of, 146
    moment of inertia of, 142–147
    of plane figures, 16–34
    second moment of, 142
    surface, 15–37
Area measures, 16
Assumed operating temperature, bearing, 283–285
Atmospheric pressure, 199, 208
Average speed, 152
Axial pitch, 249

B

Backing, 248
Ball bearings, 286–288
Bars
    hollow circular, 185–186, 190
    rectangular, 186
Base circle, 228
Base circle diameter, 241
Base diameter, spline, 266
Base time allowances, 363–364
Bearing capacity number, 283
Bearing pressure, 282–283
Bearing pressure parameter, 282
Bearings, 272, 280–289
    ball and roller, 286–288
    plain, 280–283
    thrust, 280, 283–285
Bernoulli's equation, 209–211

416 • INDEX

Bevel gear applications, 248
Bevel gear formulas, 250, 251–252
Bevel gear nomenclature, 247, 249
Bevel gears, 245–248
BHN (Brinell hardness number), 291
Bolts, working strength of, 262–263
Boyle's law, 221–224
Break-even analysis, 332–333
Brinell hardness number (BHN), 291
British thermal unit (BTU), 360
Broaching, 302
    unit power for, 305
Broaching formulas, 304
BTU (British thermal unit), 360
Budgetary cost control, 336

C

Cam displacement, 228
Cam formulas, 232–237
Cam nomenclature, 228–237
Cam problems, 237–239
Cam profile, 228
Cams, 227–239
Capital recovery factor, 334, 335
Center angle of arc, 26
Center distance, 241
Center of gravity, 132–140
Centigrade temperature scale, 221
Centimeter-gram-second (cgs) system, 87
Centrifugal force, 165–166
Centripetal force, 166
Centroids, 133–134
cgs (centimeter-gram-second) system, 85
Characteristics of logarithms, 6
Charles's law, 222–223
Chord of circular segment, 30–31
Circle sectors, center of gravity of, 136
Circle segments, center of gravity of, 136
Circles, 28–31
    area of, 29
    moment of inertia of, 146
Circular bars, hollow, 185–186, 191

Circular pitch
    gear, 241
    spline, 266
Circular sectors, 29–30
Circular segments, 30–31
Circular shafts, 184–185, 190–191
Circular thickness, 241
Circumference of circle, 28
Clearance modulus, bearing, 286
Clockwise moments, negative, 106, 110
Coefficient
    of expansion, 181–182
    of friction, 112–113, 215–216
    table of, 116–117
Coiled springs, 276–278
Colinear forces, 86
Column formulas, Gordon-Rankin, 197
Columns
    intermediate, 196–197
    parabolic, 197
    slender, 197–198
Combined gas law, 223
Common logarithms, 6
Complementary angles, 75
Components, 86
Composition of forces, 86
Compression member formulas, 193–196
Compression members, strength of, 193–200
Compression springs, 278–280
Compressive stress, 169, 171–172
Concurrent forces, 86, 93–97
Cone distance, 248
Cones
    center of gravity of, 140
    conical-surface area of, 44
    frustrum of, *see* Frustrum of cone
    volume of, 44
Conical-surface area of cone, 44
Conjugate cam, 228
Constant acceleration and rotational motion, 157–160
Constant acceleration linear motion, 154–155
Constant velocity motion, 152–153, 232–233

Continuity equation, 212
Contribution ratio, 333–334
Control chart, 330
Coordinate system, 105
Coplanar forces, 86, 93–108
Coplanar parallel forces, 101
Cosecant function, 60
Cosine function, 60
Cosines, Law of, 71–72
Cost and production rate equations, 340–349
Cost control, budgetary, 336
Cotangent function, 60
Counterclockwise moments, positive, 106, 110
Couple, 97
    forces and, 97–99
Couple forces, 86
Crossover point, 228
Crown backing, 248
Crown height, 248
Cube roots, 5
Cubes, volume of, 38
Cutting-speed formulas, 291, 294–297
Cutting-speed tables, 291, 292–294
Cutting speeds
    and feeds, 290–297
    optimized, 351–353
Cycloidal motion, 236–237
Cylinder cam, 227
Cylinders
    center of gravity of, 138
    surface area of, 42–43
    volume of, 42–43

D

Darcy formula, 214
Decagons, 27
Deceleration, 154
Decimal exponents, 12–13
Dedendum, 241
Dedendum angle, 248
Deformation, thermal, 181–182
Density, 200
    of air, 226
    of liquids, 202

Depth-of-cut factors, 296
Design stress, 180
Diagonal dimension, 16
Diameter of circle, 28
Diametral pitch, 241
Differential manometer, 208
Differential pulley, 130–131
Dimensions of plane figures, 16–34
Discharge factor, 225
Displacement diagrams, 230–231
Drilling, 300
Drilling formulas, 304
Drilling operations, 352
Drilling tool-life formulas, 353
Drum cam, 227
Dynamic hydraulic systems, 209–221
Dynamic viscosity, 204
Dynamics, 83

E

Economical order quantity, 326
Economics
    in machining, 338
    in manufacturing, 332–335
Elastic limit, 170, 180
Elastic region, 170
Elasticity, modulus of, see Modulus of elasticity
Electrical horsepower, 360
Electrical power, 359–360
Electricity, 358–360
Ellipses
    area of, 32
    perimeter of, 32
Ellipsoids, volume of, 51
Engagement, length of, 264
Equivalent standard manpower requirement, 365
Euler formula, 198
Evaluation instrument, 367
Exercises, answers to, 371–381
Expansion, coefficient of, 181–182
Exponents, 3–4, 11–12
    alphabetical, 12
    decimal, 12–13
    fractional, 12–13

negative, 12
numerical, 12
Extension springs, 278–280
External form diameter, spline, 271
External major diameter, spline, 269
External minor diameter, spline, 270

F

Face angle, 248
Face width, 241
Facilities, physical, 354–362
Factor of safety, 180–181
Fahrenheit temperature scale, 223
Feed factors, 295
Fillets
    area of, 32
    center of gravity of, 136
Fittings, friction loss in, 220–222
Flat spring formulas, 277
Flat springs, 274–278
Floor space, 355
Flow coefficient, 222
Flow losses, 228
Fluid concepts, static, 206–211
Fluid system calculations, 208–210
Fluidics, 199
Fluids, basic properties of, 199–204
Follower displacement, 228
Force(s), 86
    adding and subtracting, 89
    centrifugal, 165–166
    centripetal, 165–166
    colinear, 86
    composition of, 86
    concurrent, 86, 93–97
    coplanar, 86, 93–100
    couple, 86
    couple and, 97–99
    nonconcurrent, 86, 100–103
    noncoplanar, *see* Noncoplanar forces
    nonintersecting, 86, 100–103
    parallel, *see* Parallel forces
    parallelogram of, 86–87
    polygon of, 88
    power-and-force formulas, 314–315
    resolution of, 86
    skewed, 86
Force systems, 85–111
Foreign standard screw threads, 263
Form clearance, spline, 271
Fractional exponents, 12–13
Friction, 112–119, 217
    coefficient of, *see* Coefficient of friction
Friction horsepower, bearing, 282
Friction losses
    in fittings, 218–220
    in pipes, 216–218
Friction torque, bearing, 282
Frustrum
    of cone
        center of gravity of, 139–140
        volume of, 45
    of pyramid
        center of gravity of, 140
        volume of, 41
Fuel consumption, 361
Fundamental trigonometric identities, 75

G

Gas flow losses, 225
Gas laws, 221–224
Gauge pressure, 199–207, 208
Gay-Lussac's law, 223
Gear problems, 253
Gear ratio, 241
Gears, 240–259
    bevel, 245–249
    helical, 253
    spur, 240–245
    worm, 249, 252–255
Geometry, 15
Gordon-Rankin column formulas, 197
Graphical solutions, 86–91
Gravitational unit system, 85
Gravity
    center of, 132–141
    specific, 201
Greek letters, 11
Grinding, 300–301
Grinding operations formulas, 310–311

Guide bearings, 280
Gyration, radius of, 192–196

H

Half-angle formulas, 77–79
Haversines, 79
Head loss, 218
Heat services, 360–361
Helical gear formulas, 256
Helical gears, 253
Hemispheres, center of gravity of, 138
Heptagons, 27
Hexagons, regular, area of, 24–25
Holding costs, 327
Hollow circular bars, 185–186, 197
Horsepower, 84
  electrical, 358
Horsepower formulas, 164
Horsepower rating, 164
Human resources management, 362–365
Hydraulic system principles and equations, 209–215, 217
Hydraulic system problems, 215–220
Hydraulic systems, 200
  dynamic, 209
Hyphoid gear, 248
Hypotenuse, 20

I

Impulse, 85
Inclined-plane wedge problems, 115–119
Index, 5
Index cam, 228
Industrial organization, 319–320
Inertia, 84
  moments of, *see* Moments of inertia
  radius of, 192–195
Inspection process, 329
Instantaneous speed, 150
Intermediate columns, 196–197
Internal form diameter, spline, 270–271

Internal major diameter, spline, 269
Internal minor diameter, spline, 269
Inventory control, 325–328
Inverse trigonometric functions, 232
Involute spline, 267

J

Job evaluation, 367–369
Job rating plan, 368
Journal bearings, 280

K

Kinematic viscosity, 203–204
Kinematics, 83
Kinetic coefficient of friction, 116–117
Kinetic friction, 112
Kinetics, 83

L

Lathe turning cost basis, 339
Lathes, 298
Law
  of Cosines, 70–71
  of Sines, 68–69
  of Tangents, 71–72
Layout, plant, 355–358
Lead, 253
Lead angle, 254
Length of engagement, 263–264
Levers, 120–123
Linear measures, 16
Linear motion, 154–155
Lines, 15
  center of gravity of, 134–135
Liquids, density of, 202
Logarithmic operations, 8
Logarithmic tables, using, 7
Logarithms, 5–9
  common, 6
Lot quantity, 323

M

Machine elements, 260
Machining cost increase, 340
Machining costs, 338–340, 348–349
Machining economics, 338
Machining formulas, 302–313
Machining operations, 290
    basic, 297
    descriptions of, 298–302
Machining power requirements, 314
Major diameter fit, spline, 269
Management controls, 319
Managerial ratios, 321
Manometer
    differential, 209
    open-tube, 207
Manpower planning, 363, 365
Manpower requirement, equivalent standard, 365
Mantissas of logarithms, 6
Manufacturing economics, 332–334
Mass, 84, 85
Mathematical solutions
    coplanar forces, 93
    noncoplanar forces, 105–110
Matter, 84
Measure, units of, 16
Mechanical stress, 170
Mechanics, 83–84
    complex, 125
Merit rating, 369
Meter-kilogram-second (MKS) system, 85
Milling, 298–299
Milling formulas, 304
Milling operations, 299, 352
Milling tool-life formulas, 353
Minimum effective space width, spline, 268
MKS (meter-kilogram-second) system, 85
Modulus of elasticity, 171, 178–179
    in shear, 171
Moment(s)
    of force, 85–86, 90–91
    of inertia, 143–149
        of areas, 144–147
        polar, 142, 145–149
    second, of area, 142
Moment arm, 106
Momentum, 85
Motion
    constant acceleration and rotational, 157–159
    constant acceleration linear, 156–159, 154–157
    constant velocity, 152–153, 232–233
Mounting distance, 248
Multiple angles, 76–77

N

Negative angles, 75–76
Negative clockwise moments, 109, 110
Negative exponents, 12
Nonagons, 27
Nonconcurrent forces, 86, 100–103
Noncoplanar forces, 86, 105–110
    nonparallel, 108–110
    parallel, 105–108
Nonintersecting forces, 86, 102–103
Nonparallel noncoplanar forces, 108–110
Notation, scientific, 4
Number of teeth, 241
Numerical exponents, 12

O

Oblique angle trigonometry, 66–73
Obtuse triangles, area of, 21–22
Octagons, regular, area of, 26
Ohm's law, 358
Open-tube manometer, 207
Operation time per piece, 345–346
Optimization formulas, 350–353
Optimized cutting speeds, 351–353
Optimized tool life, 352
Optimum quantity order, 326
Origin, point of, 105
Out-of-stock costs, 327–328
Outside diameter, 241

Index • 421

P

Parabolas
    area of portion of, 33–34
    center of gravity for area of, 137–138
Parabolic columns, 197
Parabolic motion, 233–234
Paraboloids
    segments of, 51–53
    volume of, 51–52
Parallel connection, 359
Parallel forces, 86, 89–90
    coplanar, 101–102
    noncoplanar, 105–108
Parallelogram(s)
    area of, 18–19
    center of gravity of, 135
    of forces, 86–88
Pascal units, 171
Pascal's law, 205–206
Pentagons, 27
Personnel evaluation, 366–370
Physical facilities, 354–362
Pipe fittings, proportional constant for, 219
Pipes, friction losses in, 216–218
Pitch circle, 228
Pitch curve, 228
Pitch diameter
    gear, 241, 248
    spline, 268
Pitch point, 228
Plain bearings, 280–283
Plane figures, areas and dimensions of, 16–34
Plane trigonometry, 59, 60
Planes, 15
Planing and shaping, 299–300
Plant layout, 355–357
Plant services, 358–361
Plastic region, 170
Plate cam, 227
Pneumatic systems, 221–224
Point of origin, 105
Poisson's ratio, 171
Polar moments of inertia, 142, 146–149
Polygon(s)
    of forces, 88
    regular, area of, 27–28
Positive counterclockwise moments, 106, 110
Pound force, 87
Power, 84, 164
    electrical, 359–360
    unit, *see* Unit power *entries*
Power-and-force formulas, 314–315
Power formulas, 163
Power requirements, machining, 314
Powers of numbers, 3–4, 11–12
Pressure, 199–200, 206–207
    absolute, 200, 222
    atmospheric, 199, 208
Pressure angle, 228, 241
Pressure head, 206
Prime circle, 228
Prisms
    square, volume of, 38
    volume of, 39
Production control, 323–325
Production rate, 341, 346–347
Production rate equations, 340–349
Profit-to-sales ratio, 336
Proportional constant
    for pipe fittings, 219
    for smooth bends, 219
Proportional limit, 170
Pulley, differential, 130–131
Pulley systems, simple, 126–130
Pyramids
    center of gravity of, 140
    frustrum of, *see* Frustrum of pyramid
    volume of, 40

Q

Quality assurance, 329–332

R

Radial contact ball bearings, 286–287
Radial roller bearings, 286–280
Radians per second, conversion of angular velocity to, 160

Radicals, 4
Radicands, 4
Radius
   of circle, 28
   of gyration, 192–196
   of inertia, 192–196
Rankine temperature scale, 221
Reaming operations, 352
Reaming tool-life formulas, 353
Rectangles
   area of, 18
Rectangular bars, 186
Rectangular plate springs, 273, 274
Recorder point, 327
Repose, angle of, 113
Reserve factor, 180
Resolution of forces, 86
Resultants, 86
Reynolds number, 213–214
Right angle trigonometry, 59–64
Right triangles, area of, 20
Rivets, 265, 267–268
Roller bearings, 286–288
Root angle, 248
Root diameter, 241

S

Safety, factor of, 180–181
Sampling inspection tables, 331
Scalar quantities, 84
Scientific notation, 4
Screw-thread breaks, 264–265
Screw-thread nomenclature, 262
Screw-thread series, 262
Screws, 260–265
Secant function, 59–61
Section modulus, 192
Semielliptic springs, 275–276
Semiperimeters, 77
Series connection, 359
Services, plant, 358–362
Shafts, circular, 184–185, 190–191
Shaping and planing, 299–300
Shear, modulus of elasticity in, 171
Shearing strain, 177

Shearing stress, 169, 202
   torsional, 183–184
Short-compression members, 196
Side fit, spline, 267
Simple harmonic motion, 234–236
Sine function, 60, 62
Sines, Law of, 68–70
Siphon, 213
Size factor, 226
Skewed forces, 86
Slender columns, 197–198
Slenderness ratio, 192–193
Solids, 15–16
   center of gravity of, 138–140
   volumes of, 37–54
Span control, 320–323
Spandrels
   area of, 31
   center of gravity of, 137
Specific gravity, 200
Specific volume, 201
Specific weight, 200–202
Speed
   average, 150
   instantaneous, 150
Spheres
   surface area of, 46–47
   volume of, 47
Spherical sectors, 47–48
Spherical segments, 48–49
   center of gravity of, 140
Spherical trigonometry, 58
Spherical wedges, 50
Spherical zones, 49
   center of gravity of, 140
Spiral bevel gear, 247
Splines, 266–272
Springs, 272–278
   coiled, 276–278
   flat, 276–277
Spur gear formulas, 243–245
Spur gear nomenclature, 240
Spur gears, 240–245
Square prisms, volume of, 38
Square roots, 4
Squares, area of, 16–17
Static coefficient of friction, 116
Static fluid concepts, 203–208

Static friction, 112
Statics, 83
Straight bevel gear, 249
Strain, 170, 175–177
    shearing, 177
Strength
    of compression members, 191–198
    tensile, 171
    ultimate, 171
    yield, 170–171
Stress, 169
    compressive, 169, 171–173
    design, 180
    shearing, *see* Shearing stress
    tensile, 169, 171–173
    thermal, 169
    torsional shearing, 184–187
    working, 180
Stress-strain diagrams, 170
Stub pitch, spline, 268
Subtraction, vector, 89
Surface area, 15–37
    of cylinder, 43
    of sphere, 46
Surveying, 58

T

Tangent function, 60–62
Tangents, Law of, 71–73
Tapping operations, 352–353
Tapping tool-life formulas, 352–353
Taylor's equation, 350
Temperature, 223
Tensile strength, 171
Tensile stress, 169, 171–173
Thermal deformation, 181–182
Thermal stress, 169
Thin-walled closed sections, 188–189
Thin-walled open sections, 187–188
Thread, screw, *see* Screw-thread
    entries
Thrust ball bearings, 287
Thrust bearings, 280, 283–285, 286
Thrust roller bearings, 288
Tool life, optimized, 352

Torque, 183–187
    of force, 84–85
Torricelli's theorem, 212–213
Torsion, 183
Torsional shearing stress, 184–189
Torus, 53–54
Total flow of lubricant required,
    bearing, 283
Trace point, 228
Transition point, 228
Trapezium, area of, 23–24
Trapezoids
    area of, 22–23
    center of gravity of, 136
Triangles
    acute, area of, 21
    area of, 20
    moment of inertia of, 142
    obtuse, area of, 21–22
    right, area of, 20
Triangular plate spring, 274–275
Trigonometric formulas, 74–79
Trigonometric functions, 60–62
    inverse, 232
    using, 61–63
Trigonometric identities, 74–76
Trigonometric laws, 68–73
Trigonometry, 58
    oblique angle, 66–70
    plane, 58–59
    right angle, 62–63
    spherical, 58
Turning, 298
Turning formulas, 304
Turning operations, 351
Turning tool-life formulas, 353
Twist, angle of, 190–191

U

Ultimate strength, 171
Unit power for broaching, 305–306
Unit power requirements, 305
Unit strain, 170
Unit systems, 16, 85
    absolute, 85

V

Vacuum, term, 199
Vector addition, 89
Vector measures, 84
Vector subtraction, 89
Velocity, 85, 150
    angular, *see* Angular velocities
Viscosity, 203–204
    dynamic, 204
    kinematic, 203–204
Volume coefficient, 220
Volume measures, 16
Volumes
    of solid figures, 37–54
    specific, 201

W

Water services, 360–361
Wattage, 359
Watt meters, 314
Wedges
    spherical, 50–51
    volume of, 42
Weight, 80
    specific, 200–201
Weight density, 201–202
Wheel-and-pulley formulas, 125–131
Whole depth of tooth, 242

Work, 84, 163–164
Work formulas, 162
Work measurement, 363–365
Work standards, 363
Working depth of tooth, 242
Working stress, 180
Worm, 249, 252
Worm gear formulas, 249–255
Worm gears, 249–253
Worm wheel, 252

X

x axis, 105
X factor, bearing, 283

Y

y axis, 105
Yield point, 170
Yield strength, 170–171
Young's modulus, 171, 178

Z

z axis, 105
Zones, spherical, *see* Spherical zones

AUDEL®

**Over a Century of Excellence
for the Professional
and
Vocational Trades and the Crafts**

**Order now from your local bookstore
or use the convenient order form at
the back of this book.**

AUDEL

These fully illustrated, up-to-date guides and manuals mean a better job done for mechanics, engineers, electricians, plumbers, carpenters, and all skilled workers.

Contents

Electrical	II
Machine Shop and Mechanical Trades	III
Plumbing	IV
Heating, Ventilating and Air Conditioning	IV
Pneumatics and Hydraulics	V
Carpentry and Construction	V
Woodworking	VI
Maintenance and Repair	VI
Automotive and Engines	VII
Drafting	VII
Hobbies	VII
Macmillan Practical Arts Library	VIII

Electrical

House Wiring sixth edition
Roland E. Palmquist
5½ x 8¼ Hardcover 256 pp. 150 illus.
ISBN: 0-672-23404-1 $13.95

Rules and regulations of the current National Electrical Code® for residential wiring, fully explained and illustrated: • basis for load calculations • calculations for dwellings • services • nonmetallic-sheathed cable • underground feeder and branch-circuit cable • metal-clad cable • circuits required for dwellings • boxes and fittings • receptacle spacing • mobile homes • wiring for electric house heating.

Practical Electricity fourth edition
Robert G. Middleton; revised by L. Donald Meyers
5½ x 8¼ Hardcover 504 pp. 335 illus.
ISBN: 0-672-23375-4 $14.95

Complete, concise handbook on the principles of electricity and their practical application: • magnetism and electricity • conductors and insulators • circuits • electromagnetic induction • alternating current • electric lighting and lighting calculations • basic house wiring • electric heating • generating stations and substations.

Guide to the 1984 Electrical Code®
Roland E. Palmquist
5½ × 8¼ Hardcover 664 pp. 225 illus.
ISBN: 0-672-23398-3 $19.95

Authoritative guide to the National Electrical Code® for all electricians, contractors, inspectors, and homeowners: • terms and regulations for wiring design and protection • wiring methods and materials • equipment for general use • special occupancies • special equipment and conditions • and communication systems. Guide to the 1987 NEC® will be available in mid-1987.

Mathematics for Electricians and Electronics Technicians
Rex Miller
5½ x 8¼ Hardcover 312 pp. 115 illus.
ISBN: 0-8161-1700-4 $14.95

Mathematical concepts, formulas, and problem solving in electricity and electronics: • resistors and resistance • circuits • meters • alternating current and inductance • alternating current and capacitance • impedance and phase angles • resonance in circuits • special-purpose circuits. Includes mathematical problems and solutions.

Fractional Horsepower Electric Motors
Rex Miller and Mark Richard Miller
5½ x 8¼ Hardcover 436 pp. 285 ill.
ISBN: 0-672-23410-6 $15.95

Fully illustrated guide to small-to-moderate-size electric motors in home appliances and industrial equipment: • terminology • repair tools and supplies • small DC and universal motors • split-phase, capacitor-start, shaded pole, and special motors • commutators and brushes • shafts and bearings • switches and relays • armatures • stators • modification and replacement of motors.

Electric Motors
Edwin P. Anderson; revised by Rex Miller
5½ x 8¼ Hardcover 656 pp. 405 ill.
ISBN: 0-672-23376-2 $14.95

Complete guide to installation, maintenance, and repair of all types of electric motors: • AC generators • synchronous motors • squirrel-ca motors • wound rotor motors • DC motors • fractional-horsepower motors • magnetic contractors • motor testing and maintenance • motor calculations • meters • wir. diagrams • armature windings • DC armature rewinding procedu • and stator and coil winding.

Home Appliance Servicing fourth edition
Edwin P. Anderson; revised by Rex Miller
5½ x 8¼ Hardcover 640 pp. 345 ill.
ISBN: 0-672-23379-7 $15.95

Step-by-step illustrated instruction on all types of household appliance • irons • toasters • roasters and broilers • electric coffee makers • space heaters • water heaters • el tric ranges and microwave ovens • mixers and blenders • fans and ble ers • vacuum cleaners and floor polishers • washers and dryers • dis washers and garbage disposals • refrigerators • air conditioners ar dehumidifiers.

Television Service Manual

n edition
**bert G. Middleton; revised by
seph G. Barrile**
2 x 8¼ Hardcover 512 pp. 395 illus.
N: 0-672-23395-9 $15.95

actical up-to-date guide to all
ects of television transmission and
eption, for both black and white
d color receivers: • step-by-step
intenance and repair • broadcast-
g • transmission • receivers • anten-
s and transmission lines
terference • RF tuners • the video
annel • circuits • power supplies
ignment • test equipment.

Electrical Course for Apprentices and Journeymen

ond edition
land E. Palmquist
x 8¼ Hardcover 478 pp. 290 illus.
N: 0-672-23393-2 $14.95

ctical course on operational
ory and applications for training
d re-training in school or on
job: • electricity and matter
nits and definitions • electrical
nbols • magnets and magnetic
ds • capacitors • resistance • elec-
magnetism • instruments and
asurements • alternating currents
C generators • circuits • trans-
mers • motors • grounding and
und testing.

Questions and Answers for Electricians Examinations
eighth edition
land E. Palmquist
x 8¼ Hardcover 320 pp. 110 illus.
N: 0-672-23399-1 $12.95

ed on the current National
ctrical Code®, a review of exams
apprentice, journeyman, and
ster, with explanations of princi-
  underlying each test subject:
m's Law and other formulas
wer and power factors • lighting
anch circuits and feeders • trans-
mer principles and connections
ring • batteries and rectification
tage generation • motors • ground
 ground testing.

Machine Shop and Mechanical Trades

Machinists Library
fourth edition 3 vols
Rex Miller
5½ x 8¼ Hardcover 1,352 pp. 1,120 illus.
ISBN: 0-672-23380-0 $38.95

Indispensable three-volume reference for machinists, tool and die makers, machine operators, metal workers, and those with home workshops.

Volume I, Basic Machine Shop
5½ x 8¼ Hardcover 392 pp. 375 illus.
ISBN: 0-672-23381-9 $14.95

• Blueprint reading • benchwork
• layout and measurement • sheet-metal hand tools and machines
• cutting tools • drills • reamers • taps
• threading dies • milling machine cutters, arbors, collets, and adapters.

Volume II, Machine Shop
5½ x 8¼ Hardcover 528 pp. 445 illus
ISBN: 0-672-23382-7 $14.95

• Power saws • machine tool operations • drilling machines • boring
• lathes • automatic screw machine
• milling • metal spinning.

Volume III, Toolmakers Handy Book
5½ x 8¼ Hardcover 432 pp. 300 illus.
ISBN: 0-672-23383-5 $14.95

• Layout work • jigs and fixtures
• gears and gear cutting • dies and diemaking • toolmaking operations
• heat-treating furnaces • induction heating • furnace brazing • cold-treating process.

Mathematics for Mechanical Technicians and Technologists
John D. Bies
5½ x 8¼ Hardcover 392 pp. 190 illus.
ISBN: 0-02-510620-1 $17.95

Practical sourcebook of concepts, formulas, and problem solving in industrial and mechanical technology: • basic and complex mechanics
• strength of materials • fluidics
• cams and gears • machine elements
• machining operations • management controls • economics in machining • facility and human resources management.

Millwrights and Mechanics Guide
third edition
Carl A. Nelson
5½ x 8¼ Hardcover 1,040 pp. 880 illus.
ISBN: 0-672-23373-8 $22.95

Most comprehensive and authoritative guide available for millwrights and mechanics at all levels of work or supervision: • drawing and sketching • machinery and equipment installation • principles of mechanical power transmission • V-belt drives • flat belts • gears • chain drives • couplings • bearings • structural steel
• screw threads • mechanical fasteners
• pipe fittings and valves • carpentry
• sheet-metal work • blacksmithing
• rigging • electricity • welding
• pumps • portable power tools • mensuration and mechanical calculations.

Welders Guide third edition
James E. Brumbaugh
5½ x 8¼ Hardcover 960 pp. 615 illus.
ISBN: 0-672-23374-6 $23.95

Practical, concise manual on theory, operation, and maintenance of all welding machines: • gas welding equipment, supplies, and process
• arc welding equipment, supplies, and process • TIG and MIG welding
• submerged-arc and other shielded-arc welding processes • resistance, thermit, and stud welding • solders and soldering • brazing and braze welding • welding plastics • safety and health measures • symbols and definitions • testing and inspecting welds. Terminology and definitions as standardized by American Welding Society.

Welder/Fitters Guide
John P. Stewart
8½ x 11 Paperback 160 pp. 195 illus.
ISBN: 0-672-23325-8 $7.95

Step-by-step instruction for welder/fitters during training or on the job: • basic assembly tools and aids
• improving blueprint reading skills
• marking and alignment techniques
• using basic tools • simple work practices • guide to fabricating weldments • avoiding mistakes • exercises in blueprint reading • clamping devices • introduction to using hydraulic jacks • safety in weld fabrication plants • common welding shop terms.

Sheet Metal Work
John D. Bies
5½ x 8¼ Hardcover 456 pp. 215 illus.
ISBN: 0-8161-1706-3 $17.95

On-the-job sheet metal guide for manufacturing, construction, and home workshops: • mathematics for sheet metal work • principles of drafting • concepts of sheet metal drawing • sheet metal standards, specifications, and materials • safety practices • layout • shear cutting
• holes • bending and folding • forming operations • notching and clipping • metal spinning • mechanical fastening • soldering and brazing
• welding • surface preparation and finishes • production processes.

Power Plant Engineers Guide
third edition
Frank D. Graham; revised by Charlie Buffington
5½ x 8¼ Hardcover 960 pp. 530 illus.
ISBN: 0-672-23329-0 $16.95

All-inclusive question-and-answer guide to steam and diesel-power engines: • fuels • heat • combustion • types of boilers • shell or fire-tube boiler construction • strength of boiler materials • boiler calculations • boiler fixtures, fittings, and attachments • boiler feed pumps • condensers • cooling ponds and cooling towers • boiler installation, startup, operation, maintenance and repair • oil, gas, and waste-fuel burners • steam turbines • air compressors • plant safety.

Mechanical Trades Pocket Manual
second edition
Carl A. Nelson
4 x 6 Paperback 364 pp. 255 illus.
ISBN: 0-672-23378-9 $10.95

Comprehensive handbook of essentials, pocket-sized to fit in the tool box: • mechanical and isometric drawing • machinery installation and assembly • belts • drives • gears • couplings • screw threads • mechanical fasteners • packing and seals • bearings • portable power tools • welding • rigging • piping • automatic sprinkler systems • carpentry • stair layout • electricity • shop geometry and trigonometry.

Plumbing

Plumbers and Pipe Fitters Library
third edition 3 vols
Charles N. McConnell; revised by Tom Philbin
5½ x 8¼ Hardcover 952 pp. 560 illus.
ISBN: 0-672-23384-3 $34.95

Comprehensive three-volume set with up-to-date information for master plumbers, journeymen, apprentices, engineers, and those in building trades.

Volume 1, Materials, Tools, Roughing-In
5½ x 8¼ Hardcover 304 pp. 240 illus.
ISBN: 0-672-23385-1 $12.95

• Materials • tools • pipe fitting • pipe joints • blueprints • fixtures • valves and faucets.

Volume 2, Welding, Heating, Air Conditioning
5½ x 8¼ Hardcover 384 pp. 220 illus.
ISBN: 0-672-23386-x $13.95

• Brazing and welding • planning a heating system • steam heating systems • hot water heating systems • boiler fittings • fuel-oil tank installation • gas piping • air conditioning.

Volume 3, Water Supply, Drainage, Calculations
5½ x 8¼ Hardcover 264 pp. 100 illus.
ISBN: 0-672-23387-8 $12.95

• Drainage and venting • sewage disposal • soldering • lead work • mathematics and physics for plumbers and pipe fitters.

Home Plumbing Handbook third edition
Charles N. McConnell
8½ x 11 Paperback 200 pp. 100 illus.
ISBN: 0-672-23413-0 $10.95

Clear, concise, up-to-date fully illustrated guide to home plumbing installation and repair: • repairing and replacing faucets • repairing toilet tanks • repairing a trip-lever bath drain • dealing with stopped-up drains • working with copper tubing • measuring and cutting pipe • PVC and CPVC pipe and fittings • installing a garbage disposals • replacing dishwashers • repairing and replacing water heaters • installing or resetting toilets • caulking around plumbing fixtures and tile • water conditioning • working with cast-iron soil pipe • septic tanks and disposal fields • private water systems.

The Plumbers Handbook seventh edition
Joseph P. Almond, Sr.
4 x 6 Paperback 352 pp. 170 illus.
ISBN: 0-672-23419-x $10.95

Comprehensive, handy guide for plumbers, pipe fitters, and apprentices that fits in the tool box or pocket: • plumbing tools • how to read blueprints • heating systems • water supply • fixtures, valves, and fittings • working drawings • roughing and repair • outside sewage lift station • pipes and pipelines • vents, drain lines, and septic systems • lead work • silver brazing and soft soldering • plumbing systems • abbreviations, definitions, symbols, and formulas.

Questions and Answers for Plumbers Examinations second edition
Jules Oravetz
5½ x 8¼ Paperback 256 pp. 145 illus.
ISBN: 0-8161-1703-9 $9.95

Practical, fully illustrated study guide to licensing exams for apprentice, journeyman, or master plumber: • definitions, specifications, and regulations set by National Bureau of Standards and by various state codes • basic plumbing installation • drawings and typical plumbing system layout • mathematics • materials and fittings • joints and connections • traps, cleanouts, and backwater valves • fixtures • drainage, vents, and vent piping • water supply and distribution • plastic pipe and fittings • steam and hot water heating.

HVAC

Air Conditioning: Home and Commercial second edition
Edwin P. Anderson; revised by Rex Miller
5½ x 8¼ Hardcover 528 pp. 180 illus.
ISBN: 0-672-23397-5 $15.95

Complete guide to construction, installation, operation, maintenance and repair of home, commercial, and industrial air conditioning systems, with troubleshooting charts • heat leakage • ventilation requirements • room air conditioners • refrigerants • compressors • condensing equipment • evaporators • water-cooling systems • central air conditioning • automobile air conditioning • motors and motor control.

Heating, Ventilating and Air Conditioning Library second edition 3 vols
James E. Brumbaugh
5½ x 8¼ Hardcover 1,840 pp. 1,275
ISBN: 0-672-23388-6 $42.95

Authoritative three-volume reference for those who install, operate, maintain, and repair HVAC equipment commercially, industrially, or home. Each volume fully illustrated with photographs, drawings, tables and charts.

Volume I, Heating Fundamentals, Furnaces, Boilers, Boiler Conversions
5½ x 8¼ Hardcover 656 pp. 405 illus.
ISBN: 0-672-23389-4 $16.95

• Insulation principles • heating calculations • fuels • warm-air, hot water, steam, and electrical heating systems • gas-fired, oil-fired, coal-fired, and electric-fired furnaces • boilers and boiler fittings • boiler and furnace conversion.

Volume II, Oil, Gas and Coal Burners, Controls, Ducts, Piping, Valves
5½ x 8¼ Hardcover 592 pp. 455 illus.
ISBN: 0-672-23390-8 $15.95

• Coal firing methods • thermostats and humidistats • gas and oil controls and other automatic controls •

...and duct systems • pipes, pipe
...gs, and piping details • valves
...lve installation • steam and
...ater line controls.

...me III, Radiant Heating,
...r Heaters, Ventilation, Air
...ditioning, Heat Pumps,
...leaners
8¼ Hardcover 592 pp. 415 illus.
0-672-23391-6 $14.95

...iators, convectors, and unit
...rs • fireplaces, stoves, and
...neys • ventilation principles • fan
...ion and operation • air condi-
...g equipment • humidifiers and
...midifiers • air cleaners and

Burners fourth edition
n M. Field
8¼ Hardcover 360 pp. 170 illus.
0-672-23394-0 $15.95

...-date sourcebook on the
...ruction, installation, operation,
...g, servicing, and repair of all
...of oil burners, both industrial
...omestic: • general electrical
...up and wiring diagrams of
...natic control systems • ignition
...n • high-voltage transportation
...ational sequence of limit
...ols, thermostats, and various
...• combustion chambers • drafts
...neys • drive couplings • fans
...wers • burner nozzles • fuel
...s.

frigeration:
me and
mmercial second edition
n P. Anderson; revised by
...iller
8¼ Hardcover 768 pp. 285 illus.
0-672-23396-7 $17.95

...cal, comprehensive reference
...chnicians, plant engineers, and
...wners on the installation,
...tion, servicing, and repair of
...hing from single refrigeration
...to commercial and industrial
...ns: • refrigerants • compressors
...moelectric cooling • service
...ment and tools • cabinet main-
...ce and repairs • compressor
...ation systems • brine systems
...rmarket and grocery refrigera-
...locker plants • fans and blowers
...ng • heat leakage • refrigeration-
...alculations.

Pneumatics and Hydraulics

Hydraulics for Off-the-Road Equipment second edition
Harry L. Stewart; revised by
Tom Philbin
5½ x 8¼ Hardcover 256 pp. 175 illus.
ISBN: 0-8161-1701-2 $13.95

Complete reference manual for those who own and operate heavy equipment and for engineers, designers, installation and maintenance technicians, and shop mechanics: • hydraulic pumps, accumulators, and motors • force components • hydraulic control components • filters and filtration, lines and fittings, and fluids • hydrostatic transmissions • maintenance • troubleshooting.

Pneumatics and Hydraulics fourth edition
Harry L. Stewart; revised by
Tom Philbin
5½ x 8¼ Hardcover 512 pp. 315 illus.
ISBN: 0-672-23412-2 $15.95

Practical guide to the principles and applications of fluid power for engineers, designers, process planners, tool men, shop foremen, and mechanics: • pressure, work and power • general features of machines • hydraulic and pneumatic symbols • pressure boosters • air compressors and accessories • hydraulic power devices • hydraulic fluids • piping • air filters, pressure regulators, and lubricators • flow and pressure controls • pneumatic motors and tools • rotary hydraulic motors and hydraulic transmissions • pneumatic circuits • hydraulic circuits • servo systems.

Pumps fourth edition
Harry L. Stewart; revised by
Tom Philbin
5½ x 8¼ Hardcover 508 pp. 360 illus.
ISBN: 0-672-23400-9 $15.95

Comprehensive guide for operators, engineers, maintenance workers, inspectors, superintendents, and mechanics on principles and day-to-day operations of pumps: • centrifugal, rotary, reciprocating, and special service pumps • hydraulic accumulators • power transmission • hydraulic power tools • hydraulic cylinders • control valves • hydraulic fluids • fluid lines and fittings.

Carpentry and Construction

Carpenters and Builders Library
fifth edition 4 vols
John E. Ball; revised by
Tom Philbin
5½ x 8¼ Hardcover 1,224 pp. 1,010 illus.
ISBN: 0-672-23369-x $39.95
Also available in a new boxed set at no extra cost:
ISBN: 0-02-506450-9 $39.95

These profusely illustrated volumes, available in a handsome boxed edition, have set the professional standard for carpenters, joiners, and woodworkers.

Volume 1, Tools, Steel Square, Joinery
5½ x 8¼ Hardcover 384 pp. 345 illus.
ISBN: 0-672-23365-7 $10.95

• Woods • nails • screws • bolts • the workbench • tools • using the steel square • joints and joinery • cabinetmaking joints • wood patternmaking • and kitchen cabinet construction.

Volume 2, Builders Math, Plans, Specifications
5½ x 8¼ Hardcover 304 pp. 205 illus.
ISBN: 0-672-23366-5 $10.95

• Surveying • strength of timbers • practical drawing • architectural drawing • barn construction • small house construction • and home workshop layout.

Volume 3, Layouts, Foundations, Framing
5½ x 8¼ Hardcover 272 pp. 215 illus.
ISBN: 0-672-23367-3 $10.95

• Foundations • concrete forms • concrete block construction • framing, girders and sills • skylights • porches and patios • chimneys, fireplaces, and stoves • insulation • solar energy and paneling.

Volume 4, Millwork, Power Tools, Painting
5½ x 8¼ Hardcover 344 pp. 245 illus.
ISBN: 0-672-23368-1 $10.95

• Roofing, miter work • doors • windows, sheathing and siding • stairs • flooring • table saws, band saws, and jigsaws • wood lathes • sanders and combination tools • portable power tools • painting.

Complete Building Construction
second edition
John Phelps; revised by
Tom Philbin
5½ x 8¼ Hardcover 744 pp. 645 illus.
ISBN: 0-672-23377-0 $19.95

Comprehensive guide to constructing a frame or brick building from the

v

footings to the ridge; • laying out building and excavation lines • making concrete forms and pouring fittings and foundation • making concrete slabs, walks, and driveways • laying concrete block, brick, and tile • building chimneys and fireplaces • framing, siding, and roofing • insulating • finishing the inside • building stairs • installing windows • hanging doors.

Complete Roofing Handbook
James E. Brumbaugh
5½ x 8¼ Hardcover 536 pp. 510 illus.
ISBN: 0-02-517850-4 $29.95

Authoritative text and highly detailed drawings and photographs on all aspects of roofing: • types of roofs • roofing and reroofing • roof and attic insulation and ventilation • skylights and roof openings • dormer construction • roof flashing details • shingles • roll roofing • built-up roofing • roofing with wood shingles and shakes • slate and tile roofing • installing gutters and downspouts • listings of professional and trade associations and roofing manufacturers.

Complete Siding Handbook
James E. Brumbaugh
5½ x 8¼ Hardcover 512 pp. 450 illus.
ISBN: 0-02-517880-6 $29.95

Companion to *Complete Roofing Handbook*, with step-by-step instructions and drawings on every aspect of siding: • sidewalls and siding • wall preparation • wood board siding • plywood panel and lap siding • hardboard panel and lap siding • wood shingle and shake siding • aluminum and steel siding • vinyl siding • exterior paints and stains • refinishing of siding, gutter and downspout systems • listings of professional and trade associations and siding manufacturers.

Masons and Builders Library
second edition 2 vols
Louis M. Dezettel; revised by Tom Philbin
5½ x 8¼ Hardcover 688 pp. 500 illus.
ISBN: 0-672-23401-7 $23.95

Two-volume set on practical instruction in all aspects of materials and methods of bricklaying and masonry: • brick • mortar • tools • bonding • corners, openings, and arches • chimneys and fireplaces • structural clay tile and glass block • brick walks, floors, and terraces • repair and maintenance • plasterboard and plaster • stone and rock masonry • reading blueprints.

Volume 1, Concrete, Block, Tile, Terrazzo
5½ x 8¼ Hardcover 304 pp. 190 illus.
ISBN: 0-672-23402-5 $13.95

Volume 2, Bricklaying, Plastering, Rock Masonry, Clay Tile
5½ x 8¼ Hardcover 384 pp. 310 illus.
ISBN: 0-672-23403-3 $12.95

Woodworking

Woodworking and Cabinetmaking
F. Richard Boller
5½ x 8¼ Hardcover 360 pp. 455 illus.
ISBN: 0-02-512800-0 $16.95

Compact one-volume guide to the essentials of all aspects of woodworking: • properties of softwoods, hardwoods, plywood, and composition wood • design, function, appearance, and structure • project planning • hand tools • machines • portable electric tools • construction • the home workshop • and the projects themselves – stereo cabinet, speaker cabinets, bookcase, desk, platform bed, kitchen cabinets, bathroom vanity.

Wood Furniture: Finishing, Refinishing, Repairing second edition
James E. Brumbaugh
6½ x 8¼ Hardcover 352 pp. 185 illus.
ISBN: 0-672-23409-2 $12.95

Complete, fully illustrated guide to repairing furniture and to finishing and refinishing wood surfaces for professional woodworkers and do-it-yourselfers: • tools and supplies • types of wood • veneering • inlaying • repairing, restoring, and stripping • wood preparation • staining • shellac, varnish, lacquer, paint and enamel, and oil and wax finishes • antiquing • gilding and bronzing • decorating furniture.

Maintenance and Repair

Building Maintenance second edition
Jules Oravetz
5½ x 8¼ Hardcover 384 pp. 210 illus.
ISBN: 0-672-23278-2 $9.95

Complete information on professional maintenance procedures used in office, educational, and commercial buildings: • painting and decorating • plumbing and pipe fitting

• concrete and masonry • carpen • roofing • glazing and caulking • sheet metal • electricity • air conditioning and refrigeration • insect and rodent control • hea • maintenance management • custodial practices.

Gardening, Landscaping a Grounds Maintenance
third edition
Jules Oravetz
5½ x 8¼ Hardcover 424 pp. 34
ISBN: 0-672-23417-3 $15.95

Practical information for those maintain lawns, gardens, and in trial, municipal, and estate grou • flowers, vegetables, berries, an house plants • greenhouses • la • hedges and vines • flowering s and trees • shade, fruit and nut trees • evergreens • bird sanctua • fences • insect and rodent con • weed and brush control • road walks, and pavements • drainage • maintenance equipment • golf course planning and maintenan

Home Maintenance a Repair: Walls, Ceilings and Floors
Gary D. Branson
8½ x 11 Paperback 80 pp. 80 illus
ISBN: 0-672-23281-2 $6.95

Do-it-yourselfer's step-by-step to interior remodeling with pro sional results: • general mainter • wallboard installation and rep • wallboard taping • plaster repa • texture paints • wallpaper tech niques • paneling • sound contr • ceiling tile • bath tile • energy conservation.

Painting and Decorating
Rex Miller and Glenn E. Bake
5½ x 8¼ Hardcover 464 pp. 32
ISBN: 0-672-23405-x $18.95

Practical guide for painters, dec tors, and homeowners to the most up-to-date materials and niques: • job planning • tools ar equipment needed • finishing materials • surface preparation • applying paint and stains • dec ing with coverings • repairs and maintenance • color and decor principles.

Macmillan Practical Arts Library
Books for and by the Craftsman

World Woods in Color
W.A. Lincoln
7 × 10   Hardcover   300 pages
300 photos
ISBN: 0-02-572350-2 $39.95

Large full-color photographs show the natural grain and features of nearly 300 woods: • commercial and botanical names • physical characteristics, mechanical properties, seasoning, working properties, durability, and uses • the height, diameter, bark, and places of distribution of each tree • indexing of botanical, trade, commercial, local, and family names • a full bibliography of publications on timber study and identification.

The Woodworker's Bible
Alf Martensson
8 × 10   Paperback   288 pages   900 illus.
ISBN: 0-02-011940-2 $12.95

For the craftsperson familiar with basic carpentry skills, a guide to creating professional-quality furniture, cabinetry, and objects d'art in the home workshop: • techniques and expert advice on fine craftsmanship whether tooled by hand or machine • joint-making • assembling to ensure fit • finishes. Author, who lives in London and runs a workshop called Woodstock, has also written *The Book of Furnituremaking.*

Cabinetmaking: The Professional Approach
Alan Peters
8½ × 11   Hardcover   208 pages   175 illus.
(8 pp. color)
ISBN: 0-02-596200-0 $29.95

A unique guide to all aspects of professional furniture making, from an English master craftsman: • the Cotswold School and the birth of the furniture movement • setting up a professional shop • equipment • finance and business efficiency • furniture design • working to commission • batch production, training, and techniques • plans for nine projects.

The Woodturner's Art: Fundamentals and Projects
Ron Roszkiewicz
8 × 10   Hardcover   256 pages   300 illus.
ISBN: 0-02-605250-4 $24.95

A master woodturner shows how to design and create increasingly difficult projects step-by-step in this book suitable for the beginner and the more advanced student: • spindle and faceplate turning • tools • techniques • classic turnings from various historical periods • more than 30 types of projects including boxes, furniture, vases, and candlesticks • making duplicates • projects using combinations of techniques and more than one kind of wood. Author has also written *The Woodturner's Companion.*

Cabinetmaking and Millwork
John L. Feirer
7⅛ × 9½   Hardcover   992 pages
2,350 illus. (32 pp. in color)
ISBN: 0-02-537350-1 $47.50

The classic on cabinetmaking that covers in detail all of the materials, tools, machines, and processes used in building cabinets and interiors, the production of furniture, and other work of the finish carpenter and millwright: • fixed installations such as paneling, built-ins, and cabinets • movable wood products such as furniture and fixtures • which woods to use, and why and how to use them in the interiors of homes and commercial buildings • metrics and plastics in furniture construction.

Carpentry and Building Construction
John L. Feirer and Gilbert R. Hutchings
7½ × 9½   hardcover   1,120 pages
2,000 photos (8 pp. in color)
ISBN: 0-02-537360-9 $50.00

A classic by Feirer on each detail of modern construction: • the various machines, tools, and equipment from which the builder can choose • laying of a foundation • building frames for each part of a building • details of interior and exterior work • painting and finishing • reading plans • chimneys and fireplaces • ventilation • assembling prefabricated houses.

Tree Care second edition
John M. Haller
8½ x 11 Paperback 224 pp. 305 illus.
ISBN: 0-02-062870-6 $9.95

New edition of a standard in the field, for growers, nursery owners, foresters, landscapers, and homeowners: • planting • pruning • fertilizing • bracing and cabling • wound repair • grafting • spraying • disease and insect management • coping with environmental damage • removal • structure and physiology • recreational use.

Upholstering
updated
James E. Brumbaugh
5½ x 8¼ Hardcover 400 pp. 380 illus.
ISBN: 0-672-23372-x $12.95

Essentials of upholstering for professional, apprentice, and hobbyist: • furniture styles • tools and equipment • stripping • frame construction and repairs • finishing and refinishing wood surfaces • webbing • springs • burlap, stuffing, and muslin • pattern layout • cushions • foam padding • covers • channels and tufts • padded seats and slip seats • fabrics • plastics • furniture care.

Automotive and Engines

Diesel Engine Manual fourth edition
Perry O. Black; revised by William E. Scahill
5½ x 8¼ Hardcover 512 pp. 255 illus.
ISBN: 0-672-23371-1 $15.95

Detailed guide for mechanics, students, and others to all aspects of typical two- and four-cycle engines: • operating principles • fuel oil • diesel injection pumps • basic Mercedes diesels • diesel engine cylinders • lubrication • cooling systems • horsepower • engine-room procedures • diesel engine installation • automotive diesel engine • marine diesel engine • diesel electrical power plant • diesel engine service.

Gas Engine Manual third edition
Edwin P. Anderson; revised by Charles G. Facklam
5½ x 8¼ Hardcover 424 pp. 225 illus.
ISBN: 0-8161-1707-1 $12.95

Indispensable sourcebook for those who operate, maintain, and repair gas engines of all types and sizes: • fundamentals and classifications of engines · engine parts • pistons • crankshafts • valves • lubrication, cooling, fuel, ignition, emission control and electrical systems • engine tune-up • servicing of pistons and piston rings, cylinder blocks, connecting rods and crankshafts, valves and valve gears, carburetors, and electrical systems.

Small Gasoline Engines
Rex Miller and Mark Richard Miller
5½ x 8¼ Hardcover 640 pp. 525 illus.
ISBN: 0-672-23414-9 $16.95

Practical information for those who repair, maintain, and overhaul two- and four-cycle engines – with emphasis on one-cylinder motors – including lawn mowers, edgers, grass sweepers, snowblowers, emergency electrical generators, outboard motors, and other equipment up to ten horsepower: • carburetors, emission controls, and ignition systems • starting systems • hand tools • safety • power generation • engine operations • lubrication systems • power drivers • preventive maintenance • step-by-step overhauling procedures • troubleshooting • testing and inspection • cylinder block servicing.

Truck Guide Library 3 vols
James E. Brumbaugh
5½ x 8¼ Hardcover 2,144 pp. 1,715 illus.
ISBN: 0-672-23392-4 $45.95

Three-volume comprehensive and profusely illustrated reference on truck operation and maintenance.

Volume 1, Engines
5½ x 8¼ Hardcover 416 pp. 290 illus.
ISBN: 0-672-23356-8 $16.95

• Basic components · engine operating principles • troubleshooting • cylinder blocks • connecting rods, pistons, and rings • crankshafts, main bearings, and flywheels • camshafts and valve trains • engine valves.

Volume 2, Engine Auxiliary Systems
5½ x 8¼ Hardcover 704 pp. 520 illus.
ISBN: 0-672-23357-6 $16.95

• Battery and electrical systems • spark plugs • ignition systems, charging and starting systems • lubricating, cooling, and fuel systems • carburetors and governors • diesel systems • exhaust and emission-control systems.

Volume 3, Transmissions, Steering, and Brakes
5½ x 8¼ Hardcover 1,024 pp. 905 illus.
ISBN: 0-672-23406-8 $16.95

• Clutches • manual, auxiliary, and automatic transmissions • frame and suspension systems • differentials and axles, manual and power steering • front-end alignment • hydraulic, power, and air brakes • wheels and tires • trailers.

Drafting

Answers on Blueprint Reading
fourth edition
Roland E. Palmquist; revised by Thomas J. Morrisey
5½ x 8¼ Hardcover 320 pp. 275 illus.
ISBN: 0-8161-1704-7 $12.95

Complete question-and-answer instruction manual on blueprints of machines and tools, electrical systems, and architecture: • drafting scale • drafting instruments • conventional lines and representations • pictorial drawings • geometry of drafting • orthographic and working drawings • surfaces • detail drawing • sketching • map and topographical drawings • graphic symbols • architectural drawings • electrical blueprints • computer-aided design and drafting. Also included is an appendix of measurements • metric conversions • screw threads and tap drill sizes • number and letter sizes of drills with decimal equivalents • double depth of threads • tapers and angles.

Hobbies

Complete Course in Stained Glass
Pepe Mendez
8½ x 11 Paperback 80 pp. 50 illus.
ISBN: 0-672-23287-1 $8.95

Guide to the tools, materials, and techniques of the art of stained glass, with ten fully illustrated lessons: • how to cut glass • cartoon and pattern drawing • assembling and cementing • making lamps using various techniques • electrical components for completing lamps • sources of materials • glossary of terminology and techniques of stained glasswork.

Just select your books, fill out the card, and mail today.

Money-Back Guarantee

BUSINESS REPLY MAIL
FIRST CLASS   PERMIT NO. 348   NEW YORK, NY

POSTAGE WILL BE PAID BY ADDRESSEE

Macmillan Publishing Company
Audel® Library
866 Third Avenue
New York, NY 10022

Attention: Special Sales Department

NO POSTAGE
NECESSARY
IF MAILED
IN THE
UNITED STATES